Lecture Notes in Artificial Intelligence 7188

Subseries of Lecture Notes in Computer Science

Scott Sanner   Marcus Hutter (Eds.)

Recent Advances in Reinforcement Learning

9th European Workshop, EWRL 2011
Athens, Greece, September 9-11, 2011
Revised Selected Papers

 Springer

Series Editors

Randy Goebel, University of Alberta, Edmonton, Canada
Jörg Siekmann, University of Saarland, Saarbrücken, Germany
Wolfgang Wahlster, DFKI and University of Saarland, Saarbrücken, Germany

Volume Editors

Scott Sanner
NICTA and the Australian National University
7 London Circuit
Canberra, ACT 2601 Australia
E-mail: Scott.Sanner@nicta.com.au

Marcus Hutter
Australian National University
Research School of Computer Science
Canberra, ACT 0200 Australia
E-mail: marcus.hutter@anu.edu.au

ISSN 0302-9743   e-ISSN 1611-3349
ISBN 978-3-642-29945-2   e-ISBN 978-3-642-29946-9
DOI 10.1007/978-3-642-29946-9
Springer Heidelberg Dordrecht London New York

Library of Congress Control Number: 2012937733

CR Subject Classification (1998): I.2.6, I.2, F.1-2, G.3, C.1.3

LNCS Sublibrary: SL 7 – Artificial Intelligence

Typesetting: Camera-ready by author, data conversion by Scientific Publishing Services, Chennai, India

Printed on acid-free paper

Springer is part of Springer Science+Business Media (www.springer.com)

Preface

This volume contains the papers presented at EWRL 2011: the 9th European Workshop on Reinforcement Learning held in Athens, Greece, September 9–11, 2011. The workshop was co-located with the European Conference on Machine Learning and Principles and Practice of Knowledge Discovery in Databases (ECML PKDD 2011). The technical program of EWRL 2011 contained 38 plenary talks by authors of submitted papers and 4 plenary invited talks:

- Peter Auer (University of Leoben, Austria), "UCRL and Autonomous Exploration"
- Kristian Kersting (Fraunhofer IAIS and University of Bonn, Germany), "Increasing Representational Power and Scaling Inference in Reinforcement Learning"
- Peter Stone (University of Texas Austin, USA), "PRISM Practical RL: Representation, Interaction, Synthesis, and Mortality"
- Csaba Svepesvari (University of Alberta, Canada), "Towards Robust Reinforcement Learning Algorithms"

All authors of presented papers were invited to present their papers at an evening poster session held on September 10; posters were also solicited for late-breaking results with the option to submit a full paper after the workshop.

In total, there were 40 submissions to the workshop from 58 authors representing 18 countries (9 European). All papers were reviewed by at least two reviewers, with some papers receiving a third review. We had an international Program Committee of 24 members and 9 additional reviewers residing in 11 countries (7 European). Reviewing proceeded in two rounds: 24 of 39 submitted papers were accepted to the LNAI proceedings in the first round; after the workshop authors were allowed to resubmit previous submissions or submit new papers from late-breaking posters in a second round of reviewing in which 4 of 14 submissions (13 resubmissions, 1 new submission) were accepted for the LNAI proceedings. The present volume contains the texts of the 28 papers accepted for the LNAI proceedings in the first and second round.

The best paper prize of 500 Euro was awarded to the paper "Automatic Discovery of Ranking Formulas for Playing with Multi-armed Bandits" by authors Francis Maes, Louis Wehenkel and Damien Ernst (University of Liège, Belgium).

EWRL was first held in 1995 and has been held every 2-3 years since that time. Over the years, EWRL has become one of the premier events for the discussion and dissemination of recent research results in the field of reinforcement learning. This year's workshop was organized by Marcus Hutter (Australian National University, Canberra, Australia), Matthew Robards (Australian National University, Canberra, Australia), Scott Sanner (NICTA and Australian National University, Canberra, Australia), Peter Sunehag (Australian National University,

Canberra, Australia), and Marco Weiring (University of Groningen, Groningen, The Netherlands).

We would like to thank our financial sponsors: the *Artificial Intelligence Journal*, the Australian National University, NICTA, and PASCAL2. Their generous support has allowed us to continue the tradition of holding EWRL as a free event open to all interested participants. We would like to thank ECML PKDD 2011 for allowing us to co-locate with them and for advertising EWRL on their website. We are grateful that we could use the excellent conference management system EasyChair, which has been developed mainly by Andrei Voronkov and hosted at the University of Manchester; the system is cost-free. We thank Springer for their support in preparing and publishing this volume of *Lecture Notes in Artificial Intelligence*.

Last but not least, we wish to thank the organizers, Program Committee members, additional reviewers, authors, speakers and all attendees for this engaging and highly successful installment of EWRL.

February 2012   Scott Sanner
                                                      Marcus Hutter

Organization

Organizing Committee

General Workshop Chair	Marcus Hutter
Local Organizing Chair	Matthew Robards
Program Committee Chair	Scott Sanner
Treasurer	Peter Sunehag
Miscellaneous	Marco Wiering

Program Committee

Edwin Bonilla	NICTA and Australian National University, Australia
Emma Brunskill	University of California, Berkeley, USA
Peter Dayan	University College London, UK
Carlos Diuk	Princeton University, USA
Alan Fern	Oregon State University, USA
Fernando Fernandez	Universidad Carlos III de Madrid, Spain
Mohammad Ghavamzadeh	INRIA Lille, France
Marcus Hutter	Australian National University and NICTA, Australia
Kristian Kersting	Fraunhofer IAIS and University of Bonn, Germany
Shie Mannor	Technion, Israel
Ronald Ortner	University of Leoben, Austria
Joelle Pineau	McGill University, Canada
Doina Precup	McGill University, Canada
Matthew Robards	Australian National University and NICTA, Australia
Scott Sanner	NICTA and Australian National University, Australia
Guy Shani	Ben-Gurion University, Israel
David Silver	University College London, UK
Peter Sunehag	Australian National University, Australia
Prasad Tadepalli	Oregon State University, USA
Matthew Taylor	Lafayette College, USA

William Uther   NICTA and the University of New South Wales,
                                      Australia
Martijn Van Otterlo   Katholieke Universiteit Leuven, Belgium
Thomas Walsh   University of Arizona, USA
Marco Wiering   University of Groningen, The Netherlands

Additional Reviewers

Bou Ammar, Haitham   Shào, Wén
Daswani, Mayank   Van Hasselt, Hado
Kalyanakrishnan, Shivaram   Vroman, Monica
Lattimore, Tor   Visentin, Daniel
Nguyen, Phuong Minh

With Thanks to Our EWRL 2011 Sponsors

Table of Contents

Function Approximation Methods for Reinforcement Learning

Macro-actions in Reinforcement Learning

Policy Search and Bounds

Multi-Task and Transfer Reinforcement Learning

Multi-Agent Reinforcement Learning

Apprenticeship and Inverse Reinforcement Learning

Real-World Reinforcement Learning

Invited Talk:
UCRL and Autonomous Exploration

Peter Auer

University of Leoben
Leoben, Austria

Abstract. After reviewing the main ingredients of the UCRL algorithm and its analysis for online reinforcement learning — exploration vs. exploitation, optimism in the face of uncertainty, consistency with observations and upper confidence bounds, regret analysis — I show how these techniques can also be used to derive PAC-MDP bounds which match the best currently available bounds for the discounted and the undiscounted setting. As typical for reinforcement learning, the analysis for the undiscounted setting is significantly more involved.

In the second part of my talk I consider a model for autonomous exploration, where an agent learns about its environment and how to navigate in it. Whereas evaluating autonomous exploration is typically difficult, in the presented setting rigorous performance bounds can be derived. For that we present an algorithm that optimistically explores, by repeatedly choosing the apparently closest unknown state — as indicated by an optimistic policy — for further exploration.

Acknowledgements. This is joint work with Shiau Hong Lim. The research leading to these results has received funding from the European Community's Seventh Framework Programme (FP7/2007-2013) under grant agreement 231495 (CompLACS).

Invited Talk:
Increasing Representational Power and Scaling Inference in Reinforcement Learning

Kristian Kersting

Fraunhofer IAIS and University of Bonn
Bonn, Germany

Abstract. As robots are starting to perform everyday manipulation tasks, such as cleaning up, setting a table or preparing simple meals, they must become much more knowledgeable than they are today. Natural environments are composed of objects, and the possibilities to manipulate them are highly structured due to the general laws governing our relational world. All these need to be acknowledged when we want to realize thinking robots that efficiently learn how to accomplish tasks in our relational world.

Triggered by this grand vision, this talk discusses the very promising perspective on the application of Statistical Relational AI techniques to reinforcement learning. Specifically, it reviews existing symbolic dynamic programming and relational RL approaches that exploit the symbolic structure in the solution of relational and first-order logical Markov decision processes. They illustrate that Statistical Relational AI may give new tools for solving the "scaling challenge". It is sometimes mentioned that scaling RL to real-world scenarios is a core challenge for robotics and AI in general. While this is true in a trivial sense, it might be beside the point. Reasoning and learning on appropriate (e.g. relational) representations leads to another view on the "scaling problem": often we are facing problems with symmetries not reflected in the structure used by our standard solvers. As additional evidence for this, the talk concludes by presenting our ongoing work on the first lifted linear programming solvers for MDPs. Given an MDP, our approach first constructs a lifted program where each variable presents a set of original variables that are indistinguishable given the objective function and constraints. It then runs any standard LP solver on this program to solve the original program optimally.

Acknowledgements. This talk is based on joint works with Babak Ahmadi, Kurt Driessens, Saket Joshi, Roni Khardon, Tobias Lang, Martin Mladenov, Sriraam Natarajan, Scott Sanner, Jude Shavlik, Prasad Tadepalli, and Marc Toussaint.

Invited Talk:
PRISM – Practical RL: Representation, Interaction, Synthesis, and Mortality

Peter Stone

University of Texas at Austin
Austin, USA

Abstract. When scaling up RL to large continuous domains with imperfect representations and hierarchical structure, we often try applying algorithms that are proven to converge in small finite domains, and then just hope for the best. This talk will advocate instead designing algorithms that adhere to the constraints, and indeed take advantage of the opportunities, that might come with the problem at hand. Drawing on several different research threads within the Learning Agents Research Group at UT Austin, I will discuss four types of issues that arise from these contraints and opportunities: 1) Representation – choosing the algorithm for the problem's representation and adapating the representation to fit the algorithm; 2) Interaction – with other agents and with human trainers; 3) Synthesis – of different algorithms for the same problem and of different concepts in the same algorithm; and 4) Mortality – the opportunity to improve learning based on past experience and the constraint that one can't explore exhaustively.

S. Sanner and M. Hutter (Eds.): EWRL 2011, LNCS 7188, p. 3, 2012.

Invited Talk:
Towards Robust Reinforcement
Learning Algorithms

Csaba Szepesvári

University of Alberta, Edmonton, Canada

Abstract. Most reinforcement learning algorithms assume that the system to be controlled can be accurately approximated given the measurements and the available resources. However, this assumption is overly optimistic for too many problems of practical interest: Real-world problems are messy. For example, the number of unobserved variables influencing the dynamics can be very large and the dynamics governing can be highly complicated. How can then one ask for near-optimal performance without requiring an enormous amount of data? In this talk we explore an alternative to this standard criterion, based on the concept of regret, borrowed from the online learning literature. Under this alternative criterion, the performance of a learning algorithm is measured by how much total reward is collected by the algorithm as compared to the total reward that could have been collected by the best policy from a fixed policy class, the best policy being determined in hindsight. How can we design algorithms that keep the regret small? Do we need to change existing algorithm designs? In this talk, following the initial steps made by Even-Dar et al. and Yu et al., I will discuss some of our new results that shed some light on these questions.

Acknowledgements. The talk is based on joint work with Gergely Neu, Andras Gyorgy and Andras Antos.

S. Sanner and M. Hutter (Eds.): EWRL 2011, LNCS 7188, p. 4, 2012.

Automatic Discovery of Ranking Formulas for Playing with Multi-armed Bandits

Francis Maes, Louis Wehenkel, and Damien Ernst

University of Liège
Dept. of Electrical Engineering and Computer Science
Institut Montefiore, B28, B-4000, Liège - Belgium

Abstract. We propose an approach for discovering in an automatic way formulas for ranking arms while playing with multi-armed bandits.
The approach works by defining a grammar made of basic elements such as for example addition, subtraction, the max operator, the average values of rewards collected by an arm, their standard deviation etc., and by exploiting this grammar to generate and test a large number of formulas. The systematic search for good candidate formulas is carried out by a built-on-purpose optimization algorithm used to navigate inside this large set of candidate formulas towards those that give high performances when using them on some multi-armed bandit problems.
We have applied this approach on a set of bandit problems made of Bernoulli, Gaussian and truncated Gaussian distributions and have identified a few simple ranking formulas that provide interesting results on every problem of this set. In particular, they clearly outperform several reference policies previously introduced in the literature. We argue that these newly found formulas as well as the procedure for generating them may suggest new directions for studying bandit problems.

Keywords: Multi-armed Bandits, Exploration vs. exploitation, Automatic formula discovery.

1 Introduction

In the recent years, there has been a revival of interest in multi-armed bandit problems, probably due to the fact that many applications might benefit from progress in this context [5,3].

A very popular line of research that has emerged for solving these problems focuses on the design of simple policies that compute in an incremental way an index for each arm from the information collected on the arm and then play the arm with the highest ranking index. One of the most well-known such "index-based" policies is $UCB1$ proposed in [2]. It associates at every play opportunity t the value $\bar{r}_k + \sqrt{(C \ln t)/t_k}$ to every arm k where $\bar{r}_k$ is the average value of the rewards collected so far by playing this arm, t_k the number of times it has been played, and C is typically set to 2.

Simple index-based policies as $UCB1$ have two main advantages. First they can be implemented on-line with bounded memory and computing resources,

S. Sanner and M. Hutter (Eds.): EWRL 2011, LNCS 7188, pp. 5–17, 2012.

which is important in some applications. Second, due to their simplicity, they are interpretable and hence may lend themselves to a theoretical analysis that may provide insights about performance guarantees.

For designing such simple policies, researchers often use their own intuition to propose a new arm-ranking formula that they test afterwards on some problems to confirm or refute it. We believe that by proceeding like this, there may be interesting ranking formulas that have little chances to be considered. Therefore we propose a systematic approach for discovering such ranking formulas in the context of multi-armed bandit problems.

Our approach seeks to determine automatically such formulas and it works by considering a rich set of candidate formulas in order to identify simple ones that perform well. This is achieved based on the observation that a formula can be grown using a sequence of operations. For example the simple formula $\sqrt{\bar{r}_k}$ could be seen as the result of two operations: select $\bar{r}_k$ then apply $\sqrt{\cdot}$ to $\bar{r}_k$. From there, the search for a well-performing combination of operations of a given complexity is determined by using a specific search procedure where the constants that may appear in a formula are separately optimized using an estimation of distribution algorithm. Using an optimization procedure to identify in a large set of candidates policies well-performing ones has already been tested with success in a previous paper where we were focusing on the identification of linear index formulas over a high-dimensional feature space, with the sole goal of minimizing the regret. Here we focus on the discovery of simple interpretable formulas within a set of non-linear formulas generated by a richer grammar, with the goal of providing further insight into the nature of the multi-armed bandit problem [7].

The rest of this paper is organized as follows. In the next section, we remind the basics about multi-armed bandit problems. In Section 3, we present our procedure for identifying simple formulas for index-based policies performing well on a set of bandit problems. Section 4 reports simulation results and discusses seven such formulas that have been discovered by our procedure. Finally, Section 5 concludes and presents some future research directions.

2 Multi-armed Bandit Problem and Policies

We now describe the (discrete) multi-armed bandit problem and briefly introduce some index-based policies that have already been proposed to solve it.

2.1 The K-armed Bandit Problem

We denote by $i \in \{1, 2, \ldots, K\}$ the ($K \geq 2$) arms of the bandit problem, by ν_i the reward distribution of arm i, and by μ_i its expected value; b_t is the arm played at round t, and $r_t \sim \nu_{b_t}$ is the obtained reward. $H_t = [b_1, r_1, b_2, r_2, \ldots, b_t, r_t]$ is a vector that gathers the history over the first t plays, and we denote by $\mathcal{H}$ the set of all possible histories of any length t.

A policy $\pi : \mathcal{H} \to \{1, 2, \ldots, K\}$ is an algorithm that processes at play t the vector H_{t-1} to select the arm $b_t \in \{1, 2, \ldots, K\}$:

$$b_t = \pi(H_{t-1}). \tag{1}$$

The regret of the policy π after a horizon of T plays is defined by:

$$R_T^\pi = T\mu^* - \sum_{t=1}^{T} r_t, \tag{2}$$

where $\mu^* = \max_k \mu_k$ denotes the expected reward of the best arm. The expected regret is therefore given by

$$E[R_T^\pi] = \sum_{k=1}^{K} E[t_k(T)](\mu^* - \mu_k), \tag{3}$$

where $t_k(T)$ is the number of times arm k is used in the first T rounds.

The K-armed bandit problem aims at finding a policy π^* that for a given K minimizes the expected regret (or, in other words, maximizes the expected reward), ideally for any horizon T and any $\{\nu_i\}_{i=1}^{K}$.

2.2 Index-Based Bandit Policies

Index-based bandit policies are based on a ranking *index* that computes for each arm k a numerical value based on the sub-history of responses H_{t-1}^k of that arm gathered at time t. These policies are sketched in Algorithm 1 and work as follows. During the first K plays, they play sequentially the machines $1, 2, \ldots, K$ to perform initialization. In all subsequent plays, these policies compute for every machine k the score $index(H_{t-1}^k, t) \in \mathbb{R}$ that depends on the observed sub-history H_{t-1}^k of arm k and possibly on t. At each step t, the arm with the largest score is selected (ties are broken at random).

Here are some examples of popular index functions:

$$index^{\text{UCB1}}(H_{t-1}^k, t) = \overline{r}_k + \sqrt{\frac{C \ln t}{t_k}} \tag{4}$$

$$index^{\text{UCB1-Bernoulli}}(H_{t-1}^k, t) = \overline{r}_k + \sqrt{\frac{\ln t}{t_k} \min\left(1/4, \overline{\sigma}_k + \sqrt{\frac{2 \ln t}{t_k}}\right)} \tag{5}$$

$$index^{\text{UCB1-Normal}}(H_{t-1}^k, t) = \overline{r}_k + \sqrt{16 \frac{t_k \overline{\sigma}_k^2}{t_k - 1} \frac{\ln(t - 1)}{t_k}} \tag{6}$$

$$index^{\text{UCB-V}}(H_{t-1}^k, t) = \overline{r}_k + \sqrt{\frac{2\overline{\sigma}_k^2 \zeta \ln t}{t_k}} + c\frac{3\zeta \ln t}{t_k} \tag{7}$$

where $\overline{r}_k$ and $\overline{\sigma}_k$ are the mean and standard deviation of the rewards so far obtained from arm k and t_k is the number of times it has been played.

Algorithm 1. Generic index-based discrete bandit policy

1: Given scoring function $index : \mathcal{H} \times \{1, 2, \ldots, K\} \to \mathbb{R}$,
2: **for** $t = 1$ to K **do**
3:      Play bandit $b_t = t$ ▷ Initialization: play each bandit once
4:      Observe reward r_t
5: **end for**
6: **for** $t = K$ to T **do**
7:      Play bandit $b_t = \text{argmax}_{k \in \{1,2,\ldots,K\}} index(H_{t-1}^k, t)$
8:      Observe reward r_t
9: **end for**

Policies UCB1, UCB1-BERNOULLI[1] and UCB1-NORMAL[2] have been proposed by [2]. UCB1 has one parameter $C > 0$ whose typical value is 2. Policy UCB-V has been proposed by [1] and has two parameters $\zeta > 0$ and $c > 0$. We refer the reader to [2,1] for detailed explanations of these parameters.

3   Systematic Search for Good Ranking Formulas

The ranking indexes from the literature introduced in the previous section depend on t and on three statistics extracted from the sub-history H_{t-1}^k : $\bar{r}_k$, $\bar{\sigma}_k$ and t_k. This section describes an approach to systematically explore the set of ranking formulas that can be build by combining these four variables, with the aim to discover the best-performing one for a given set of bandit problems.

We first formally define the (infinite) space of candidate formulas $\mathcal{F}$ in Section 3.1. Given a set of bandit problems P, a time horizon T, and $\mathcal{F}$, the aim is to solve:

$$f^* = \underset{f \in \mathcal{F}}{\text{argmin}} \, \Delta_{P,T}(f) \qquad (8)$$

where $\Delta_{P,T}(f) \in \mathbb{R}$ is a loss function which is equal to the average expected regret over the set of bandit problems P when those are played with a time horizon T. Note that in our simulation this loss function will be estimated by running for every element of P 1000 simulations over the whole time horizon. Since we consider formulas that may contain constant parameters, solving Equation 8 is a mixed structure/parameter optimization problem. Section 3.2 describes the procedure to search for formula structures and Section 3.3 focuses on parameter optimization. Note that in order to avoid overfitting, some form of regularization or complexity penalty can be added to the objective function $\Delta(\cdot)$. This is however not necessary in our study since we purposely restrict our search algorithm to rather small formulas.

---

[1] The original name of this policy is UCB-TUNED. Since this paper mostly deals with policies having parameters, we changed the name to UCB1-BERNOULLI to make clear that no parameter tuning has to be done with this policy.
[2] Note that this index-based policy does not strictly fit inside Algorithm 1 as it uses an additional condition to play bandits that were not played since a long time.

3.1 A Grammar for Generating Index Functions

Sets of mathematical formulas may be formalized by using grammars. In fields such as compiler design or genetic programming [6], these formulas are then described by parse trees which are labeled ordered trees. We use this approach for our index formulas, by using the grammar given in Figure 1: an *index* formula F is either a binary expression $B(F, F)$, a unary expression $U(F)$, an atomic variable V or a constant cst. The set of binary operators B includes the four elementary mathematic operations $(+, -, \times, \div)$ and the *min* and *max* operators. The three unary operators U are the logarithm ln, the square root $sqrt$ and the inverse function $inverse$. As mentioned previously, we consider four elementary variables V, which are the empirical reward mean $\bar{r}_k$ and standard deviation $\bar{\sigma}_k$ of arm k, the number t_k of times it has been played so far and the current time step t. For example, two formulas in $\mathcal{F}$ are given in the form of parse trees in the right part of Figure 1.

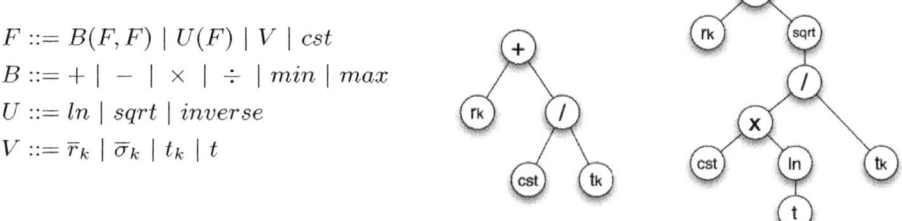

$$F ::= B(F, F) \mid U(F) \mid V \mid cst$$
$$B ::= + \mid - \mid \times \mid \div \mid min \mid max$$
$$U ::= ln \mid sqrt \mid inverse$$
$$V ::= \bar{r}_k \mid \bar{\sigma}_k \mid t_k \mid t$$

Fig. 1. The grammar used for generating candidate index functions and two example formula parse trees corresponding to $\bar{r}_k + C/t_k$ and $\bar{r}_k + \sqrt{Cln(t)/t_k}$.

In the following, the formula *structure* refers to its parse tree with constants labeled as cst. Each constant is a *parameter* with values $\in \mathbb{R}$. We denote formula structures by f, formula parameters by $\theta \in \mathbb{R}^C$ where C is the number of constants and parameterized formulas by $f_\theta \in \mathcal{F}$.

3.2 Generation of Candidate Formula Structures

We now describe the iterative depth-limited approach that we adopt to generate candidate formula structures f. We rely on a search space $\mathcal{S}$ the elements s of which are ordered lists of "current" formulas: $s = (f^1, \ldots, f^S)$. The elementary search step consists in expanding a state by appending to it a new formula f^{S+1} constructed from its current formulas $f^1, \ldots, f^S$, by using one of the following operations:

- Create a constant: $f^{S+1} = cst$
- Create a variable: $f^{S+1} = \bar{r}_k, \bar{\sigma}_k, t_k$ or t
- Create an unary operation: $f^{S+1} = U(f^i)$ where $i \in [1, S]$
- Create a binary operation: $f^{S+1} = B(f^i, f^j)$ where $i \in [1, S], j \in [i, S]$

Algorithm 2. Formula structure search algorithm

Given loss function $\Delta^{structure} : \mathcal{S} \to \mathbb{R}$,
Given max depth $d_{max} \geq 1$,

1: $s_0 \leftarrow ()$
2: **for** $\ell \in [0, \infty[$ **do**
3:    $s_\ell^* \leftarrow$ DepthLimitedSearch $(s_\ell, \Delta^{structure}, d_{max})$
4:    **if** $\ell \geq 1$ and $\Delta^{structure}(s_\ell^*) \geq \Delta^{structure}(s_{\ell-1}^*)$ **then**
5:       **return** $s_{\ell-1}^*$
6:    **end if**
7:    $s_{\ell+1} \leftarrow$ firstSuccessorState(s_ℓ, s_ℓ^*)
8:    $t \leftarrow \ell + 1$
9: **end for**

In this search space, starting from the empty initial state $s_0 = ()$, the formula $\bar{r}_k + cst/t_k$ can for example be built with the following operation sequence: $(f^1 = \bar{r}_k) \to (f^2 = cst) \to (f^3 = t_k) \to (f^4 = f^2/f^3) \to (f^5 = f^1 + f^4)$.

We use the function $\Delta^{structure}(s) \in \mathbb{R}$ to evaluate the loss associated to a state $s \in \mathcal{S}$. This loss only depends on the last formula f^S of state s. If f^S is a parameter-free formula (i.e., without constants), $\Delta^{structure}(s)$ is equal to the loss $\Delta_{P,H}(f^S)$. Otherwise, $\Delta^{structure}(s)$ is equal to the best achieved loss $\Delta_{P,H}(f_{\theta^*}^S)$ during the optimization of parameters $\theta \in \mathbb{R}^C$ with the procedure described in the next section.

Our iterative depth-limited search approach is specified by Algorithm 2 and works in the following way. We start with the initial state $s_0 = ()$. Each search iteration uses a depth-limited search to select the next formula that will be added to the current state. At iteration ℓ, depth-limited search works by exploring exhaustively the search space, starting from s_ℓ with a maximum exploration depth of d_{max}. During this exploration, the loss function $\Delta^{structure}$ is computed for every visited state, and the state that minimizes this function is returned[3]. s_ℓ^* is thus the best state that can be reached with the depth-limited search procedure when starting from s_ℓ. Once $s_\ell^* = (f_1, \ldots, f_\ell, f_{\ell+1}^*, \ldots, f_{\ell+d_{max}}^*)$ is identified, the formula $f_{\ell+1}^*$ is added to the current state s_ℓ to form the new state $s_{\ell+1} = (f_1, \ldots, f_\ell, f_{\ell+1}^*)$. The search procedure stops as soon as an iteration ℓ does not lead to an improved value of $\Delta^{structure}(s_\ell^*)$ compared to its previous iteration.

Our search procedure may evaluate the same formula structures multiple times. To avoid this, we implement $\Delta^{structure}(\cdot)$ with a cache that stores all the values $\Delta^{structure}(\cdot)$ that have already been computed since the beginning of search. To further reduce the number of calls to $\Delta^{structure}(\cdot)$, this cache relies on a normal version of the formula, which is computed by iteratively applying the following rules until the formula remains unchanged:

- Commutativity: fix the order of the operands of commutative operators: $+, \times, max$ and min

---

[3] Ties are broken by giving the preference to smaller formulas.

- Multiplication $\forall x \in \mathcal{F}, 1 \times x \rightarrow x$
- Division: $\forall x \in \mathcal{F}, x \div x \rightarrow 1, x \div 1 \rightarrow x,$
- Subtraction: $\forall x \in \mathcal{F}, x - x \rightarrow 0$
- Inverse function: $\forall x \in \mathcal{F}, y \in \mathcal{F}, x \times inverse(y) \rightarrow x \div y, x \div inverse(y) \rightarrow x \times y, 1 \div x \rightarrow inverse(x), inverse(inverse(x)) \rightarrow x$

Note that our search procedure in rather naive and could be improved in several ways. However, it proves to be efficient enough to already discover several interesting new bandit policies, as shown in the next section. A simple way to improve the efficiency of our approach would be to extend the set of rewriting rules with advanced mathematical rules such as factorization and division simplification.

3.3   Optimization of Constants

Each time a formula structure f with a number $C \geq 1$ of parameters is evaluated, we run a derivative-free global optimization algorithm to determine the best values $\theta^* \in \mathbb{R}^C$ for these parameters. In this work, we rely on a powerful, yet simple, class of metaheuristics known as *Estimation of Distribution Algorithms* (EDA) [4]. EDAs rely on a probabilistic model to describe promising regions of the search space and to sample good candidate solutions. This is performed by repeating iterations that first *sample* a population of N candidate values of $\theta \in \mathbb{R}^C$ using the *current* probabilistic model and then *fit* a *new* probabilistic model given the $b < N$ best candidates. Many different kinds of probabilistic models may be used inside an EDA. The most basic form of EDAs uses one marginal distribution per variable to optimize and is known as the *univariate marginal distribution algorithm* [8]. We have adopted this approach that, while simple, proved to be effective to solve our general parameter optimization problem.

Our EDA algorithm thus proceeds as follows. There is one Normal distribution per parameter to optimize $c \in \{1, 2, \ldots, C\}$. At the first iteration, these distributions are initialized as standard Normal distributions. At each subsequent iteration, N candidates $\theta \in \mathbb{R}^C$ are sampled using the current distributions and evaluated; then the p distributions are re-estimated using the $b < N$ best candidates of the current iteration. Optimization stops when the loss $\Delta_{P,H}(f_\theta)$ has not decreased during the last i_{stop} iterations and then returns the parameters θ^* for which we observed the minimum of the loss function $\Delta_{P,H}(f_{\theta^*})$.

4   Numerical Experiments

We now apply our formula discovery approach on a set of twelve discrete bandit problems with different kinds of distributions and different numbers of arms. We will focus on the analysis of seven interesting formulas that were discovered.

4.1   Experimental Setup

We considered the following 12 discrete bandit problems:

- **2-Bernoulli arms**: PROBLEM-1, PROBLEM-2 and PROBLEM-3 are two-arms bandit problems with Bernoulli distributions whose expectations (μ_1, μ_2) are respectively $(0.9, 0.6)$, $(0.9, 0.8)$ and $(0.55, 0.45)$.
- **10-Bernoulli arms**: PROBLEM-4 – PROBLEM-6 correspond to the three first problems, in which the arm with lowest reward expectation is duplicated 9 times. For example, the reward expectations of PROBLEM-4 are $(0.9, 0.6, 0.6, 0.6, 0.6, 0.6, 0.6, 0.6, 0.6, 0.6)$.
- **2-Gaussian arms**: PROBLEM-7 – PROBLEM-9 are two-arms bandit problems with Normal distributions. As previously the expectations μ_1 and μ_2 of these distributions are respectively $(0.9, 0.6)$, $(0.9, 0.8)$ and $(0.55, 0.45)$. We use the same standard deviation $\sigma = 0.5$ for all these Normal distributions.
- **2-Truncated Gaussian arms**: PROBLEM-10 – PROBLEM-12 correspond to the three previous problems with Gaussian distributions truncated to the interval $[0, 1]$ (that means that if you draw a reward which is outside of this interval, you throw it away and draw a new one).

We have first applied our index formula discovery algorithm with loss functions $\Delta_{P,T}$ where P contains only one of the twelve problems and with $T = 100$. The values of d_{max} considered with these twelve different loss functions are 1,2 and 3. We then performed a set of experiments by optimizing the loss $\Delta_{P,T}$ with P containing the twelve problems so as to look for policies performing well on average for this set of problems. The population size of the EDA optimizer is $N = 12C$ and $b = 3C$ is used as the number of best solutions used to fit new distributions, where C is the number of parameters. EDA optimization iterations are stopped after $i_{stop} = 5$ iterations without improvement.

In this setting, with our C++ based implementation and a 1.9 Ghz processor, performing the search with $d_{max} = 1$ takes $\simeq 10$ seconds, with $d_{max} = 2$ it takes a few minutes, and with $d_{max} = 3$ it takes some weeks of CPU time.

4.2   Discovered Policies

Table 1 reports the best formulas that were discovered for each set P used to define the loss function and each value d_{max}. Since we could not always complete the search with d_{max}, we only report the best formulas found after 2 search iterations and 3 search iterations.

When $d_{max} = 1$, Algorithm 2 aims at performing purely greedy search. As could be expected the best formula that can be found in a greedy way, when starting from an empty formula, is always $\text{GREEDY}(H^k_{t-1}, t) = \bar{r}_k$. To discover more interesting formulas, the search algorithm has to perform a deeper exploration. With $d_{max} = 2$, the algorithm discovers medium-performing index functions, without any recurrent pattern between the found solutions. The results with $d_{max} = 3$ are more interesting. In this last setting, the automatic discovery process most of the time converges to one of the following formulas:

Table 1. Discovered index formulas for horizon $T = 100$ on twelve bandit problems. For each problem, the best performing index formula is shown in bold. - denotes results that we could not compute within 10 days.

P	$d_{max} = 1$	$d_{max} = 2$	$d_{max} = 3$	
			2 steps	3 steps
PROBLEM-1	$\overline{r}_k$	$t_k(\overline{\sigma}_k - C_1)(\overline{\sigma}_k - C_2)$	$max(\overline{r}_k, C)$	-
PROBLEM-2	$\overline{r}_k$	$(\overline{\sigma}_k - \overline{r}_k - 1)\overline{\sigma}_k$	$max(\overline{r}_k, C)$	$t_k(\overline{r}_k - C)$
PROBLEM-3	$\overline{r}_k$	$\sqrt{2\overline{r}_k + ln\overline{r}_k} - ln\overline{r}_k$	$\overline{r}_k(\overline{r}_k - C)$	$t_k(\overline{r}_k - C)$
PROBLEM-4	$\overline{r}_k$	$ln(\overline{r}_k - \overline{\sigma}_k)$	$max(\overline{r}_k, C)$	-
PROBLEM-5	$\overline{r}_k$	$\frac{1}{\overline{\sigma}_k}$	$\overline{\sigma}_k(\overline{\sigma}_k - C)$	-
PROBLEM-6	$\overline{r}_k$	$\overline{r}_k - \overline{\sigma}_k$	$\overline{r}_k(\overline{r}_k - C)$	-
PROBLEM-7	$\overline{r}_k$	$t_k(\overline{r}_k^2 - \overline{\sigma}_k)$	$max(\overline{r}_k, C)$	-
PROBLEM-8	$\overline{r}_k$	$2\overline{r}_k - \overline{\sigma}_k/\overline{r}_k$	$max(\overline{r}_k, C)$	$max(\overline{r}_k, \frac{C}{\overline{r}_k})$
PROBLEM-9	$\overline{r}_k$	$t_k(\overline{r}_k - \overline{\sigma}_k) + \overline{\sigma}_k$	$max(\overline{r}_k, C)$	$t_k(\overline{r}_k - C)$
PROBLEM-10	$\overline{r}_k$	$\overline{r}_k^2 - \overline{r}_k$	$max(\overline{r}_k, C)$	-
PROBLEM-11	$\overline{r}_k$	$\overline{r}_k\overline{\sigma}_k + ln\overline{r}_k$	$max(\overline{r}_k, C)$	$max(\overline{r}_k, C)$
PROBLEM-12	$\overline{r}_k$	$\overline{r}_k(\sqrt{\overline{r}_k} - 1)$	$\overline{r}_k + 1/t_k$	$t_k(\overline{r}_k - C)$
PROBLEM-1–12	$\overline{r}_k$	$\overline{r}_k - \overline{\sigma}_k + ln(\overline{r}_k - \overline{\sigma}_k)$	$\overline{r}_k + 1/t_k$	$\overline{r}_k + C/t_k$

$$\text{FORMULA-1}(H_{t-1}^k, t) = max(\overline{r}_k, C) \quad \text{FORMULA-2}(H_{t-1}^k, t) = t_k(\overline{r}_k - C)$$

$$\text{FORMULA-3}(H_{t-1}^k, t) = \overline{r}_k(\overline{r}_k - C) \quad \text{FORMULA-4}(H_{t-1}^k, t) = \overline{\sigma}_k(\overline{\sigma}_k - C)$$

$$\text{FORMULA-5}(H_{t-1}^k, t) = \overline{r}_k + C/t_k$$

In addition to these five formulas, we consider two additional formulas that we manually derived from $d_{max} = 2$ solution to PROBLEM-9 and PROBLEM-7:

$$\text{FORMULA-6}(H_{t-1}^k, t) = t_k(\overline{r}_k - C\overline{\sigma}_k) \quad \text{FORMULA-7}(H_{t-1}^k, t) = t_k(\overline{r}_k^2 - C\overline{\sigma}_k)$$

4.3   Evaluation of the Discovered Ranking Formulas

Policies Used for Comparison. We have compared our discovered policies against various policies that have been proposed in the literature. These are the ϵ_n-GREEDY policy [9] as described in [2], the policies introduced by [2]: UCB1, UCB1-BERNOULLI, UCB1-NORMAL and UCB2, and the policy UCB-V proposed by [1]. UCB1-BERNOULLI and UCB1-NORMAL are parameter-free policies respectively designed for bandit problems with Bernoulli distributions and for problems with Normal distributions.

Results with Optimized Constants. Table 2 reports the losses achieved by our discovered policies when their constants are optimized as well as the losses of the set of policies used for comparison. The parameters of the latter policies have also been optimized for generating these results, except for UCB2 which is used with a value for its parameter α equal to 0.001, as suggested in [2]. As we can see, five out of the seven formulas discovered outperform all the policies used for comparison. FORMULA-2 seems to be the best since it outperforms all the other

Table 2. Loss of all policies on the 12 discrete bandit problems with horizon $T = 1000$. The scores in bold indicate learned or discovered policies that outperform all baseline policies.

Policy	2-Bernoulli			10-Bernoulli			2-Gaussian			2-Tr. Gaussian		
	1	2	3	4	5	6	7	8	9	10	11	12
Baseline policies												
UCB1	4.3	7.5	9.8	25.0	34.7	49.2	6.2	10.9	10.9	17.6	23.3	22.8
UCB1-BERNOULLI	4.4	9.1	9.8	39.1	57.0	60.9	6.2	10.9	10.9	34.9	31.2	30.8
UCB1-NORMAL	47.7	35.4	30.6	248.8	90.6	85.6	36.0	30.7	30.7	73.3	40.5	39.2
UCB2(0.001)	6.9	10.1	11.4	61.6	67.0	67.8	7.4	11.4	11.4	39.3	31.5	30.7
UCB-V	4.6	5.8	9.5	26.3	34.1	50.8	5.3	10.0	10.0	14.4	21.4	22.0
ϵ_n-GREEDY	5.9	7.5	12.8	32.9	37.5	54.7	8.2	11.2	11.2	20.0	23.6	21.5
Discovered policies												
GREEDY	31.9	23.1	36.6	69.4	46.1	71.6	57.2	38.1	38.1	94.9	44.8	41.2
FORMULA-1	**1.6**	**3.3**	**7.2**	**14.8**	**26.7**	**43.7**	**2.1**	**8.0**	**8.0**	**7.5**	**18.2**	**19.3**
FORMULA-2	**1.1**	**1.9**	**5.7**	**10.2**	**21.0**	**38.3**	**1.9**	**4.9**	**4.9**	**8.2**	**14.4**	**15.1**
FORMULA-3	4.8	11.2	16.6	28.6	46.1	59.4	9.6	17.0	17.0	27.5	27.3	26.9
FORMULA-4	14.6	12.0	47.3	60.4	46.1	88.3	139.8	46.6	46.6	91.8	42.4	45.9
FORMULA-5	**2.2**	**3.8**	**7.0**	**18.7**	**28.6**	**45.1**	**3.6**	**7.5**	**7.5**	**10.1**	**17.4**	**18.1**
FORMULA-6	**1.1**	**2.0**	**5.3**	**10.6**	**20.8**	**38.1**	**3.3**	**11.1**	**7.8**	**10.5**	**20.7**	**21.4**
FORMULA-7	**1.2**	**2.0**	**5.6**	**10.7**	**19.6**	**39.0**	**2.4**	**7.1**	**5.6**	**7.8**	**17.7**	**17.7**

policies on 9 among the 16 test problems. Another formula that behaves very well on all the problems is FORMULA-7. Though FORMULA-3 and FORMULA-4 may work well on very specific problems with the horizon $T = 100$ used for search, these formulas give poor loss values on all problems with horizon $T = 1000$ and are thus not further investigated.

Influence of the Parameters. For a parameter-dependent formula to be good, it is important to be able to identify default values for its parameters that will lead to good results on the bandit problem to be played, given that no or very little a priori knowledge (e.g., only the type of distribution and the number of arms is known) on this problem is available. Figure 2 illustrates the performances of the formulas that appeared in Table 2 to be the most promising, with respect to the evolution of the value of their parameter. We see that FORMULA-2 always leads to a loss which very rapidly degrades when the parameter moves away from its optimal value. However, its optimal parameter value is always located in the interval [*expected reward of the second best arm, expected reward of the optimal arm*]. Actually, we could have anticipated – at least to some extent – this behavior by looking carefully at this formula. Indeed, we see that if the value of its parameter is in this interval, the policy will be more and more likely to play the best arm as the number of plays goes on, while otherwise, it is not necessarily the case.

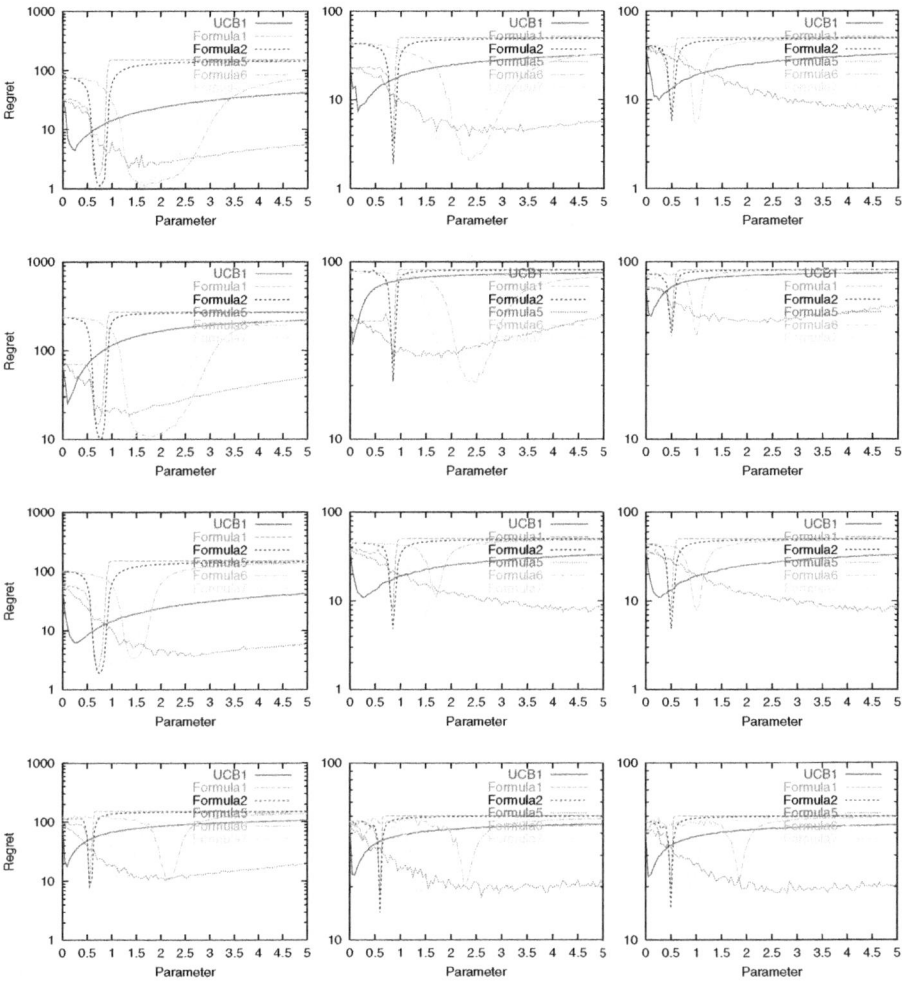

Fig. 2. Loss (average expected regret) as a function of each formula numerical parameter. From top to bottom: 2-Bernoulli problems, 10-Bernoulli problems, 2-Gaussian problems and 2-Truncated Gaussian problems.

For FORMULA-7, which was the second best formula when optimized, the loss degrades less rapidly when moving away from the optimal parameter value. However, it is difficult to explain here the location of its optimum. For FORMULA-5, it is quite easy to find a value for its parameter that works well on each test problem. For example, with its parameter chosen equal to 3.5, performances which are always better than those obtained with the *optimized* version of UCB1 are observed. It is worth noticing that this FORMULA-5 is actually very close to UCB1. Indeed, it also associates to an arm an index value which is the sum of $\bar{r}_k$ plus a term that could also been seen as an upper confidence bias. Contrary to UCB1, this term decreases in t_k rather than in $\sqrt{t_k}$ but, more importantly, it does not increase explicitly with t.

5   Conclusions

We have proposed in this paper an approach for automatically identifying simple arm ranking formulas for playing with multi-armed bandits. This approach has lead us to discover several well performing formulas. Interestingly, some of them share similarities with those previously proposed in the literature. However, several formulas found to perform very well are behaving in a completely different way. For example, they yield indexes that are not guaranteed to converge (in probability) to a monotonic transformation of the expected rewards of the arms.

These newly found formulas are quite intriguing and we believe that it would be worth investigating their analytical properties. In particular, it would be interesting to better understand on which class of bandit problems they may be advantageous to use.

It is likely that one key issue that should be addressed in this analysis is the influence of the parameter values inside these formulas on the performance of the policy. Indeed, those that were found to be the best when their parameter was optimized were also found to be those for which we could not find a parameter value leading to a policy working well on every bandit problem of our test set.

Investigating how our approach for discovering policies could be adapted so as to discover policies that could have satisfactory performances on every problem of a candidate set of problems for default values of their parameters is also an interesting research direction. This could be done for example by scoring the policies in a different way inside our search algorithm, for example by using as score the worst regret that the policy achieves over the set rather than the average regret obtained on a set of problems.

Acknowledgements. Damien Ernst acknowledges the financial support of the Belgian National Fund of Scientific Research (FNRS) of which he is a Research Associate. This paper presents research results of the European excellence network PASCAL2 and of the Belgian Network BIOMAGNET, funded by the Interuniversity Attraction Poles Programme, initiated by the Belgian State, Science Policy Office.

References

1. Audibert, J.-Y., Munos, R., Szepesvári, C.: Tuning Bandit Algorithms in Stochastic Environments. In: Hutter, M., Servedio, R.A., Takimoto, E. (eds.) ALT 2007. LNCS (LNAI), vol. 4754, pp. 150–165. Springer, Heidelberg (2007)
2. Auer, P., Fischer, P., Cesa-Bianchi, N.: Finite-time analysis of the multi-armed bandit problem. Machine Learning 47, 235–256 (2002)
3. Chakrabarti, D., Kumar, R., Radlinski, F., Upfal, E.: Mortal multi-armed bandits. In: Neural Information Processing Systems, pp. 273–280 (2008)

4. Gonzalez, C., Lozano, J.A., Larrañaga, P.: Estimation of Distribution Algorithms, A New Tool for Evolutionary Computation. Kluwer Academic Publishers (2002)
5. Lee, C.S., Wang, M.H., Chaslot, G., Hoock, J.B., Rimmel, A., Teytaud, O., Tsai, S.R., Hsu, S.C., Hong, T.P.: The computational intelligence of MoGo Revealed in Taiwan's computer Go tournaments. IEEE Transactions on Computational Intelligence and AI in Games (2009)
6. Looks, M.: Competent Program Evolution. PhD thesis, Washington University in St. Louis (2006)
7. Maes, F., Wehenkel, L., Ernst, D.: Learning to play k-armed bandit problems (2011) (submitted)
8. Pelikan, M., Mühlenbein, H.: Marginal distributions in evolutionary algorithms. In: Proceedings of the International Conference on Genetic Algorithms Mendel (1998)
9. Sutton, R.S., Barto, A.G.: Reinforcement Learning: An Introduction. MIT Press (1998)

Goal-Directed Online
Learning of Predictive Models

Sylvie C.W. Ong, Yuri Grinberg, and Joelle Pineau

School of Computer Science
McGill University, Montreal, Canada

Abstract. We present an algorithmic approach for integrated learning and planning in predictive representations. The approach extends earlier work on predictive state representations to the case of online exploration, by allowing exploration of the domain to proceed in a goal-directed fashion and thus be more efficient. Our algorithm interleaves online learning of the models, with estimation of the value function. The framework is applicable to a variety of important learning problems, including scenarios such as apprenticeship learning, model customization, and decision-making in non-stationary domains.

Keywords: predictive state representation, online learning, model-based reinforcement learning.

1 Introduction

Reinforcement learning (RL) is the problem of learning how to behave so as to maximize long-term reward through interactions with a dynamic environment. In recent years, there has been an increased interest in the case of *batch* reinforcement learning, whereby the data used for learning how to behave are collected *a priori*. However in many domains, the case of *online* reinforcement learning is more relevant than the batch case. An additional complication arising in the online case is that the agent must explore its environment efficiently. Hence, one of the central issues in online reinforcement learning is the trade-off between exploration and exploitation. Generally speaking, to ensure efficiency in learning, there is a need for approaches that can achieve *goal-directed learning* from interaction [23].

Much of the prior work on reinforcement learning in unknown domains has focused on domains with full state observability [23]. The problem typically becomes much harder in partially observable environments, where the state of a dynamical system is in general a function of the entire history of actions and observations. Methods have been proposed for online learning in the partially observable Markov decision process (POMDP) [17,20]. However these methods require a clear definition of the latent state variables of the system; this is not always easy to formalize in some decision-theoretic systems. For example, in

S. Sanner and M. Hutter (Eds.): EWRL 2011, LNCS 7188, pp. 18–29, 2012.

human-robot interaction tasks, we typically need to handle partial state observability, yet it is often unclear what set of latent variables are appropriate for capturing the human's intention, cognitive state, or attentional level, to name just a few relevant factors.

The predictive state representation (PSR) is a promising approach that forgoes the notion of underlying latent variables, and instead represents system state by the occurrence probabilities of future observation sequences, conditioned on future action sequences [13]. Since the representation is based entirely on observable quantities, in principle, predictive models should be easier to learn from data. PSRs have also been shown to have greater representational power than POMDPs [21]. Prior work on learning predictive models have mostly taken the approach of learning a complete model of the system with the objective of obtaining good predictive accuracy for all possible future behaviors, then planning with the learned model. Thus they either assume the training data is acquired through interaction with the environment using a purely exploratory policy [4], or else side-step the exploration problem by learning from a batch of pre-sampled action-observation trajectories [2].

Such approaches have several practical limitations. First, they assume an arbitrary division between the learning phase and acting phase, which implies that the cost of learning is unimportant. Second, when dealing with large domains (i.e. many observations and actions) and limited data, they tend to suffer from sparse parameter estimation and numerical instability (with some exception, e.g. [2]). Third, they don't naturally extend to handle non-stationary environments, unless a separate mechanism is involved to detect environment changes and initialize a new learning phase.

In this paper, we propose an online algorithm for learning predictive models of dynamical systems through *goal-directed* interaction. To our knowledge, this is the first approach for learning predictive models that integrates learning and planning in a way that explicitly tackles the problem of exploration and exploitation. Our approach is (loosely) based on the actor-critic framework. Model parameters are estimated in an online fashion, interleaved with steps of data gathering and policy optimization. The agent's behavior policy for interacting with the environment is derived from the most recently optimized policy, which in turn is obtained from planning with the current model and the data gathered thus far.

The contributions of this paper are primarily algorithmic and empirical (rather than theoretical). In contrast to previous approaches to predictive model learning, the objective is not to learn the complete model, but to only learn relevant aspects of the model for obtaining a good policy. Compared to prior work, our approach should therefore be more data efficient and thus more easily scalable to large action/observation domains, while having less empirical regret while learning. Since learning and planning are integrated, our approach should also more easily handle non-stationary environments. In preliminary investigations, we demonstrate the performance of our method on a robot navigation domain.

2   Predictive State Representations

A controlled, discrete-time, finite dynamical system generates observations from
a set $\mathcal{O}$ in response to actions from a set $\mathcal{A}$. At each time step t, an agent
interacting with the system takes an action $a_t \in \mathcal{A}$ and receives an observation
$o_t \in \mathcal{O}$. A history, $h = a_1 o_1 a_2 o_2 \cdots a_t o_t$, is a sequence of past actions and
observations, while a test, $\tau = a^1 o^1 a^2 o^2 \cdots a^k o^k$, is a sequence of actions and
observations that may occur in the future. The prediction for test τ given history
h, $p(\tau|h)$, is the conditional probability that the observation sequence in τ occurs,
if the actions specified in τ are executed. The null test ε is the zero-length test.
By definition, the prediction for ε is always 1[13].

Let T be the set of all tests and H the set of all histories. Given an ordering
over H and T, we define the system-dynamics matrix, $\mathcal{D}$, with rows corresponding to histories and columns corresponding to tests. Matrix entries are the pre-
dictions of tests, i.e., $D_{i,j} = p(\tau_j|h_i)$ [21]. Suppose the matrix has a finite number
of linearly independent columns, and we denote the set of tests corresponding to
those columns as the core tests, Q. Then, the prediction of any test is the linear
combination of the predictions of Q. Let $p(Q|h)$ be a vector of predictions for Q
given history h, and m_τ be a vector of weights for test τ, then, for all possible
tests τ, there exists a set of weights m_τ, such that $p(\tau|h) = m_\tau^\mathsf{T} p(Q|h)$. Thus,
the vector $p(Q|h)$ is a sufficient statistic for history, i.e. it represents the system
state, b. The set of all possible states form the PSR state-space, $\mathcal{B}$, and we refer
to the state $b = p(Q|h)$, as the projection of history h on the PSR state-space.
Define M_{ao} as the matrix with rows m_{aoq}^T, for all $q \in Q$. Given a history h, a
new action a, and subsequent observation o, the PSR state vector is updated by,

$$p(Q|hao) = \frac{p(aoQ|h)}{p(ao|h)} = \frac{M_{ao}p(Q|h)}{m_{ao}^\mathsf{T}p(Q|h)} \ . \tag{1}$$

The projection of any history can be calculated by repeatedly applying the above
equation, starting from the prediction vector for the empty history, $p(Q|\phi)$, which
represents the initial system state, m_0.

A PSR model is completely specified by the core tests Q, weight vectors m_{aol},
for all $a \in A$, $o \in O$, $l \in Q \cup \{\varepsilon\}$, and m_0,. We denote these model parameters
collectively as $\mathbf{M}$. The usual approach to model learning is to approximate the
system-dynamics matrix, $\mathcal{D}$, through interactions with the system, i.e. by sampling action-observation trajectories, then use $\mathcal{D}$ to estimate the model $\mathbf{M}$[21].

3   Planning in PSRs

An agent taking action $a \in \mathcal{A}$ in a dynamic environment receives not just an
observation $o \in \mathcal{O}$ but also a reward $r \in \mathcal{R}$. (Rewards are real numbers, however
we assume that there is a finite set $\mathcal{R}$ of possible values for immediate rewards.)
Denoting $R(b, a)$ as the expected immediate reward for action a at system state b,
the goal of planning in PSRs for an infinite horizon problem is to find an optimal
policy that maximizes the expected cumulative rewards $\mathrm{E}[\sum_t \gamma^t R(b_t, a_t)]$. Here,

γ is a discount factor that constrains the expected sum to be finite, $b_t = p(Q|h_t)$ is the system state and a_t the action, at time t.

In general, there are two approaches to planning in PSRs. One approach is to extend POMDP planning algorithms that exploit the fact that V^*, the value function for an optimal policy, can be approximated arbitrarily closely by a piecewise-linear, convex function of the probability distribution (or belief) over latent state variables. This approach represents a value function as a set of linear functions and performs value iteration on the set, generating a new set of linear functions at each iteration. The prediction vector $p(Q|h)$ is the state representation in PSRs that is equivalent to the belief state in POMDPs. It has been shown that for each action $a \in \mathcal{A}$, the expected immediate reward $R(b, a)$ is a linear function of the prediction vector $b = p(Q|h)$ and thus, just as in POMDPs, the optimal value function V^* is also a piecewise linear function of the PSR system state[12]. Previous work has used this idea for planning in PSRs, for both exact[12], as well as approximate planning[10,11,3].

An alternative approach is to extend function approximation algorithms for RL that were originally developed for fully observable systems. The system state is compactly represented by a vector of features instead of the state variable, and a mapping is learned from features to the value function, Q-function, or actions. The features are usually prespecified (for example, using domain knowledge) and the mapping function is usually parametric, so planning reduces to learning the parameter values of the mapping function. The prediction vector $p(Q|h)$ is the state representation in PSRs that is equivalent to the feature vector in fully observable systems. Previous work has extended RL methods such as Q-learning[12], SARSA[18] and Natural Actor-Critic[1], by learning mapping functions as described above, with the PSR system state taking the place of feature vectors. We adopt this latter approach due to its flexibility and scalability.

4   Online Reinforcement Learning with Predictive Models

We now present our algorithmic approach to model-based online RL in Algorithm 1. Just as in the PSR planning methods mentioned in Section 3, the algorithm does planning in the PSR state-space, and aims to learn a good policy as a function of the PSR system state.

4.1   Algorithm Overview

Given a PSR model, $\mathbf{M}$, any history can be projected to the corresponding PSR state-space. To learn such a model, the usual steps mentioned in Section 2 are performed – sampling action-observation trajectories from the environment, approximating the system-dynamics matrix $\mathcal{D}$, and estimating the model $\mathbf{M}$. Unlike previous approaches however, our objective is online RL, not offline planning, thus model learning in our algorithm is goal-directed – trajectories are sampled so as to balance exploration of the environment with exploitation to obtain good rewards. To accomplish this, the algorithm interleaves optimizing

---

Algorithm 1. Goal-directed Online Model Learning

1: **input:** Initial behavior policy, π_{init}^e.
2: **initialization:** $i \leftarrow 1$; $\pi_i^e \leftarrow \pi_{init}^e$.
3: **repeat**
4:　　Sample $\bar{\mathcal{G}}_i$, a set of N_i trajectories from the environment (Algorithm 2, which requires as input, behaviour policy π_i^e and model $\mathbf{M}_{i-1}$.) Let $\mathcal{G}_i$ denote the set of trajectories sampled up to iteration i, i.e. $\mathcal{G}_i = \mathcal{G}_{i-1} \bigcup \bar{\mathcal{G}}_i$.
5:　　From $\mathcal{G}_i$, approximate the system-dynamics matrix $\mathcal{D}_i$, and estimate $\mathbf{M}_i$ (described in Section 4.2).
6:　　Optimize policy π_i with $\mathbf{M}_i$ and $\mathcal{G}_i$ (Algorithm 3).
7:　　$i \leftarrow i + 1$.
8:　　Update the behaviour policy π_i^e using π_{i-1}.
9: **until** stopping conditions are reached.
10: **output:** Predictive model $\mathbf{M}$, and its policy π.

---

a policy π using the current model $\mathbf{M}$ (line 6), and, sampling trajectories (line 4) using a behavior policy π^e derived from π (line 8). The model $\mathbf{M}$ is then updated based on the samples (line 5), and the process repeats. A key feature of the algorithm is that learning the model and planning with the model are integrated. If the process is interrupted or stopping conditions reached (line 9), the algorithm outputs both a model $\mathbf{M}$ and its policy π (line 10). This is in contrast to previous approaches which learn the model and then plan with it, as two separate and distinct processes.

The algorithm is a general framework with different possible choices for behaviour policies, and model learning and planning methods. For concreteness, we describe a particular implementation of these three steps.

The behaviour policy π^e is a stochastic policy which maps the system state b to a probability $\pi^e(b, a)$ for each action $a \in \mathcal{A}$. Policy π^e repeatedly samples trajectories of the form $a_1, o_1, r_1, a_2, o_2, r_2 \ldots$, starting from the initial system state (Algorithm 2). At the first iteration of the algorithm, before any model learning and policy optimization have occurred, π_1^e is set to the initial behaviour policy π_{init}^e (line 2, Algorithm 1), a blind/open-loop policy such as a uniform random action policy. Thus, action sampling in line 6 of Algorithm 2 is not conditioned on the system state, $a_t \sim \pi_i^e(a)$, and the state update step in line 7 is skipped. In subsequent iterations, $i \geq 2$, π_i^e is updated based on π_{i-1}, the policy optimized in the previous iteration (line 8, Algorithm 1). To balance exploration and exploitation, possible choices for π^e include softmax and ϵ-greedy policies. In our experiments, we implemented the latter –

$$\pi_i^e(b, a) = \begin{cases} 1 - \epsilon & \text{for } a = \pi_{i-1}(b) \\ \frac{\epsilon}{|\mathcal{A}| - 1} & \text{for all other actions } a \end{cases}$$

– where $\pi_{i-1}(b)$ is the optimal action at system state b, and ϵ is akin to the exploration constant.

The model learning and policy optimization steps are evoked every N_i sampled trajectories. We describe these steps in the next two sections.

Algorithm 2. Adaptive Goal-Directed Sampling

1: **input:** Behaviour policy π^e, model $\mathbf{M}$, number of trajectories to sample N, and maximum number of time steps T.
2: **repeat**
3:    $t \leftarrow 1$.
4:    Set b_t to the initial system state, m_0.
5:    **repeat**
6:       Sample $a_t \sim \pi_i^e(b_t, a)$. Execute a_t, get observation o_t and reward r_t from the environment.
7:       Update the system state b_{t+1} using equation (1) and model $\mathbf{M}$.
8:       $t \leftarrow t + 1$.
9:    **until** $t = T$.
10: **until** N trajectories have been sampled.
11: **output:** Sampled trajectories $\bar{\mathcal{G}}$.

4.2   Online Model Learning

Our model learning method is based on an online linear compression of the observed trajectories. In particular, we use an online singular value decomposition (SVD) method to update model parameters from streaming data.

To obtain a compact, low dimensional model representation, we learn a transformed PSR (TPSR) model [19]. TPSRs are a generalization of PSRs where the model parameters, m_0', M_{ao}' and m_{ao}', are transformed versions of PSR parameters, m_0, M_{ao} and m_{ao}, respectively, and give equivalent predictions for any tests τ. The TPSR model's compactness arises from its state being a linear combination of predictions for a potentially large set of PSR core tests. The state update proceeds as the PSR update in equation (1), with the transformed parameters substituting the PSR parameters.

Our model learning method relies on recent progress in learning TPSRs using spectral methods, as introduced in [2]. In this approach, instead of learning the system-dynamics matrix $\mathcal{D}$, only the thin SVD decomposition of $\mathcal{D}$ is maintained. New data is incorporated directly into the SVD decomposition using the online update method described in [5]. The computational complexity of the online update in the i-th iteration of our algorithm is $O(s_i \times |\mathbf{M}|^3)$, where $|\mathbf{M}|$ is the TPSR dimension, and s_i is the effective number of updated entries in $\mathcal{D}_i$ due to the additional data incorporated from newly sampled trajectories $\bar{\mathcal{G}}_i$. (In contrast, regular (batch) SVD has complexity $O(n^3)$, for $\mathcal{D}_i$ of size $n \times n$, i.e., the complexity is dependent on the size of $\mathcal{D}$, which could grow very quickly with the size of the action/observation spaces.) Further advantages of this model learning method are that it avoids actually building and storing $\mathcal{D}$, and it is amenable to tracking of non-stationary environments.

4.3   Policy Optimization

Our policy optimization method is based on fitted value iteration [9], an approach which formulates function approximation as a sequence of supervised learning

Algorithm 3. Fitted Q Iteration [7] for Predictive Models

1: **Input:** A set of tuples, $\mathcal{F}$, of the form (b_t, a_t, r_t, b_{t+1}), constructed from a set of sampled trajectories $\mathcal{G}$ and using a predictive model **M**.
2: **Initialization:**
3:    $k \leftarrow 0$.
4:    Set $\hat{Q}_k(b, a)$ to be zero for all $b \in \mathcal{B}$, $a \in \mathcal{A}$.
5: **Iterations:**
6: **repeat**
7:    $k \leftarrow k + 1$.
8:    Build training set, $\mathcal{T} = \{(i^l, o^l), l = 1, \cdots, |\mathcal{F}|\}$, where:
9:        $i^l = (b_t^l, a_t^l)$, and, $o^l = r_t^l + \gamma \max_a \hat{Q}_{k-1}(b_{t+1}^l, a)$.
10:    Apply a regression algorithm to learn the function $\hat{Q}_k(b, a)$ from training set $\mathcal{T}$.
11: **until** stopping conditions are reached.
12: **Output:** policy π, where $\pi(b) \leftarrow \text{argmax}_a \hat{Q}(b, a)$.

---

problems. In particular, our algorithm performs something akin to batch mode Q-learning with function approximation, by applying an ensemble of trees algorithm to a sequence of regression problems. This fitted Q iteration approach has been shown to give good performance in fully observable systems and has good convergence properties [7]. We extend the algorithm to partially observable systems in Algorithm 3.

Given a set of trajectories $\mathcal{G}$ and a predictive model **M**, the goal is to learn $\hat{Q}(b, a)$, an estimation of the expected cumulative reward as a result of executing action a from system state b. Given $\hat{Q}(b, a)$, the optimal action at b is $\pi(b) \leftarrow \text{argmax}_a \hat{Q}(b, a)$. The input to the algorithm is a set of tuples, (b_t, a_t, r_t, b_{t+1}), constructed from one-step system transitions in sampled trajectories of the form, $a_1, o_1, r_1, a_2, o_2, r_2, \cdots$. This requires tracking the system state b_t at every time step: for each trajectory, b_1 is set to the initial system state m_0, then for each subsequent time step, b_{t+1} is updated from b_t, a_t, and o_t, using equation (1). The set of tuples is used for solving a sequence of regression problems. At iteration k in the sequence, the regression function to be learned is the mapping from (b, a) to $\hat{Q}_k(b, a)$. The training set for fitting the regression function has as input the pairs (b_t, a_t), and target outputs $r_t + \gamma \max_a \hat{Q}_{k-1}(b_{t+1}, a)$, where $\hat{Q}_{k-1}(b, a)$ is the regression function learned in the previous iteration $k - 1$, and γ is a discount factor. Repeated applications of the regression algorithm in successive iterations result in increasingly better approximations to the true Q-function.

We use extremely randomized trees (Extra-Trees)[8] as the regression algorithm. This predicts the value of a regression function using an ensemble of trees. The tree building procedure requires specification of the number of trees to build M_{tree}, and the minimal leaf size n_{min} (refer to [7] and [8] for more details).

5   Experimental Results

We evaluated our proposed algorithm on two Gridworld problems of different sizes, depicted in Figure 1. The agent does not sense its position directly, only

(a)   (b)

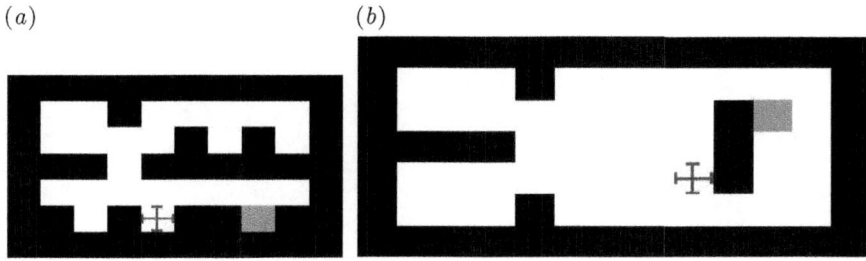

Fig. 1. Maps for Gridworld problems, GW-25 (a) and GW-47 (b). The cross is the starting position and the lightly shaded cell is the goal. Refer to the text for details.

the presence of adjacent walls in the four cardinal directions, hence there are 16 possible observations. In each time step, the agent takes a step in one of the four directions and the action succeeds with probability 0.8, otherwise, the agent moves in one of the perpendicular directions. The agent remains in place if it bumps into a wall. We refer to the two problems as GW-25 and GW-47, their respective environments contain 25 and 47 free cells. The agent receives zero reward everywhere except at the goal where it receives a reward of 1.

Our algorithm samples trajectories of fixed lengths (11 and 13 for GW-25 and GW-47, respectively). In the first iteration, 100 such trajectories are collected for GW-25 and 150 for GW-47, followed by 10 trajectories in each subsequent iteration. The trajectories are used to estimate probabilities of histories of up to length 6 and 8 for GW-25 and GW-47, respectively, and tests of up to length 4. The online TPSR learning process incorporates probability estimations from newly acquired data of each iteration and updates a 3-dimensional state-space TPSR model . We ran the fitted Q algorithm with 100 iterations and a discount factor of 0.95. The extremely randomized trees in the algorithm built single trees across all actions, in an ensemble consisting of 25 trees (M_{tree} parameter) and with minimum leaf size set at 15 data points (n_{min} parameter).

To illustrate the advantages of our approach we compare our algorithm with the usual (regular) approach in PSR literature which learns a model using a purely exploratory policy (uniform random action selection) for trajectory sampling, followed by planning only after model learning is completed [3].

Results of the policy performance with increasing number of sampled trajectories are presented in Figure 2. Plots (a), (b) and (c) compare the policies learned by our algorithm (adaptive) and the regular algorithm on GW-25. We show results for different settings of the exploration constant ϵ in our algorithm. Plot (d) shows performances on GW-47. Policies were evaluated by generating 100 trajectories and logging the proportion of trajectories that reached the goal.

Figure 2 shows that given the same number of sampled trajectories, our algorithm improves its policy faster than the usual PSR model learning and planning approach, and that the performance gap increases with increased problem size (compare plots (a) for GW-25 with (d) for GW-47.) Plots (a) to (c) show that the difference between the results for $\epsilon = 0.2$ and $\epsilon = 0.3$ are minimal. However,

(a) (b)

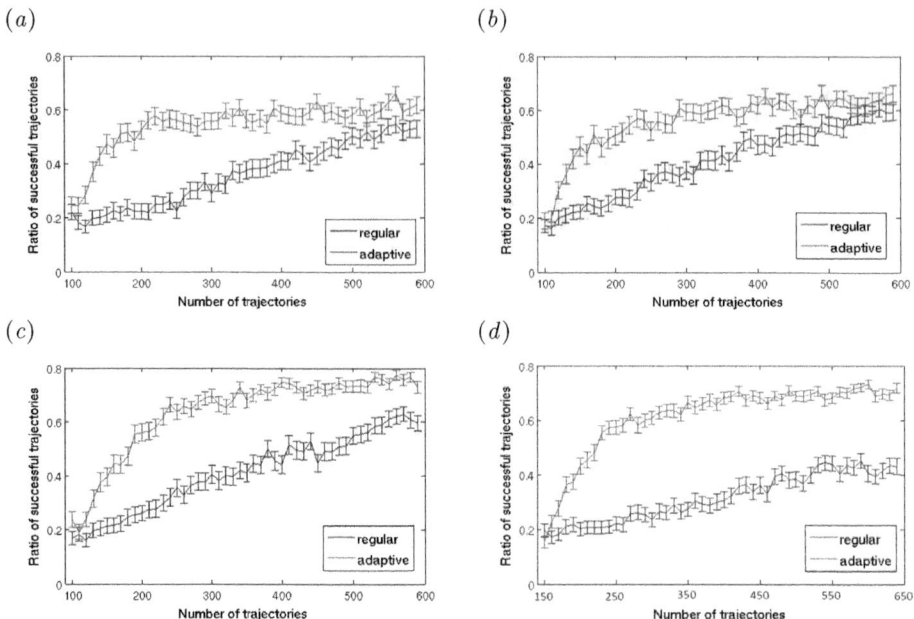

(c) (d)

Fig. 2. Experimental results on the Gridworld problems depicted in Figure 1. Plots (a), (b) and (c) show policy performances on GW-25, with exploration constant ϵ set at 0.2, 0.3, and 0.5 respectively, in our algorithm. Plot (d) shows policy performances on GW-47, with $\epsilon = 0.2$ in our algorithm. All results were averaged over 100 runs. Error bars indicate standard deviations.

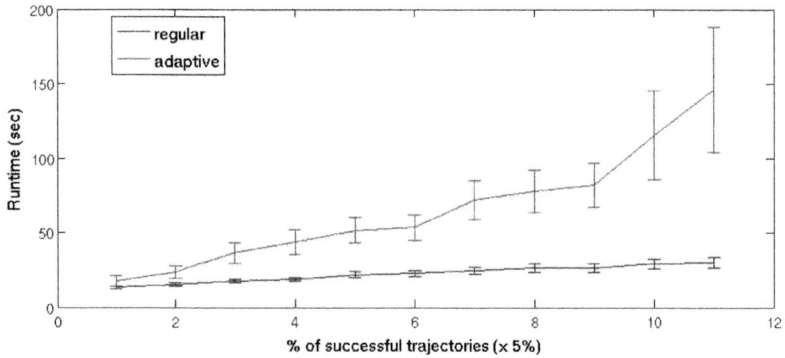

Fig. 3. Run time comparisons on Gridworld problem GW-25, with $\epsilon = 0.5$ in our algorithm. Results were averaged over 40 runs. Error bars indicate standard deviations. The experiments were carried on a PC with two 3.3GHz CPUs and 7GB of RAM, running Ubuntu 10.03 OS.

for $\epsilon = 0.5$ two differences are noticeable. Firstly, for the initial 250 sampled trajectories, the policies for $\epsilon = 0.5$ improve slightly slower as compared to when ϵ is set to lower values. This is expected – with the bigger exploration factor, the optimal policy is invoked less. The second, more interesting observation is that, starting from the 350th sampled trajectory onwards, the performance of these policies continues to grow beyond the maximum achieved by the other policies. A possible explanation is that more exploration results in a more accurate model, which in turn is used to find a better policy. Yet, the majority of exploration is performed "on the way to the goal", leading to significantly better policies compared to those learned from completely random exploration. In a sense, it can be seen as a type of exploration-exploitation tradeoff which is different from the usual case in RL involving estimation of action values. Here, lack of exploration leads to inaccurate estimates of system dynamics, but exploring uniformly at random results in slow convergence to the optimal policy.

In Figure 3 we compare the execution times required for each algorithm to reach the same levels of performance on GW-25 (with $\epsilon = 0.5$ in our algorithm). Our algorithm essentially trades off data efficiency with computational complexity. Data efficiency is often crucial in real world settings – experience is more expensive than computation and the aim is to learn in as few trials or with as little experience as possible. Another important advantage is that our algorithm is 'anytime' – at any stage, the optimal policy (based on the most recently learned model) is immediately available to the agent. This is not the case for the usual approach of planning only after model learning is completed.

6   Related Work

There has been few prior work that address online learning of PSR models. McCracken and Bowling [15], and Aberdeen et al. [1], employ a constrained-gradient algorithm for online PSR learning. Boots and Gordon [2] learn TPSR models online by utilizing low-rank modifications of thin SVDs. However, none of these approaches address the problem of integrated learning and planning. In particular, although the work in [2] uses a similar model learning method as ours, it focuses on learning alone, while in [1], planning proceeds together with learning but the resultant policy from planning does not inform the behaviour policy for model learning.

The idea of not learning a complete model that can predict all possible futures has been explored in approximate PSRs ([6],[22],[24]), where the objective is to learn models that accurately predict only specific quantities of interest. The relevant predictions are specified beforehand and efficient use of sampled trajectories is achieved by aggregating tests and histories according to these predictions. In contrast, our approach achieves data efficiency by biasing sampling towards goal-directed trajectories, with sampling behaviour adaptively and automatically learned from system interaction.

7   Discussion and Conclusion

This work proposes a new integrated learning and planning approach in PSRs. We show in preliminary experiments the advantages over doing learning and planning separately – given the same amount of data for learning, our method was able to learn a model and a value function that gives better policies, and the policy performance gap increases with increased problem size.

Our approach is suitable for domains where the initial system conditions are such that a good policy can be obtained without learning the dynamics of the whole state space or, equivalently, without searching in the whole space of policies. For example, in navigation tasks, the initial system state would be concentrated on a small portion of the environment if for example, the robot is learning how to navigate in a building, starting from a particular room within the building. Another necessary assumption is that the goal is reachable within the limits we impose on the trajectory length, starting from the initial system state. In practice, we make sure that the trajectories are of sufficient length in relation to the path to the goal.

There are some interesting and important learning problems, which are particularly amenable to solution with our approach. In apprenticeship learning, the goal is to learn a policy that is at least as good or better than an expert. We can apply our method to apprenticeship learning in an unknown domain, by deriving the initial behaviour policy from the expert policy, and then iteratively improving upon it. Our method could also be applied to model customization. Given a basic model that is not customized to the current environment, our algorithm iteratively learns a better suited model. Lastly, given the online nature of our algorithm, and the integration of learning and planning, our approach can be extended to learning in non-stationary domains. We aim to investigate the applicability of our method to these classes of problems in our future work.

It would be interesting to compare empirically our algorithm with history-based approaches such as U-Tree [14], Monte-Carlo AIXI [25] and ΦMDPI [16]. These algorithms also integrate learning and planning, while attempting to balance exploration and exploitation. Their approach differs from ours in representing system state with histories rather than predicted probabilities of future experiences. We leave this also to future work.

Acknowledgments. The authors wish to thank Doina Precup (McGill University) for helpful discussions regarding this work. Funding was provided by the National Institutes of Health (grant R21 DA019800) and the NSERC Discovery Grant program.

References

1. Aberdeen, D., Buffet, O., Thomas, O.: Policy-gradients for PSRs and POMDPs. In: AISTATS (2007)
2. Boots, B., Gordon, G.J.: An online spectral learning algorithm for partially observable nonlinear dynamical systems. In: Proceedings AAAI (2011)

3. Boots, B., Siddiqi, S., Gordon, G.: Closing the learning-planning loop with predictive state representations. In: Proceedings of Robotics: Science and Systems (2010)
4. Bowling, M., McCracken, P., James, M., Neufeld, J., Wilkinson, D.: Learning predictive state representations using non-blind policies. In: Proceedings ICML (2006)
5. Brand, M.: Fast low-rank modifications of the thin singular value decomposition. Linear Algebra and its Applications 415, 20–30 (2006)
6. Dinculescu, M., Precup, D.: Approximate predictive representations of partially observable systems. In: Proceedings ICML (2010)
7. Ernst, D., Geurts, P., Wehenkel, L.: Tree-based batch mode reinforcement learning. Journal of Machine Learning (2005)
8. Geurts, P., Ernst, D., Wehenkel, L.: Extremely randomized trees. Machine Learning 63, 3–42 (2006)
9. Gordon, G.J.: Approximate Solutions to Markov Decision Processes. Ph.D. thesis, School of Computer Science, Carnegie Mellon University (1999)
10. Izadi, M.T., Precup, D.: Point-Based Planning for Predictive State Representations. In: Bergler, S. (ed.) Canadian AI. LNCS (LNAI), vol. 5032, pp. 126–137. Springer, Heidelberg (2008)
11. James, M.R., Wessling, T., Vlassis, N.: Improving approximate value iteration using memories and predictive state representations. In: AAAI (2006)
12. James, M.R., Singh, S., Littman, M.L.: Planning with predictive state representations. In: International Conference on Machine Learning and Applications, pp. 304–311 (2004)
13. Littman, M., Sutton, R., Singh, S.: Predictive representations of state. In: Advances in Neural Information Processing Systems, NIPS (2002)
14. McCallum, A.K.: Reinforcement Learning with Selective Perception and Hidden State. Ph.D. thesis, University of Rochester (1996)
15. McCracken, P., Bowling, M.: Online discovery and learning of predictive state representations. In: Neural Information Processing Systems, vol. 18 (2006)
16. Nguyen, P., Sunehag, P., Hutter, M.: Feature reinforcement learning in practice. Tech. rep. (2011)
17. Poupart, P., Vlassis, N.: Model-based bayesian reinforcement learning in partially observable domains. In: Tenth International Symposium on Artificial Intelligence and Mathematics, ISAIM (2008)
18. Rafols, E.J., Ring, M., Sutton, R., Tanner, B.: Using predictive representations to improve generalization in reinforcement learning. In: IJCAI (2005)
19. Rosencrantz, M., Gordon, G.J., Thrun, S.: Learning low dimensional predictive representations. In: Proceedings ICML (2004)
20. Ross, S., Pineau, J., Chaib-draa, B., Kreitmann, P.: A Bayesian approach for learning and planning in partially observable Markov decision processes. Journal of Machine Learning Research 12, 1655–1696 (2011)
21. Singh, S., James, M., Rudary, M.: Predictive state representations: A new theory for modeling dynamical systems. In: Proceedings UAI (2004)
22. Soni, V., Singh, S.: Abstraction in predictive state representations. In: AAAI (2007)
23. Sutton, R.S., Barto, A.G.: Reinforcement Learning: An Introduction. The MIT Press (1998)
24. Talvitie, E., Singh, S.: Simple local models for complex dynamical systems. In: Advances in Neural Information Processing Systems, NIPS (2008)
25. Veness, J., Ng, K.S., Hutter, M., Uther, W., Silver, D.: A Monte-Carlo AIXI approximation. JAIR 40, 95–142 (2011)

Gradient Based Algorithms with Loss Functions and Kernels for Improved On-Policy Control

Matthew Robards and Peter Sunehag*

Australian National University Nicta

Abstract. We introduce and empirically evaluate two novel online gradient-based reinforcement learning algorithms with function approximation – one model based, and the other model free. These algorithms come with the possibility of having non-squared loss functions which is novel in reinforcement learning, and seems to come with empirical advantages. We further extend a previous gradient based algorithm to the case of full control, by using generalized policy iteration. Theoretical properties of these algorithms are studied in a companion paper.

1 Introduction

The ability to learn online is an important trait for reinforcement learning (RL) algorithms. In recent times there have been significant focus on using stochastic gradient descent to enable online reinforcement learning, [1], [4], [9], [10] with significant theoretical advances.

In this paper we introduce two new algorithms – one model based, and one model free – for the control case of online reinforcement learning with function approximation with a novel extension to (possibly) non-squared loss functions. We show both algorithms to perform well on the mountain-car and cart-pole domains with the model free algorithm being faster with complexity $O(d)$ compared to the model-based algorithm which is $O(d^2)$. We further show, however, that our model based algorithm is the only algorithm (in our comparison set) which performs well in domains where observational noise makes the process non-Markov. We show this on the noisy observation mountain-car, and noisy observation cart-pole domains. We extend the gradient temporal difference (GTD) algorithms of [4], [9] with generalized policy iteration in order to make them comparable in the case of full control (note that we do not try to claim that this upholds the theory of this algorithm, however it does perform well empirically). Our algorithms are presented with a novel theoretical analysis in the companion paper [6].

The novel contributions of this paper can be summed up as follows:

- Introduction and evaluation of non-squared loss functions in reinforcement learning,

---

* This author was supported by ARC grant DP0988049.

S. Sanner and M. Hutter (Eds.): EWRL 2011, LNCS 7188, pp. 30–41, 2012.

- Introduction of a model free and a model based algorithm which performs well on standard reinforcement learning benchmark problems,
- Introduction of a model based algorithm which performs well under noisy observations (even when the noise makes the process non-Markov),
- Extension to full control and evaluation of GTD algorithms.

1.1 Related Work

The classical methods SARSA(λ) and Q-estimation were introduced [8] in the tabular reinforcement learning setting and were heuristically extended to linear function approximation. These methods, however, are known to have convergence issues in this more general setting. To address this issue the first gradient descent method was then introduced aiming to minimize the Bellman error [1]. A further series of gradient descent methods were proposed for temporal difference learning and optimal control [4], [9], [10]. Unlike the present work, however, these methods minimize a *projected* Bellman error. Further these algorithms have little empirical analysis for optimal control. In this paper we give an empirical analysis of this gradient-based algorithm along with ours.

1.2 Outline

Section 2 presents Markov decision processes and the two key background algorithms. We then present our algorithms residual gradient Q-estimation in Section 3, and model based Q-estimation in Section 4. In Section 5 we provide our experimental analysis, and we conclude in Section 6.

2 Preliminaries and Stochastic Gradient TD Algorithms

2.1 Markov Decision Processes

We assume a (finite, countably infinite, or even continuous) Markov decision process (MDP) [5] given by the tuple $\{\mathcal{S}, \mathcal{A}, T, R, \gamma\}$. Here, we have states $s \in \mathcal{S}$, actions $a \in \mathcal{A}$, $T : \mathcal{S} \times \mathcal{A} \times \mathcal{S} \to [0,1]$ is a transition function with $T(s,a,s')$ defining the probability of transitioning from state s to s' after executing action a. $R : \mathcal{S} \times \mathcal{A} \times \mathcal{S} \to \mathbb{R}$ is a reward function where $r_t = R(s_t, a_t, s_{t+1})$ (with $R(s_t, a_t) = \mathbb{E}[R(s_t, a_t, \cdot)]$) is the (possibly stochastic) reward received for time t after observing the transition from state s_t to s_{t+1} on action a_t. Finally, $0 \leq \gamma < 1$ is a discount factor.

In this paper we will aim to find optimal state action value functions $Q : \mathcal{S} \times \mathcal{A} \to \mathbb{R}$ which best approximate the infinite discounted expected reward sum through $Q(s_t, a_t) = \sum_{i=0}^{\infty} \gamma^i \mathbb{E}[r_{t+i}]$ where we take action a_t and follow the optimal policy afterwards. This is often solved using the Bellman equation

$$Q(s_t, a_t) = \mathbb{E}[r_t | s_t, a_t] + \gamma \max_{a_{t+1}} \mathbb{E}[Q(s_{t+1}, a_{t+1}) | s_t, a_t]. \tag{1}$$

2.2   Residual Gradient TD

The original stochastic gradient descent (SGD) based reinforcement learning algorithm [1] was introduced for a state value function approximator of the form $V_t(s) = \langle w_t, \phi(s) \rangle$ for $w, \phi \in \mathbb{R}^d$ with the objective

$$w^* = \arg\min_w \mathbb{E}\left[\left(V^\pi(s) - R(s,a) - \gamma V^\pi(s')\right)^2\right] \qquad (2)$$

and stochastic gradient descent updates of the form

$$w_{t+1} = w_t - \eta_t(\phi(s_t) - \gamma\phi(s_{t+1}))(V_t(s_t) - r_t - \gamma V_t(s_{t+1})). \qquad (3)$$

2.3   GTD and Derivatives

More recently a series of algorithms [4], [9], [10] has been introduced which minimizes the MSPBE given by

$$\text{MSPBE}(w) = \left\| Q_w - \Pi T^{\pi_w} Q_w \right\|_\mu^2 \qquad (4)$$

where $Q_w \in \mathbb{R}^{|S \times A|}$ is a vector in which each element corresponds to Q evaluated at a unique state action pair with respect to the current parameters w. Furthermore μ is the stationary distribution from which samples are drawn and Π is the projection operator projecting $\Pi F = \arg\min_{f \in \text{span}(\Phi)} \|F - f\|_\mu$. The authors managed to provide some breakthrough guarantees for this algorithm.

3   Residual Gradient Q-Estimation

In this section we introduce our online model-free *residual gradient Q-estimation (RGQ)* algorithm with linear function approximation. That is, we represent our value function as $Q(s,a) = \langle \theta, \Phi(s,a) \rangle$, where $\theta, \Phi(s,a) \in \mathbb{R}^d$, and θ corresponds to the w in the previous section.

We evaluate the performance of a sequence of value function parameter estimates $\{\theta_t\}_{t=1...m}$ through

$$\sum_{t=1}^{m} C_t(\theta_t), \qquad (5)$$

for a given sequence of functions $\{C_t\}_{t=1,...,m}$, where Q_t is chosen before time observing C_t. Here we have a sequence of regularized cost functions $\{C_t\}_{t=1,...,m}$ given by

$$C_t(\theta_t) = l\left(Q_t(s_t, a_t) - \gamma Q_t(s_{t+1}, a_{t+1}), r_t\right) + \frac{\lambda}{2}\|\theta_t\|_2^2, \qquad (6)$$

and $l : \mathbb{R} \times \mathbb{R} \to \mathbb{R}^+$ is some convex loss function.

3.1   Linear Updates

We optimize our objective through gradient descent with the following update

$$\theta_{t+1} = \Pi\left[(1 - \eta_t\lambda)\theta_t - \eta_t\nabla_{\theta_t}l\left(Q_t(s_t, a_t) - \gamma Q_t(s_{t+1}, a_{t+1}), r_t\right)\right]. \qquad (7)$$

One can easily plug various loss function into this update equation. For example, for the ϵ-insensitive loss $l(x, y) = \max(0, |x - y| - \epsilon)$, we get

$$\theta_{t+1} = \Pi\Bigg[(1 - \eta_t\lambda)\theta_t$$
$$- \eta_t\left(\phi(s_t, a_t) - \gamma\phi(s_{t+1}, a_{t+1})\right)sign\left(Q_t(s_t, a_t) - r_t - \gamma Q_t(s_{t+1}, a_{t+1})\right)\Bigg], \tag{8}$$

when $\left|Q_t(s_t, a_t) - r_t - \gamma Q_t(s_{t+1}, a_{t+1})\right| - \epsilon > 0$, and no update otherwise.

3.2   Reproducing Kernel Hilbert Space Updates

We now assume our value function is of a non-linear form. Rather, we assume that it is linear in a reproducing kernel Hilbert space (RKHS) $\mathcal{H}_k$ (which is induced by the kernel $k : \mathcal{S} \times \mathcal{A} \times \mathcal{S} \times \mathcal{A} \to \mathbb{R}$) such that $Q_t(s, a) = \langle Q_t(\cdot), k((s, a), \cdot)\rangle_{\mathcal{H}_k}$ where,

$$Q_t(\cdot) = \sum_i \alpha_i k((s_i, a_i), \cdot). \tag{9}$$

Our gradient descent updates then become

$$Q_{t+1} = \Pi\left[(1 - \eta_t\lambda)Q_t - \eta_t\nabla_{Q_t}l\left(Q_t(s_t, a_t) - \gamma Q(s_{t+1}, a_{t+1}), r_t\right)\right]$$
$$= \Pi\left[(1 - \eta_t\lambda)Q_t - \eta_t\frac{\partial l\left(Q_t(s_t, a_t) - \gamma Q(s_{t+1}, a_{t+1}), r_t\right)}{\partial\left(Q_t(s_t, a_t) - \gamma Q(s_{t+1}, a_{t+1})\right)}\right.$$
$$\left. \left(k((s_t, a_t), \cdot) - \gamma k((s_{t+1}, a_{t+1}), \cdot)\right)\right], \tag{10}$$

where we have used $\nabla_f l(f(x), y) = \frac{\partial l(f(x), y)}{\partial f(x)}\frac{\partial f(x)}{\partial f} = \frac{\partial l(f(x), y)}{\partial f(x)}\frac{\partial\langle f, k(x, \cdot)\rangle_{\mathcal{H}_k}}{\partial f} = \frac{\partial l(f(x), y)}{\partial f(x)}k(x, \cdot)$. One can easily plug various loss function into this update equation.

4   Model Based Q-Estimation

In this section we introduce our online model based algorithm. We begin by formulating the objective, and discussing the way in which we optimize linear function approximators for the given objective. We then propose a kernel version of the algorithm.

4.1   The Objective

We aim to find a state-action value function Q^* such that

$$Q^*(\phi(s), a) \approx \mathbb{E}_{r|s,a}[r|s, a] + \gamma Q^*(\mathbb{E}_{s'|s,a}[\phi(s')|s, a], a'). \tag{11}$$

We attempt to learn the value function by learning two models (one for the expected reward M^R, and one for the expected next state's features M^T) and perform function approximation on the overall Q function. As an approximation to this we set the objective to minimize the expected loss $Q^* = \arg\min_Q C(Q)$ where

$$C(Q) := \sum_{t=1}^{m} L\Big(Q(\phi(s_t), a_t), M_t^R(s_t, a_t) + \gamma Q\big(M_t^T(s_t, a_t), a_{t+1}\big) \Big) + \frac{\lambda}{2} \|Q\|_2^2. \tag{12}$$

where $L : \mathbb{R} \times \mathbb{R} \to \mathbb{R}^+$ is a general loss function, for example squared or ϵ-insensitive loss given by

$$L(x, y) = max(0, |x - y| - \epsilon). \tag{13}$$

Furthermore, $M^R(s, a)$ is the predictor of the rewards, and $M^T(s, a)$ is a predictor of the next state's features. We will save the details of the optimization of Q for later. Firstly, we must discuss our transition and reward models.

4.2   Optimizing the Approximators

The optimization problem is threefold: firstly estimating the expectation over possible next state's features, secondly estimating the expected reward and thirdly, with this knowledge, optimizing the overall objective function.

Estimating the Next State Expectation. We begin by discussing how to approximate the next state. For this, we introduce the function approximator $M^T : \mathcal{S} \times \mathcal{A} \to \mathbb{R}^{d_S}$, represented by $M^T(s, a) = W\Phi(s, a)$ where W is a parameter matrix of dimension $d_S \times d$. We measure the performance of a sequence of $\{M_t^T\}_{t=1,\ldots,m}$ through

$$\sum_{t=1}^{m} \left[L\Big(M_t^T(s_t, a_t), \phi(s_{t+1}) \Big) + \lambda^T \|W_t\|_2^2 \right] \tag{14}$$

where λ^T is a regularization parameter associated with the reward model, $\|W\|_2^2 = \sum_{i,j} W_{i,j}^2$, and $L : \mathbb{R}^{d_S} \times \mathbb{R}^{d_S} \to \mathbb{R}$ is the summed component-wise loss. For example, the squared loss in this situation is given by $L(X, Y) = \|X - Y\|_2^2$. We perform gradient descent through

$$W_{t+1} = (1 - \eta_t^T \lambda^T)W_t - \eta_t^R \Big(M^T(s_t, a_t) - \phi(s_{t+1}) \Big) \Phi(s, a)^\top. \tag{15}$$

Estimating the Expected Reward. To estimate the reward we use yet another function approximator. This is a function $M^R : \mathcal{S} \times \mathcal{A} \to \mathbb{R}$ which is represented by $M^R(s,a) = \langle \psi, \Phi(s,a) \rangle$. We measure the performance of a sequence $\{M_t^R\}_{t=1,\ldots,m}$ through

$$\sum_{t=1}^{m} \left[l\left(M_t^R(s_t, a_t) - r_t \right) + \lambda^R \|\psi_t\|_2^2 \right] \tag{16}$$

where λ^R is another regularization parameter. We perform a gradient descent of the form

$$\psi_{t+1} = (1 - \eta_t^R \lambda^R)\psi_t - \eta_t^R \Phi(s_t, a_t)\left(M_t^R(s_t, a_t) - r_t \right). \tag{17}$$

4.3   Reproducing Kernel Hilbert Space Extension

We now briefly describe the reproducing kernel hilbert space (RKHS) version of the model approximations for our algorithm

Approximating the State Transition Model. We have $d_\mathcal{S}$ independent kernel regressors $(M_i^T : \mathcal{S} \times \mathcal{A} \to \mathbb{R})$, each of which estimates a single feature $(\phi_i(s'))$ of the next state s'. We want to find sequences of these functions which minimize

$$\sum_{t=1}^{m} \left[\left(M_{t,i}^T(s_t, a_t) - \phi_i(s_{t+1}) \right)^2 + \frac{\lambda}{2} \|M_{t,i}^T\|_{\mathcal{H}_k}^2 \right] \tag{18}$$

such that

$$\|M_i^T\|_{\mathcal{H}_k} \le C^T \quad \forall i \tag{19}$$

where each M_i^T is represented by

$$M_i^T(\cdot) = \sum_j \beta_j^i k((s_j, a_j), \cdot). \tag{20}$$

Note that each $M_{t,i}^T$ must be chosen before s_{t+1} is observed. We solve this using the gradient descent algorithm previously introduced getting

$$M_{t+1,i}^T = \Pi^T \left((1 - \eta_t^T \lambda^T)M_{t,i}^T - 2\eta_t^T \left(M_{t,i}^T(s_t, a_t) - \phi_i(s_{t+1}) \right) k((s_t, a_t), \cdot) \right) \tag{21}$$

where η_t^T is the step size parameter associated with the transition model at time t, and Π^T is a projection operator projecting it's argument onto the set $\{f | f \in \mathcal{H}_k, \|f\|_{\mathcal{H}_k} \le C^T\}$.

Approximating the Reward Model. To estimate the expected reward we use a function approximator of the form

$$M^R = \sum_i \alpha_i^R k((s_i, a_i), \cdot) \tag{22}$$

with the objective of finding the sequence $\{M_t^R\}_{t=1,...,m}$ which minimizes

$$\sum_{t=1}^{m} \left[\left(M_t^R(s_t, a_t) - r_t\right)^2 + \frac{\lambda^R}{2} \|M_t^R\|_2^2 \right] \tag{23}$$

where λ^R is the regularization parameter associated with the reward model. Note that each M_t^R must be chosen before r_t is observed. We then optimize using gradient descent:

$$M_{t+1}^R = \Pi^R \left((1 - \eta_t^R \lambda^R) M_t^R - \eta_t^R \left(M_t^R(s_t, a_t) - r_t \right) k((s_t, a_t, \cdot) \right) \tag{24}$$

where Π^R is a projection operator which projects its argument onto the set $\{f, \|f\| \leq C^R\}$.

4.4   Optimizing the Value Function

We optimize for the problem given by the cost function C_t at time t, and this is given by

$$C_t(\theta_t) = l(Q_t(s_t, a_t) - \gamma Q_t(M_t^T(s_t, a_t), a_{t+1}), M_t^R(s_t, a_t)) + \frac{\lambda}{2} \|\theta_t\|_2^2, \tag{25}$$

by performing gradient descent steps through

$$\theta_{t+1} = \Pi \left(\theta_t - \eta_t \nabla_{\theta_t} C_t(\theta_t) \right). \tag{26}$$

where Π is a projection back onto a convex set.

Table 1. The optimal parameters found for each algorithm on the mountain-car (MC), cart-pole (CP), and double-cart-pole (DCP)

	RGQ	KRGQ	GTD		MBFA					SARSA(λ)		kernel SARSA		
	η	η	η_w	η_θ	η	η^R	η^T	λ^R	λ^T	η	λ	λ	η	ϵ
MC	0.1	0.1	10^{-4}	0.01	1.0	0.1	0.1	10^{-4}	10^{-3}	0.1	0.7	0.6	0.5	$5 \cdot 10^{-5}$
CP	0.01	0.1	10^{-4}	0.01	0.001	0.001	0.001	10^{-4}	10^{-3}	0.005	0.6	0.7	0.1	$5 \cdot 10^{-7}$
DCP	-	0.1	-	-	-	-	-	-	-	-	-	0.7	0.1	$5 \cdot 10^{-7}$

5   Experimental Results

We compare RGQ, kernel-RGQ (KRGQ) and MBQ with SARSA, kernel SARSA (λ) [7] and GTD on five experimental domains. Namely we test on mountain-car, cart-pole, noisy-observation mountain-car, noisy-observation cart-pole, and deterministic double-cart-pole. The noisy-observation variants of these environments involve adding noise to the agents observations which renders the environment non-Markov. We use these examples to show that MBQ is especially useful in such situations as it is robust under noisy observations.

Mountain-Car. involves driving an underpowered car up a steep hill. We use the state/action space, and transition/reward dynamics as defined in [8]. The agent receives reward -1 at each step until success when reward 1 is received. We capped episodes in the mountain-car problem to 1000 time steps. The car was initialized to a standing start (zero velocity) at a random place on the hill in each episode. We also evaluate under noisy observations.

Pole Balancing (Cart-Pole). requires the agent to balance a pole hinged atop a cart by sliding the cart along a frictionless track [8]. Rewards are zero except for -1, which is received upon failure (if the cart reaches the end of the track, or the pole exceeds an angle of ± 12 degrees). At the beginning of each episode we drew the initial pole angle uniformly from $[\pm 2]$ degrees. Further, we cap episode lengths at 1000 time steps. We report not only on this MDP, but also on a noisy version in which we add noise to the agents observations.

Double Pole Balancing (Double-Cart-Pole). This domain is similar in nature to the cart-pole domain, but much harder to solve. In the double-cart-pole domain, the agent is required to slide a cart along a frictionless track in order to balance *two* poles (of different length) hinged atop the cart. Note that this is different to the slightly easier version of the double-cart-pole task in which one pole is hinged atop another. In our version, both poles are hinged directly to the cart. The failure conditions are the same as for the cart-pole task, but the poles were initialised deterministically to vertical in each episode. The state space for this task is 6-dimensional.

5.1   Setup

We used a process of trial-and-error to find the best setup for each algorithm on the domains. The best setup found for each is given in Table 1 on the mountain-car and cart-pole domains. This setup was then used without any additional tuning on the noisy-observation variants.

For each of the linear algorithms we used RBF-coding as described in [8], with Gaussian RBFs. In all cases the widths of these RBFs was either 5% or 10% of the diameter of a normalized state-action space. In the kernel method we used a Gaussian kernel with width 10% of the state space.

5.2   Discussion of Results

On the cart-pole task we see that all our algorithms work quite well. Whilst MBQ is one of the slowest to learn, we also note that it is the only algorithm which performs well under noisy observations on this task (Figure 3). In fact we illustrate in Figure 2 that the performance of MBQ is unchanged when noise is added to the observations in the cart-pole task. The reason this occurs is because the noise in the observations is "averaged out" by our model estimates. On the mountain-car task we, again, see that all algorithms perform quite well, however in this case MBQ is more sensitive to noisy observations. In fact the most consistent algorithm on the mountain car task was GTD.

It was the case in all comparisons of our algorithms that the tradition (in RL) squared loss was not the best performer, with the exception of KRGQ with squared loss on mountain car (in which case all loss functions performed well). Hence we conclude that non-squared loss functions are beneficial in reinforcement learning, and deserve further study.

KRGQ performed competitively on both non-noisy problems, and hence is a good contribution to the kernel reinforcement learning literature. It is, however, extremely slow and prone to overfitting since it naively adds samples. This method requires further work (like that of [3], [2], [7]) to incorporate some more intelligence in the way it adds samples. The ϵ-insensitive loss provides some sparsification, however is a very simplistic method for doing so.

Finally we show in Figure (7) that KRGQ performs well on the double cart pole task. We highlight that the performance is much better than random. Note that this is the only algorithm from our selection which we could get to perform better than random.

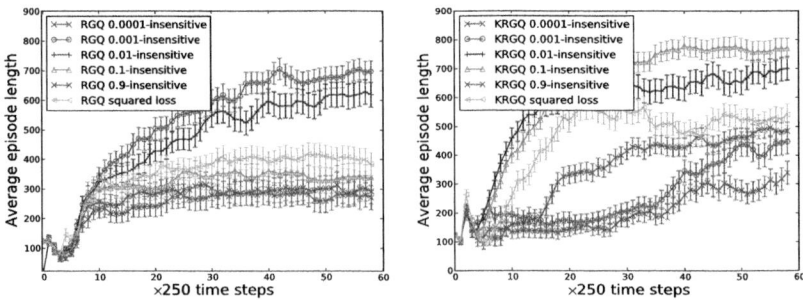

Fig. 1. RGQ (left) and KRGQ (right) applied to the cart-pole task. We show various losses.

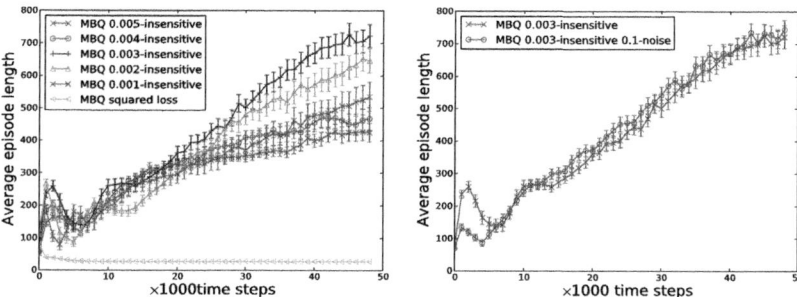

Fig. 2. Comparison of our model based algorithm (MBQ) with various losses (left). The best setup found for MBQ on noise free cart-pole, and the same setup applied to noisy observation cart-pole. Note that there is no deterioration in the performance of MBQ under noisy observations.

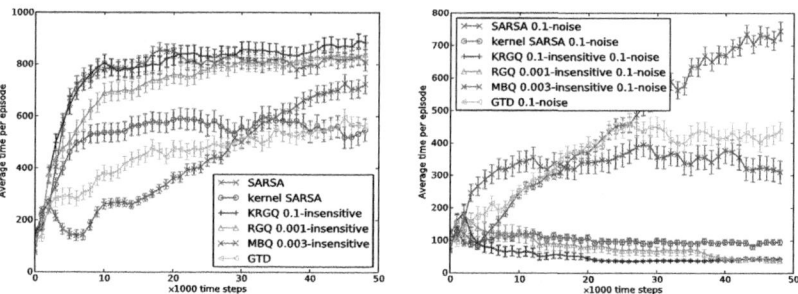

Fig. 3. The best setup found for all comparison algorithms applied to the noise free cart-pole task (left), and the noisy-observation cart-pole task (right)

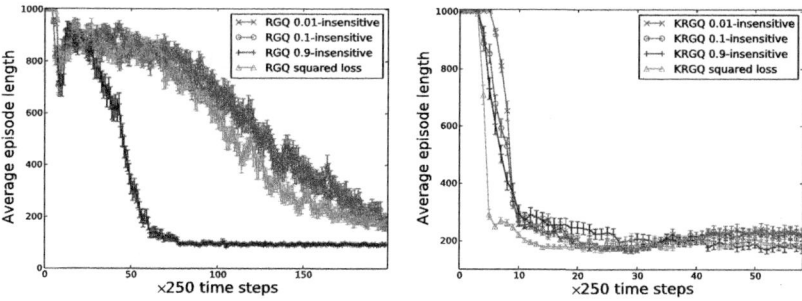

Fig. 4. RGQ (left) and KRGQ (right) applied to the mountain-car task. We show various losses.

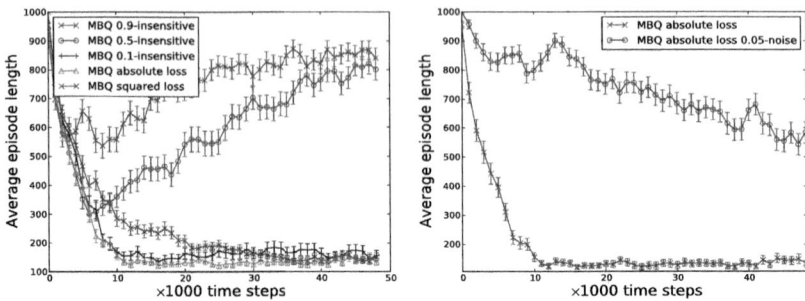

Fig. 5. MBQ applied to the mountain-car task showing various ϵ-insensitive losses as well as the squared loss (left), and the optimal setup found for MBQ on the mountain-car task compared with the same setup applied to the noisy-observation mountain-car task (right)

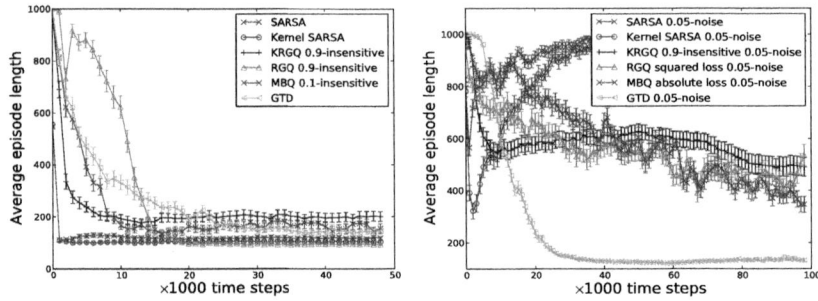

Fig. 6. The best setup found for all comparison algorithms applied to the noise free mountain-car task (left), and the noisy-observation mountain-car task (right)

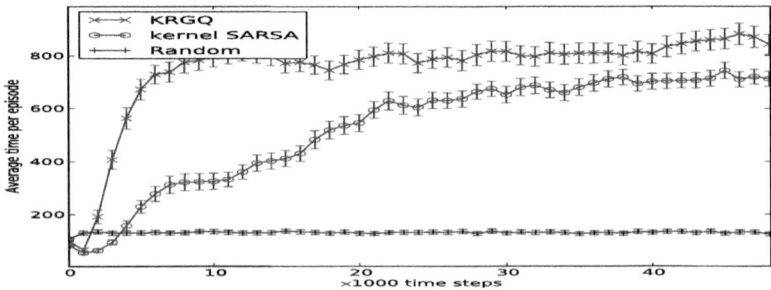

Fig. 7. The best setup found for KRGQ and kernel SARSA (λ) applied to the double-cart-pole task

6   Conclusion

We have introduced two algorithms – one model based and one model free –
with kernel variants (although we do not test the kernel variant of the model
based algorithm due to its complexity), which come with theoretical guarantees
[6]. These algorithms allow for the introduction of non-squared loss functions in
reinforcement learning which we empirically justify by showing that the squared
loss is rarely the best choice.. We further extend the GTD algorithm to the case
of full control using generalized policy iteration.

References

1. Baird, L., Moore, A.: Gradient descent for general reinforcement learning. In: Neural Information Processing Systems, vol. 11, pp. 968–974. MIT Press (1998)
2. Engel, Y., Mannor, S., Meir, R.: Reinforcement learning with Gaussian processes.
   In: 22nd International Conference on Machine Learning (ICML 2005), Bonn, Germany, pp. 201–208 (2005)
3. Engel, Y., Mannor, S., Meir, R.: Bayes meets bellman: The gaussian process approach to temporal difference learning. In: Proc. of the 20th International Conference on Machine Learning, pp. 154–161 (2003)
4. Maei, H., Szepesvri, C., Bhatnagar, S., Sutton, R.: Toward off-policy learning control with function approximation. In: Proceedings of the 27th International Conference on Machine Learning (2010)
5. Puterman, M.L.: Markov Decision Processes: Discrete Stochastic Dynamic Programming. Wiley, New York (1994)
6. Robards, M., Sunehag, P.: Online convex reinforcement learning. In: Submitted to
   9th EWRL (2011)
7. Robards, M., Sunehag, P., Sanner, S., Marthi, B.: Sparse Kernel-SARSA(λ) with
   an Eligibility Trace. In: Gunopulos, D., Hofmann, T., Malerba, D., Vazirgiannis,
   M. (eds.) ECML PKDD 2011. LNCS, vol. 6913, pp. 1–17. Springer, Heidelberg
   (2011)
8. Sutton, R., Barto, A.: Reinforcement Learning. The MIT Press (1998)
9. Sutton, R., Maei, H., Precup, D., Bhatnagar, S., Silver, D., Szepesvri, C., Wiewiora,
   E.: Fast gradient-descent methods for temporal-difference learning with linear function approximation. In: Proceedings of the 26th International Conference on Machine Learning (2009)
10. Sutton, R., Szepesvári, C., Maei, H.: A convergent o(n) temporal-difference al-
    gorithm for off-policy learning with linear function approximation. In: NIPS, pp.
    1609–1616. MIT Press (2008)

Active Learning of MDP Models

Mauricio Araya-López, Olivier Buffet,
Vincent Thomas, and François Charpillet

Nancy Université / INRIA
LORIA – Campus Scientifique – BP 239
54506 Vandoeuvre-lès-Nancy Cedex – France
`firstname.lastname@loria.fr`

Abstract. We consider the active learning problem of inferring the transition model of a Markov Decision Process by acting and observing transitions. This is particularly useful when no reward function is *a priori* defined. Our proposal is to cast the active learning task as a utility maximization problem using Bayesian reinforcement learning with belief-dependent rewards. After presenting three possible performance criteria, we derive from them the belief-dependent rewards to be used in the decision-making process. As computing the optimal Bayesian value function is intractable for large horizons, we use a simple algorithm to approximately solve this optimization problem. Despite the sub-optimality of this technique, we show experimentally that our proposal is efficient in a number of domains.

1 Introduction

Learning in Markov Decision Processes (MDPs) is usually seen as a means to maximize the total utility for a given problem. Nevertheless, learning the transition model of an MDP independently of the utility function—if it exists—can be a very useful task in some domains. For example, this can be used for learning the transition model in a batch process, where in a first stage we are interested in choosing the good actions for optimizing the information gathering process, and afterwards in a second stage, we are interested in earning rewards [5]. Moreover, there are some cases where we do not have access to the utility function, such as models for simulations or model refinement, where we want only to learn a good model, no matter which task the model will be used for.

Learning stochastic MDP models is an easy task if an exploration policy is given. In this case, the history of transitions can be seen as the data, and the problem of finding the optimal parameters for the selected distribution over the models can be solved by using likelihood maximization.

Here, we are concerned with actively learning the transition model, what raises a control problem. This amounts to finding a policy that explores optimally an MDP in order to acquire the best distribution over possible models. This differs from active supervised learning, where any sample can be queried at any time. In our setting, a sequence of actions is needed to reach a specific state from which

S. Sanner and M. Hutter (Eds.): EWRL 2011, LNCS 7188, pp. 42–53, 2012.

to acquire a new sample. This is a complex problem since one has to reason on an imperfect model in order to improve that same model.

To our knowledge, there is not much research in active learning for arbitrary stochastic MDP models [19]. Indeed, one of the few works in this domain is about learning Dynamic Bayesian Networks (DBNs) for representing factored MDPs [9], where the authors conclude that actively learning transition models is a challenging problem and new techniques are needed to address this problem properly.

Our proposal is to use the Bayesian Reinforcement Learning (BRL) machinery with belief-dependent rewards to solve this active learning task. First we cast the learning problem as a utility maximization problem by using rewards that depend on the belief that is being monitored. Then, we define some performance criteria to measure the quality of distributions produced by different policies. Using these criteria, we derive the belief-dependent rewards that will be used to find exploration policies. Due to the intractability of computing the optimal Bayesian value function, we solve this problem sub-optimally by using a simple myopic technique called EXPLOIT.

Belief-dependent rewards have been used as heuristic methods for POMDPs. For example, in coastal navigation [15], convergence is sped up by using reward shaping based on an information-based criterion. Moreover, POMDPs with belief-dependent rewards have been recently studied in [1], where classical POMDP algorithms are applied to this type of rewards only with little modifications. Unfortunately, these techniques cannot be applied, for the same reason why POMDP algorithms are not used for standard BRL: the special type of beliefs used are not suitable for these algorithms.

The remainder of the paper is organized as follows. In Section 2 we give a short review of BRL and the algorithms that have been proposed so far. Then, in Section 3 we introduce the methodology used to solve this active learning problem as a BRL problem with belief-dependent rewards, including the selected performance criteria and their respective derived rewards. In Section 4 we present the results of several experiments over some MDP models taken from the state of the art. Finally, in Section 5 we close with the conclusion and future work.

2   Background

2.1   Reinforcement Learning

A *Markov Decision Process* (MDP) [13] is defined by a tuple $\langle S, A, T, R \rangle$ where S is a finite set of system *states*, A is a finite set of possible *actions*, the *transition* function T indicates the probability to transition from one state s to another s' when some action a is performed: $T(s, a, s') = Pr(s'|s, a)$, and $R(s, a, s')$ is the instant scalar *reward* obtained during this transition. Reinforcement Learning (RL) [18] is the problem of finding an optimal decision policy—a mapping $\pi : S \rightarrow A$—when the model (T and R) is unknown but while interacting with the system. A typical performance criterion is the expected return

$$V_H^\pi(s) = E_\pi \left[\sum_{t=0}^{H} R(S_t, A_t, S_{t+1}) | S_0 = s \right],$$

where H is the planning horizon[1]. Under an optimal policy, this state value function verifies the Bellman optimality equation [3] (for all $s \in \mathcal{S}$):

$$V_H^*(s) = \max_{a \in \mathcal{A}} \sum_{s' \in \mathcal{S}} T(s, a, s') \left[R(s, a, s') + V_{H-1}^*(s') \right],$$

and computing this optimal value function allows to derive an optimal policy by behaving in a greedy manner, i.e., by picking actions in $\arg\max_{a \in \mathcal{A}} Q^*(s, a)$, where a state-action value function Q_π is defined as

$$Q_H^\pi(s, a) = \sum_{s' \in \mathcal{S}} T(s, a, s') \left[R(s, a, s') + V_{H-1}^\pi(s') \right].$$

Typical RL algorithms either (i) directly estimate the optimal state-action value function Q^* (model-free RL), or (ii) learn T and R to compute V^* or Q^* (model-based RL). Yet, in both cases, a major difficulty is to pick actions so as to make a compromise between exploiting the current knowledge and exploring to acquire more knowledge.

2.2   Model-Based Bayesian Reinforcement Learning

We consider here *model-based Bayesian Reinforcement Learning* [17] (BRL), i.e., model-based RL where the knowledge about the model—now a random vector **b**—is represented using a—generally structured—probability distribution over possible transition models. An initial distribution $Pr(\boldsymbol{b}_0)$ has to be specified, which is then updated using the Bayes rule after each new transition (s, a, s'):

$$Pr(\boldsymbol{b}_{t+1} | \boldsymbol{b}_0, h_{t+1}) = Pr(\boldsymbol{b}_{t+1} | \boldsymbol{b}_t, s_t, a_t, s^{t+1}) Pr(\boldsymbol{b}_t | \boldsymbol{b}_0, h_t),$$

where $h_t = s_0, a_0, \cdots, s_{t-1}, a_{t-1}, s_t$ is the state-action history until t. This random variable is usually known as the *belief* over the model, and therefore defines a belief-MDP with an infinite state space. Solving optimally this belief-MDP is intractable due to the increasing complexity along the planning horizon, but formulating the reinforcement learning problem using a Bayesian approach provides a sound way of dealing with the exploration-exploitation dilemma. Even though POMDP algorithms deal with belief-MDPs, one cannot directly benefit from classical POMDP algorithms because of the particular type of belief space. Other—offline or online—approximate approaches have therefore been introduced, allowing in a number of cases to prove theoretical properties. Several approaches and approximation techniques have been proposed for BRL and, as presented in [2], most approaches belong to one of the three following classes,

---

[1] For simplicity, in this paper we are focused on undiscounted finite horizon problems. However, a similar technique can be applied to the discounted infinite horizon case.

from the simplest to the most complex: *undirected approaches, myopic approaches* and *belief-lookahead approaches*.

Undirected approaches do not consider the uncertainty about the model to select the next action, and therefore do not reason about the possible gain of information. They often rely on picking random actions occasionally, e.g., using an ϵ-greedy or softmax exploration strategy, the computed Q-value being based on the average model. These algorithms usually converge to the optimal value function in the limit, but with no guarantee on the convergence speed.

Myopic approaches select the next action so as to reduce the uncertainty about the model. Some of them solve the current average MDP with an added exploration reward which favors transitions with lesser known models, as in R-MAX [4], BEB [10], or with variance based rewards [16]. Another approach, used in BOSS [2], is to solve, when the model has changed sufficiently, an optimistic estimate of the true MDP (obtained by merging multiple sampled models). For some of these algorithms, such as BOSS and BEB, there is a guarantee that, with high probability, the value function is close to some optimum (Bayesian or PAC-MDP) after a given number of samples. Yet, they may stop exploring after some time, preventing the convergence to the optimal value function.

Belief-lookahead approaches aim at optimally compromising between exploration and exploitation. One can indeed [7] reformulate BRL as the problem of solving a POMDP where the current state is a pair $\omega = (s, b)$, where s is the current observable state of the BRL and b is the belief on the hidden model. Each transition (s, a, s') is an observation that provides new information to include to b. BEETLE [12] is one of the few such approaches, one reason for their rarity being their overwhelming computational requirements. Other option is to develop the tree of beliefs and use branch-and-bound to prune the infinite expansion [6].

2.3 Chosen Family of Probability Distributions

Among various possible representations for the belief b over the model, we use here one independent Dirichlet distribution per state-action pair. We denote one of them at time t by its sufficient statistic: a positive integer vector $\boldsymbol{\theta}_{s,a}^t$ where $\boldsymbol{\theta}_{s,a}^t(s')$ is the number of observations of transition (s, a, s'), including $\boldsymbol{\theta}_{s,a}^0(s')$ a *priori* observations. The complete belief state of the system can thus be written $\omega = (s, \boldsymbol{\theta})$, where $\boldsymbol{\theta} = \{\boldsymbol{\theta}_{s,a}, \forall s, a\}$. This is called a Belief-Augmented MDP (BAMDP), a special kind of belief-MDP where the belief-state is factored into the system state and the model. A triplet (s, a, s') leads to a Bayesian update of the model, $\boldsymbol{\theta}'$ differing from $\boldsymbol{\theta}$ only in that $\boldsymbol{\theta}'_{s,a}(s') = \boldsymbol{\theta}_{s,a}(s') + 1$. Moreover, due to the properties of Dirichlet distributions, the transition function of the BAMDP $T(\omega, a, \omega')$ is given by: $Pr(\omega'|\omega, a) = \frac{\boldsymbol{\theta}_{s,a}(s')}{\|\boldsymbol{\theta}_{s,a}\|_1}$.

To sum up, BRL transforms the problem of facing an unknown model into that of making decisions when the state contains unobserved system parameters. The problem of finding a sound compromise between exploration and exploitation becomes that of solving a BAMDP given an initial set of belief-parameters $\boldsymbol{\theta}^0$.

3   Active Learning of MDP Models Using BRL

In this paper we are interested in learning the hidden model of an MDP by observing state transitions online, under an active exploration strategy. From this arises a decision-making problem, where the best policy of actions must be selected in order to optimize the learning process. For a given policy, the learning process is straightforward using the Bayesian setting, because the likelihood maximization for the joint Dirichlet distribution corresponds to the sequential Bayes update of the $\boldsymbol{\theta}$ parameter described in Section 2.2. Therefore, the optimal policy will depend on the criterion used to compare two joint Dirichlet distributions produced from different policies. Among the possible options, we have selected three performance criteria that will be described in Section 3.2.

For finding the optimal policy, we can cast the active learning task as a BRL problem with *belief-dependent rewards*, where these rewards can be derived from the performance criterion. In other words, we extend the classical definition of BRL to rewards that depend on the $\boldsymbol{\theta}$ parameter, where the Bellman equation takes the form:

$$V_H(\boldsymbol{\theta}, s) = \max_a \left[\sum_{s'} Pr(s'|s, a, \boldsymbol{\theta})(\rho(\boldsymbol{\theta}, a, \boldsymbol{\theta}') + V_{H-1}(\boldsymbol{\theta}', s')) \right], \qquad (1)$$

with $\boldsymbol{\theta}'$ the posterior parameter vector after the Bayes update with (s, a, s'), and $\rho(\boldsymbol{\theta}, a, \boldsymbol{\theta}') = \rho(s, a, s', \boldsymbol{\theta})$ the immediate belief-dependent reward. Within this formulation the problem of actively learning MDP models can be optimally solved using a dynamic programming technique. Yet, as in normal BRL, computing the exact value function is intractable because of the large branching factor of the tree expansion, so approximation techniques will be needed to address this problem.

3.1   Derived Rewards

In order to define the belief-dependent rewards needed for Equation 1, we will use the analytical expressions of the performance criteria to derive analytical expressions for immediate reward functions. As our problem has a finite horizon, one can say that the performance criteria could be used directly as a reward in the final step, whereas the rewards would be zero for the rest of the steps. Yet, this type of reward functions forces to develop the complete tree expansion in order to obtain non-zero rewards, which turns out to be extremely expensive for large horizons.

Therefore, we need a way of defining substantial immediate rewards at each step. As rewards are defined over a transition (s, a, s') and the current belief-parameter $\boldsymbol{\theta}$, we will use the Bayesian update for computing the performance difference between the current belief and the posterior belief. This is a standard reward shaping technique, which allows decomposing a potential function—here the performance criteria—in immediate rewards for each step, with the property of preserving the optimality of the generated policy [11].

Let $D(\boldsymbol{\theta}^t, \boldsymbol{\theta}^0)$ be a distance between the initial prior and the posterior parameters after t Bayes updates such that maximizing this distance amounts to maximizing the gain of information. From this we define the derived reward as follows,

$$\rho(s, a, s', \boldsymbol{\theta}^t) = D(\boldsymbol{\theta}^{t+1}, \boldsymbol{\theta}^0) - D(\boldsymbol{\theta}^t, \boldsymbol{\theta}^0),$$

where $\boldsymbol{\theta}^{t+1}$ is the set of parameters after the transition (s, a, s') from $\boldsymbol{\theta}^t$. Please recall that the Bayes update only modifies one state-action pair per update, meaning that only one state-action pair component of our distribution will change per update. This provides important simplifications in computing the performance of one transition.

In some cases, the performance criterion complies with the triangular *equality*, meaning that the derived rewards can be simply computed as

$$\rho(s, a, s', \boldsymbol{\theta}^t) = D(\boldsymbol{\theta}^t, \boldsymbol{\theta}^{t+1}), \tag{2}$$

removing the dependency from the initial prior.

3.2 Performance Criteria

Assuming that the real model is unknown, we must define a way to compare two distributions produced by different policies. As there is no *a priori* best criterion, we have selected three information-based criteria under the constraint that they can be computed analytically: the *variance difference*, the *entropy difference* and the *Bhattacharyya distance*.

Variance Difference. The first criterion is based on the simple intuition that we are seeking those policies that produce low-variance distributions. The variance for the multivariate distribution over the models corresponds to a heavily sparse matrix of size $|\mathcal{S}|^2|\mathcal{A}| \times |\mathcal{S}|^2|\mathcal{A}|$, but here we will consider that the sum of marginal variances (the trace of the matrix) is enough as a metric. The variance of the i-th element of a Dirichlet distribution is given by $\sigma^2(X_i|\boldsymbol{\alpha}) = \frac{\alpha_i(\|\boldsymbol{\alpha}\|_1 - \alpha_i)}{\|\boldsymbol{\alpha}\|_1^2(\|\boldsymbol{\alpha}\|_1 + 1)}$. Then, we define the *variance difference* as follows,

$$D_V(\boldsymbol{\theta}^t, \boldsymbol{\theta}^0) = \sum_{s,a} \sum_{s'} (\sigma^2(X_{s'}|\boldsymbol{\theta}^0_{s,a}) - \sigma^2(X_{s'}|\boldsymbol{\theta}^t_{s,a})).$$

Entropy Difference. An other common measure for probability distributions is the entropy, which measures the uncertainty of a random variable. Computing the uncertainty of beliefs seems to be a natural way of quantifying the quality of distributions. The entropy of a multivariate random variable distributed as a Dirichlet distribution with parameters $\boldsymbol{\alpha}$ is given by $H(\boldsymbol{\alpha}) = \log(B(\boldsymbol{\alpha})) + (\|\boldsymbol{\alpha}\|_1 - N)\psi(\|\boldsymbol{\alpha}\|_1) - \sum_{j=1}^{N}((\alpha_j - 1)\psi(\alpha_j)$, where $B(\cdot)$ is the generalized beta function, N is the dimensionality of the vector $\boldsymbol{\alpha}$, and $\psi(\cdot)$ is the digamma function. Then, we define the *entropy difference* as

$$D_H(\boldsymbol{\theta}^t, \boldsymbol{\theta}^0) = \sum_{s,a} (H(\boldsymbol{\theta}^0_{s,a}) - H(\boldsymbol{\theta}^t_{s,a})).$$

Bhattacharyya Distance. The two measures described above attempt to quantify how much information the distribution contains. In this context, information theory provides several notions of information such as Chernoff's, Shannon's, Fisher's or Kolmogorov's. As stated in [14], the last three are inappropriate for Dirichlet distributions, because an analytical solution of the integral does not exist or due to the non-existence for some specific values. Using the result in [14] for computing the Chernoff information between two Dirichlet distributions $C_\lambda(\alpha, \alpha')$, and fixing the parameter λ to $1/2$, we obtain the *Bhattacharyya distance* between two belief-states as follows,

$$D_B(\theta^t, \theta^0) = \sum_{s,a} -\log\left(\frac{B(\frac{\theta^t_{s,a}}{2} + \frac{\theta^0_{s,a}}{2})}{\sqrt{B(\theta^t_{s,a})B(\theta^0_{s,a})}}\right).$$

3.3   From Criteria to Rewards

It can be shown that the first two criteria presented in Section 3.2 comply with the triangular *equality*, so we can use Equation 2 to compute their respective derived reward functions. Even though this is not always true for the *Bhattacharyya* distance, we will also use Equation 2 for simplicity, knowing that we are not preserving optimality for this specific case.

Therefore, after some trivial but tedious algebra, we can define the *variance*, *entropy* and *Bhattacharyya* instant rewards. For presenting these expressions we use the helper variables $x = \theta_{s,a}(s')$ and $y = \|\theta_{s,a}\|_1$ for all the rewards, and also $z = \|\theta_{s,a}\|_2^2$ for the variance reward:

$$\rho_V(s, a, s', \theta) = \frac{1}{y+1} - \frac{z}{y^2(y+1)} + \frac{2x - y + z}{(y+1)^2(y+2)},$$

$$\rho_H(s, a, s', \theta) = \log\left(\frac{y}{x}\right) + \frac{|S|+1}{y} - \sum_{j=x}^{y}\frac{1}{j},$$

$$\rho_B(s, a, s', \theta) = \log\left[\frac{\Gamma(x)\sqrt{x}}{\Gamma(x+1/2)}\right] - \log\left[\frac{\Gamma(y)\sqrt{y}}{\Gamma(y+1/2)}\right].$$

Also, we would like to introduce a simple reward function, motivated by the exploration bonus of BEB [10], which only focuses on the difference of information from one state-action pair to another. This *state-action count* reward can be simply defined as

$$\rho_S(s, a, s', \theta) = \frac{1}{\|\theta_{s,a}\|_1} = \frac{1}{y}.$$

This reward is easier to compute than the other three, but preserves the same principle of quantifying the information gain of a Bayes update. In fact, this reward function optimizes the performance criterion

$$D_S(\theta^t, \theta^0) = \sum_{s,a}(\psi(\|\theta^t_{s,a}\|) - \psi(\|\theta^0_{s,a}\|)),$$

which turns out to be a quantity appearing in both the *entropy difference* and the *Bhattacharyya distance*.

A key property of the rewards presented above is that they tend to zero for an infinite belief evolution, meaning that there is no more to learn at this stage. Specifically, the two last rewards ρ_B and ρ_S are always positive and decreasing functions with the belief evolution, while the two first ρ_V and ρ_H can have negative values, but their absolute values are always decreasing, all of them converging to zero in the limit.

3.4   Solving BRL with Belief-Dependent Rewards

It is clear that the algorithms that have been introduced in Section 2.2 will require some modifications to work with belief-dependent rewards. For example, BEETLE uses a polynomial representation of the value function, where rewards are scalar factors multiplying monomials. In our setup, rewards will be functions multiplying monomials, which makes offline planning even more complex.

EXPLOIT, which is one of the simplest online algorithms for BRL, consists in solving the MDP corresponding to the current average model or, in other words, iterating over the Bayesian value function without performing the Bayes update. Then, EXPLOIT executes the best action for this simple MDP, updates its belief by observing the arrival state, and starts again by solving the MDP for the new average model. Please note that EXPLOIT does not guarantee converging to an optimal policy (as in ϵ-greedy Q-learning).

For the belief-dependent reward scenario, EXPLOIT takes the form

$$V_H(\boldsymbol{\theta}, s) = \max_a \left[\sum_{s'} Pr(s'|s, a, \boldsymbol{\theta}) \left(\rho(s, a, s', \boldsymbol{\theta}) + V_{H-1}(\boldsymbol{\theta}, s') \right) \right],$$

where the MDP to solve is defined by a transition model $T(s, a, s') = \boldsymbol{\theta}_{s,a}(s')/\|\boldsymbol{\theta}_{s,a}\|$ and reward function $R(s, a, s') = \rho(s, a, s', \boldsymbol{\theta})$.

If we consider now the belief-dependent rewards presented in Section 3.1, solving this MDP will not provide a lower bound—neither an upper bound—of the Bayesian value function, but only an approximation. This simple algorithm will exploit the current information about the belief to explore parts of the model where the information is still weak, and despite the sub-optimality of the approximation, this algorithm exhibits a fair exploration behavior where all state-action pairs are visited infinitely often in the limit.

4   Experiments

4.1   Experimental Setup

We have selected three small MDP models to learn, taken from the BRL state of the art: the classic *Bandit* [8] problem, and the *Chain* and *Loop* problems [17]. For each problem, we have tested several transition models, varying the

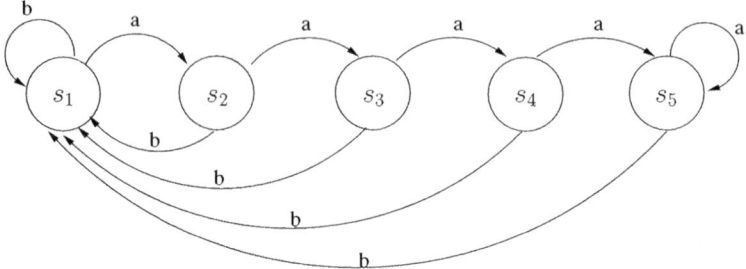

Fig. 1. Chain MDP model used for the experiments (without transition probabilities)

transition probabilities of the arcs to test our technique under different scenarios. However, due to space constraints we will present here only the results for the *Chain* problem, with a short discussion of the other problems.

In the 5-state **Chain MDP** (Figure 1), every state is connected to the state s_1 by taking action b and every state s_i is connected to the next state s_{i+1} with action a, except state s_5 that is connected to itself. In the *normal* case, the agent can "slip" at each time step with a probability of 0.2, performing the opposite action as intended. We have also tested the *deterministic* case when the probability of slipping is zero. Finally, we have tested a *mixed* version, where the probability of slipping is zero for action b, but 0.5 for action a. These two variations decrease the chances to arrive "by luck" to the states at the right of the chain.

The initial conditions of the problem are that we always start at state s_1 with the uniform distribution over the transition models as an initial prior. Other priors can be used—such as informative or structured priors—but for simplicity we will consider for this paper only the uniform one.

For evaluating the behavior of EXPLOIT, we have considered two other policies, namely the RANDOM and GREEDY policies. The RANDOM policy chooses homogeneously a random action at each step, while the GREEDY policy selects the action with largest expected immediate reward.

The three performance criteria have been tested, where the rewards for EXPLOIT and GREEDY are the respective derived rewards of Section 3.1 ρ_V, ρ_H and ρ_B depending on the evaluated performance criterion. Also, we have tested the *state-action count* reward ρ_S for each criterion and experiment.

For approximately solving the finite horizon MDPs within EXPLOIT, we have truncated the planning horizon to a small value $h = \min(2|\mathcal{S}|, H)$. Increasing h will provide better results, but to be similar in execution time with RANDOM and GREEDY, we have selected this small horizon.

4.2   Results

We have tested all the strategies on each problem for the first 100 to 1000 steps, and for each performance criterion. Figure 2 shows the average performance over 100 trials plotted with their respective 95% confidence interval for the *Chain* problem.

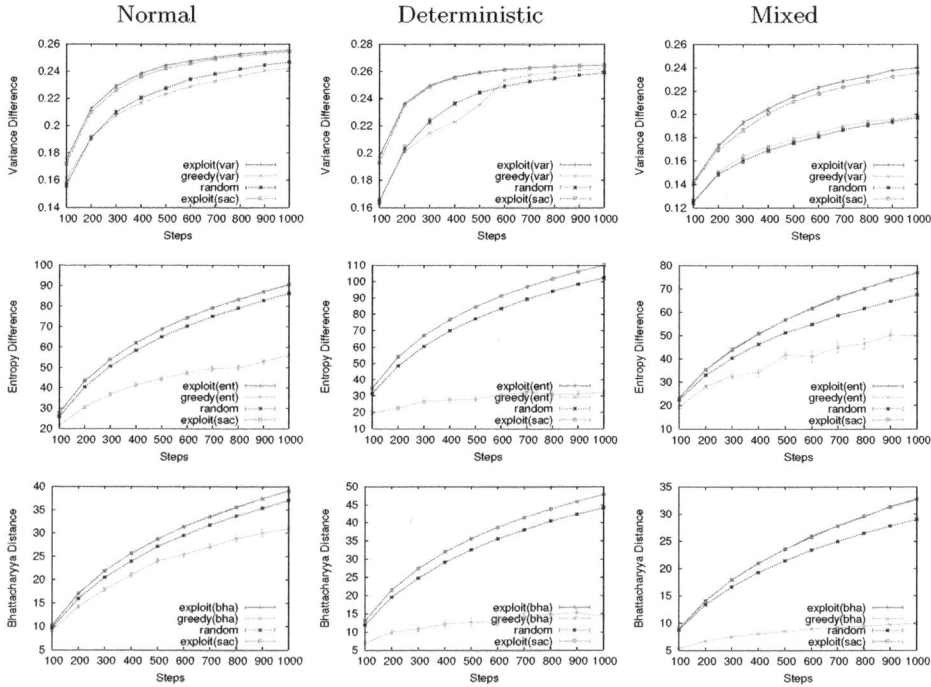

Fig. 2. Mean performance over 100 trials versus time steps, for the *Chain* problem with different models and the three performance criteria. For each plot, the RANDOM strategy (in blue ∗), the GREEDY strategy (in green ×), and the EXPLOIT algorithm with the derived reward (in red +) and with the *state-action count* reward (in magenta □) are shown.

Even though *Chain* is a small problem, it is interesting because an intelligent exploration behavior is needed to learn the model. This can be seen along the three criteria, where the GREEDY policy behaves very poorly because a lookahead strategy is needed to arrive to some states. Even though the RANDOM strategy behaves fairly well, information-based strategies outperform this simple technique in all three criteria: for a desired solution quality, several hundred steps more are needed by RANDOM to achieve the same results. Even more, for the variance criterion, it seems to be very difficult for RANDOM to achieve the same quality in the short-term.

An other important result is that the *state-action count* reward behaves similarly well as the respective derived rewards. This means that the derived rewards can be replaced by this computationally cheap reward with no much performance loss. Indeed, performing a cross-experiment for the rewards and criteria shows that all information-based rewards behave similarly well for all criteria.

For the *Bandit* problem there is not much difference between the algorithms, and the behavior is the same through the different criteria. This is because the optimal policy for exploring a fully connected MDP corresponds to fairly selecting the available actions, which resembles the RANDOM policy.

On the *Loop* problem the results resemble to the ones presented for the *Chain*: information-based rewards outperform the two simple algorithms. Yet, the improvements of our approach compared to RANDOM are milder than in the *Chain*, because a simple exploration strategy is sufficient for this problem.

5   Conclusion and Future Work

We have presented a sound and original way of modeling the problem of actively learning a stochastic MDP model with arbitrary dynamics, by casting the problem as a BRL utility maximization problem with belief-dependent rewards. To that end, we have employed three performance criteria that are commonly used to compare probability distributions, namely the variance, the entropy and the Bhattacharyya distance. For each performance criterion, we have derived a belief-dependent reward such that, in the first two cases, the accumulated rewards correspond exactly to the performance criterion. Also, we have presented a simple reward function—the *state-action count*—based on previous work on normal BRL. Even though the formulation—in theory—allows solving the problem optimally, the intractability of computing the optimal Bayesian value function leads to using sub-optimal algorithms such as EXPLOIT. Our experiments show that this simple technique produces better results than selecting actions randomly, which is the baseline technique for exploring unknown MDP models. Also, our experiments show that there is no need for selecting complex derived rewards (at least for EXPLOIT) in order to obtain good results; the *state-action count* behaves nearly as well as the theoretically derived rewards.

However, this work leaves several open questions about the possibilities of modeling the active learning of MDP models using BRL. For instance, deepening the analysis on the relationship between the *state-action count* criterion and the other criteria might help defining a more advanced reward shaping technique to derive computationally inexpensive rewards. Also, exploring other techniques used for normal BRL could improve the results as they approach the optimal solution. For example, belief-lookahead techniques can be used for refining the myopic policies proposed here, or maybe some other myopic technique could produce better results.

A natural extension is to encode prior knowledge as a structured prior, such as with DBNs in [9], or with parameter tying in [12]. This would dramatically speed up the learning process by making a much more efficient use of data, while not involving major modifications in the solution techniques.

References

1. Araya-López, M., Buffet, O., Thomas, V., Charpillet, F.: A POMDP extension with belief-dependent rewards. In: Advances in Neural Information Processing Systems 23 (NIPS 2010) (2010)
2. Asmuth, J., Li, L., Littman, M., Nouri, A., Wingate, D.: A Bayesian sampling approach to exploration in reinforcement learning. In: Proceedings of the Twenty-Fifth Conference on Uncertainty in Artificial Intelligence (UAI 2009) (2009)

3. Bellman, R.: The theory of dynamic programming. Bull. Amer. Math. Soc. 60, 503–516 (1954)
4. Brafman, R., Tennenholtz, M.: R-max - a general polynomial time algorithm for near-optimal reinforcement learning. Journal of Machine Learning Research 3, 213–231 (2003)
5. Şimşek, O., Barto, A.G.: An intrinsic reward mechanism for efficient exploration. In: Proceedings of the 23rd International Conference on Machine Learning, ICML 2006, pp. 833–840. ACM, New York (2006)
6. Dimitrakakis, C.: Tree exploration for Bayesian RL exploration. In: CIMCA/IAWTIC/ISE, pp. 1029–1034 (2008)
7. Duff, M.: Optimal learning: Computational procedures for Bayes-adaptive Markov decision processes. Ph.D. thesis, University of Massachusetts Amherst (2002)
8. Gittins, J.C.: Bandit processes and dynamic allocation indices. Journal of the Royal Statistical Society 41(2), 148–177 (1979)
9. Jonsson, A., Barto, A.G.: Active Learning of Dynamic Bayesian Networks in Markov Decision Processes. In: Miguel, I., Ruml, W. (eds.) SARA 2007. LNCS (LNAI), vol. 4612, pp. 273–284. Springer, Heidelberg (2007)
10. Kolter, J., Ng, A.: Near-Bayesian exploration in polynomial time. In: Proceedings of the Twenty-Sixth International Conference on Machine Learning, ICML 2009 (2009)
11. Ng, A.Y., Harada, D., Russell, S.: Policy invariance under reward transformations: Theory and application to reward shaping. In: Proceedings of the Sixteenth International Conference on Machine Learning, pp. 278–287. Morgan Kaufmann (1999)
12. Poupart, P., Vlassis, N., Hoey, J., Regan, K.: An analytic solution to discrete Bayesian reinforcement learning. In: Proceedings of the Twenty-Third International Conference on Machine Learning (ICML 2006) (2006)
13. Puterman, M.: Markov Decision Processes: Discrete Stochastic Dynamic Programming. Wiley-Interscience (April 1994)
14. Rauber, T., Braun, T., Berns, K.: Probabilistic distance measures of the Dirichlet and Beta distributions. Pattern Recognition 41(2), 637–645 (2008)
15. Roy, N., Thrun, S.: Coastal navigation with mobile robots. In: Advances in Neural Information Processing Systems 12, pp. 1043–1049 (1999)
16. Sorg, J., Singh, S., Lewis, R.: Variance-based rewards for approximate Bayesian reinforcement learning. In: Proceedings of the Twenty-Sixth Conference on Uncertainty in Artificial Intelligence (2010)
17. Strens, M.J.A.: A Bayesian framework for reinforcement learning. In: Proceedings of the International Conference on Machine Learning (ICML 2000), pp. 943–950 (2000)
18. Sutton, R., Barto, A.: Reinforcement Learning: An Introduction. MIT Press (1998)
19. Szepesvári, C.: Reinforcement learning algorithms for MDPs – a survey. Tech. Rep. TR09-13, Department of Computing Science, University of Alberta (2009)

Handling Ambiguous Effects in Action Learning

Boris Lesner and Bruno Zanuttini

GREYC, Université de Caen Basse-Normandie, CNRS UMR 6072, ENSICAEN
{boris.lesner,bruno.zanuttini}@unicaen.fr

Abstract. We study the problem of learning stochastic actions in propositional, factored environments, and precisely the problem of identifying STRIPS-like effects from transitions in which they are ambiguous. We give an unbiased, maximum likelihood approach, and show that maximally likely actions can be computed efficiently from observations. We also discuss how this study can be used to extend an RL approach for actions with independent effects to one for actions with correlated effects.

Keywords: stochastic action, maximum likelihood, factored MDP.

1 Introduction

Learning how one's actions affect the environment is a central issue in Artificial Intelligence, and especially in Reinforcement Learning (RL) problems. This task is far from being trivial because of many factors. In particular, actions may affect the environment differently depending on the state where they are taken. They may also be stochastic, that is, have different possible outcomes, each of which occurs with some probability each time we take the action.

A third difficulty, which is our focus in this article, is that the effects of an action are typically *ambiguous* in some states. Precisely, if some fact is true in the state where we took the action *and* in the resulting state, then it is ambiguous whether this fact was set to true by the action, or was simply unaffected by it (and persisted true). This of course makes no difference in the starting state in question, but may matter a lot if we want to generalize the corresponding observation to other states, where the fact in question is not true.

To give a concrete example, assume that you want to learn the effects of the action "play the National Lottery" by observing your neighbours. If one of them suddenly has an expensive car parked in front of his house, then he must have won (effect "become rich" occurred). On the other hand, assume another one always has had an expensive car parked in front of his huge villa. Then you will see no change when he wins, nor when he looses. Technically, for this neighbour (state) both winning and losing (effects) provoke no change (self-transition). Then, this ambiguity must be resolved for generalizing from a few "poor" neighbours and a few "rich" neighbours to a generic player.

We address here this specific problem for environments and actions described in factored form over propositional variables, with STRIPS-like effects. We provide an in-depth study of ambiguity and of maximum likelihood approaches to

S. Sanner and M. Hutter (Eds.): EWRL 2011, LNCS 7188, pp. 54–65, 2012.

learning in the presence of ambiguous effects. Though the obvious motivation for this work is RL, we place our study in a general learning context, where some transitions between states are observed, from which the effects of the (unique) stochastic action which provoked them must be learnt together with their probabilities. In this setting, we propose a linear programming approach for computing the most likely effect distributions, and use it to propose a generalization of SLF-Rmax [12] to deal with ambiguous effects.

Ambiguous effects are discussed in several works [13, in particular], and implicitly dealt with in approaches to *relational RL* [5,10,11]. Still, to the best of our knowledge they have never been investigated in depth. For instance, propositional actions are often learnt as dynamic bayesian networks (DBNs) which assume independent effects on variables [6,4,12]. Then ambiguity is bypassed by learning the effects on x independently in states satisfying x and $\overline{x}$. But such DBNs (without synchronic arcs) cannot represent effect distributions as simple as $\{(x_1 x_2, .5), (\emptyset, .5)\}$, while STRIPS-like descriptions, as we study here, are fully expressive (see the discussion by Boutilier *et al.* [1]).

Closest to ours is the work by Walsh *et al.* [13]. Our work can in fact be seen as using their linear regression approach (for known effects) with all 3^n possible effects, but introducing new techniques for reducing the exponential number of unknowns and dealing with ambiguity.

Section 2 introduces our formal setting. In Sections 3–5 we study the problem of computing most likely actions given observed transitions. We give an application to RL and conditional actions in Section 6, and conclude in Section 7.

2   Formal Setting

We consider propositional environments, described through n Boolean variables $x_1, \ldots, x_n$. A *state* s is an assignment to $x_1, \ldots, x_n$. There is a hidden stochastic action a, which can be described as a probability distribution on a finite set of effects (we identify the action and the distribution to each other). We write $\{(p_i, e_i) \mid i = 1, \ldots, \ell\}$ (with $\sum_i p_i = 1$) for this distribution, meaning that each time a is taken, exactly one of the effects e_i occurs, as sampled i.i.d. according to the p_i's. Effects are STRIPS-like, that is, each effect is a consistent term on $x_1, \ldots, x_n$. When a is taken in some state s and effect e_i occurs, the resulting state s' is obtained from s by changing all affected variables to their value in e_i. We write $s' = apply(s, e_i)$. For convenience, states and terms are also seen as sets of literals, and 3^X denotes the set of all terms/effects.

For instance, for $n = 3$, $s = 000$ and $s' = 010$ are two states, and $e_1 = \overline{x}_1 x_2$, $e_2 = \overline{x}_3$ are two effects. The distribution $\{(e_1, .7), (e_2, .2), (\emptyset, .1)\}$ defines an action a. Each time a is taken in a state, e_1 occurs (with probability .7), or e_2 occurs (.2), or nothing occurs (.1). If effect e_1 occurs in state s above, the resulting state is s', while effects e_2 and $\emptyset$ provoke a transition from s to itself.

Ambiguous Effects and Compact Representation. Ambiguity is a consequence of the following straightforward result.

Proposition 1. *Let s, s' be two states. Then the effects $e \in 3^X$ which satisfy apply$(s, e) = s'$ are exactly those for which $s' \setminus s \subseteq e \subseteq s'$ holds.*

It follows that the effects which may have occurred when a transition from s to s' is observed, can be *compactly* represented, as the set interval $[s' \setminus s, s']$. As an example, the effects which provoke a transition from $s = 000$ to $s' = 010$ are $x_2, \overline{x}_1 x_2, x_2 \overline{x}_3$, and $\overline{x}_1 x_2 \overline{x}_3$. That is, we are sure that the (atomic) effect x_2 occurred, but any other literal in s' may have been set to true, or *left* true, by the action. These effects are compactly represented by the set interval $[x_2, \overline{x}_1 x_2 \overline{x}_3]$.

Set intervals obey the rule $[e_1, e_2] \cap [e_3, e_4] = [e_1 \cup e_3, e_2 \cap e_4]$. For instance, $[x_2, \overline{x}_1 x_2 \overline{x}_3] \cap [\emptyset, x_1 x_2 \overline{x}_3]$ is $[x_2, x_2 \overline{x}_3]$, and $[x_2, \overline{x}_1 x_2 \overline{x}_3] \cap [\overline{x}_2 \overline{x}_3, x_1 x_2 \overline{x}_3]$ is empty, since $x_2 \overline{x}_2 \overline{x}_3 \not\subseteq x_2 \overline{x}_3$ holds (here, because x_2 and $\overline{x}_2 \overline{x}_3$ are inconsistent together).

Sampled Effects and Induced Observations. We consider an agent which must learn the effect distribution of a unique action a, from transitions provoked by this action in the environment. Remember that (except in Section 6) we assume a to be unconditional, i.e., to have the same effect distribution D in all states s.

For generality, we assume nothing about the states s in which transitions are observed; what we only require is that the environment samples the effects fairly to D. This setting arises naturally in many contexts. For instance, a robot may train in the factory, so as to learn accurate models of its actuators, but the situations s which it can experiment may be restricted by the facilities in the factory. In an RL setting, the agent may lose control on the visited states if other agents or exogenous events can intervene at any moment in the environment.

Formally, the examples sampled by the environment form a multiset of pairs state/effect, of the form $\mathcal{T} = \{(s_i, e_i) \mid i = 1, \ldots, m\}$, and the learner sees the corresponding multiset of transitions between states $\{(s_i, s'_i) \mid i = 1, \ldots, m\}$ (with $s'_i = \text{apply}(s_i, e_i)$). For simplicity of presentation, we define the *observations* induced by $\mathcal{T}$ to be the multiset of intervals $O_{\mathcal{T}} = \{[s'_i \setminus s_i, s'_i] \mid i = 1, \ldots, m\}$, and assume that this is the only data available to the learner. That this information is equivalent to the multiset of transitions is a consequence of Proposition 1, together with the observation that s_i can be easily retrieved from $s'_i \setminus s_i$ and s'_i. Importantly, observe that e_i is always in the set interval $[s'_i \setminus s_i, s'_i]$.

Example 1. Let $D = \{(x_1, .5), (x_1 \overline{x}_2, .3), (\emptyset, .2)\}$, and let

$$\mathcal{T} = \{(01, x_1), (00, x_1), (00, \emptyset), (00, x_1 \overline{x}_2), (01, x_1)\}$$

in which, for instance, the first element corresponds to state $s = 01$ having been chosen, and effect x_1 having been sampled. The observations induced by $\mathcal{T}$ are

$$O = \{[x_1, x_1 x_2], [x_1, x_1 \overline{x}_2], [\emptyset, \overline{x}_1 \overline{x}_2], [x_1, x_1 \overline{x}_2], [x_1, x_1 x_2]\}$$

The *frequency* of an observation $o = [s' \setminus s, s']$ in $O_{\mathcal{T}}$, written $f_{o, O_{\mathcal{T}}}$ (or f_o), is the proportion of indices i such that $[s'_i \setminus s_i, s'_i]$ is precisely o in $O_{\mathcal{T}}$. In Example 1, $f_{[x_1, x_1 \overline{x}_2]}$ is $1/5 + 1/5 = .4$. Observe that this corresponds in fact to the manifestation of two different effects (but this is hidden to the learner).

3   Most Likely Actions

We now characterize the maximally likely effect distributions D (i.e., actions) given observations $O = \{[s_i' \setminus s_i, s_i'] \mid i = 1, \ldots, m\}$. First recall from Bayes rule that the likelihood of D given O satisfies $Pr(D|O) = Pr(D) \, Pr(O|D) / Pr(O)$. In particular, if all distributions D have the same *prior* likelihood $Pr(D)$, the most likely distribution D given O is the one which maximizes $Pr(O|D)$.

Likelihood and Fairness. We first need a few straightforward lemmas. Write O for a multiset of observations induced by states/effects in $\mathcal{T}$, and $D = \{(p_i, e_i) \mid i \in I\}$, $D' = \{(p_j', e_j') \mid j \in J\}$ for two effect distributions. Moreover, write $\|D - D'\|_1$ for the L1-distance between D, D', that is, $\|D - D'\|_1 = \sum_{e \in 3^X} |p_e - p_e'|$, with $p_e = p_i$ for $e = e_i$ in D and $p_e = 0$ otherwise, and similarly for p_e'. For instance, the L1-distance from the distribution in Example 1 to the distribution $\{(x_1, .6), (\emptyset, .4)\}$ is $(.6 - .5) + (.3 - 0) + (.4 - .2) = 0.6$.

Lemma 1. *If for some $o \in O$, there is no $i \in I$ with $e_i \in o$, then $Pr(O|D)$ is 0.*

This follows from the fact that when a transition from s to s' is observed, the effect which provoked it must be in the interval $[s' \setminus s, s']$.

Now write $D_{\mathcal{T}} = \{(e, p_e)\}$ for the distribution of effects in $\mathcal{T} = \{(s', e')\}$, that is, p_e is given by $p_e = |\{(s', e') \in \mathcal{T} \mid e' = e\}| \,/\, |\mathcal{T}|$.

Lemma 2. *$Pr(O|D) > Pr(O|D')$ is equivalent to $\|D_{\mathcal{T}} - D\|_1 < \|D_{\mathcal{T}} - D'\|_1$.*

This follows, e.g., from the following bound [14], which extends Chernoff's bound to multivalued random variables:

$$Pr(\|D_{\mathcal{T}} - D\|_1 \geq \epsilon) \leq (2^\ell - 2)\mathrm{e}^{-m\epsilon^2/2} \tag{1}$$

where m is the number of samples observed (size of $\mathcal{T}$) and ℓ is the number of effects (with nonzero probability) in D. Note that other distances could be used. We use L1-distance because approximating the transition probabilities $T(\cdot|s, D)$ in an MDP with respect to it provides guarantees on the resulting value functions, but the infinite norm, for instance, gives similar results [7, "simulation lemma"]).

These observations lead us to introduce the following definition.

Definition 1. *Let $D = \{(p_i, e_i) \mid i \in I\}$ be an effect distribution. A multiset of observations O is said to be ϵ-fair to D if*

1. *for all intervals $o \in O$, there is an $i \in I$ with $p_i \neq 0$ and $e_i \in o$, and*
2. *there is a multiset of states/effects $\mathcal{T}$ such that O is induced by $\mathcal{T}$ and the distribution of effects in $\mathcal{T}$ is at L1-distance at most ϵ to D.*

Hence (Lemmas 1–2) the effect distributions D to which O is 0-fair are perfectly likely given O, and otherwise, the smaller ϵ, the more likely D given O.

Clearly enough, a multiset O is always 0-fair to at least one distribution D. Such D can be built from one effect e_o in each interval $o \in O$, with the frequency

f_o as its probability. However, there are in general many such "perfectly likely" distributions. In fact, each possible choice of effects in the construction above leads to a different one. Further, in some given observed interval it may be the case that two different effects provoked the corresponding transitions, increasing still more the number of candidate distributions. Still worse, the different most likely distributions may be arbitrarily distant from each other (hence inducing arbitrarily different planning problems, if learnt actions are used for planning).

Example 2. Let $o_1 = [x_1, x_1 x_2], o_2 = [x_2, x_1 x_2]$, and $O = \{o_1 \times 5, o_2 \times 5\}$ (o_1, o_2 observed 5 times each). Then O is 0-fair to $D_1 = \{(x_1, .5), (x_2, .5)\}$, to $D_2 = \{(x_1 x_2, 1)\}$, and to $D_3 = \{(x_1, .4), (x_1 x_2, .3), (x_2, .3)\}$. As to the last point, D_3 indeed induces O if $x_1 x_2$ is sampled once in $s = 01$ and twice in $s = 10$.

Now despite O is 0-fair to D_1 and to D_2, their L1-distance is maximal (2), and so are the transition functions $T(\cdot|00, \cdot)$ which they induce in $s = 00$.

Summarizing, we have that (with high confidence) O is always ϵ-fair to the hidden effect distribution D which indeed generated it, for some small ϵ. Nevertheless, O can also be ϵ-fair, for some small ϵ, to some distribution which is very different from D. Another way to state this is that fairness gives a lower bound on the distance to the real, hidden D. Anyway, this lower bound is a correct, unbiased measure of likelihood (Lemma 2).

Computing fairness. If some additional constraints have to be satisfied (like accounting for other observations at the same time), in general there will not be any effect distribution to which the observations are 0-fair. Hence we investigate how to compute in general how fair a given multiset of observations is to a given effect distribution.

To that aim, we give an alternative definition of fairness. To understand the construction, observe the following informal statements for a sufficiently large multiset of observations O generated by a distribution $D = \{(p_i, e_i) \mid i \in I\}$:

- if e_i induced $o \in O$, then the observed frequency of o should be close to p_i,
- it is possible that some $e_i \in D$ with a sufficiently small p_i has never been observed, i.e., that O contains no interval o with $e_i \in o$,
- every $o \in O$ *must* be induced by some effect e_i with $p_i \neq 0$,
- if $o_1, o_2 \in O$ intersect, it may be that only one $e_i \in D$ induced both (in different states s), and then f_{o_1}, f_{o_2} should approximately sum up to p_i,
- dually, if $o \in O$ contains e_i, e_j, it may be that both participated in inducing o, and then p_i, p_j should approximately sum up to f_o.

Generalizing these observations, we can see that one effect may induce several observed intervals, and dually that one interval may be induced by several effects. We take these possibilities into account by introducing values which we call *contributions of effects to intervals*, written $c_{e,o}$, and reflecting how often, among the times when e was sampled, it induced interval o.

We are now ready to characterize ϵ-fairness. The value to be minimized in Proposition 2 is the L1-distance between D and the frequencies of the realizations of effects. A particular combination of $c_{e,o}$'s indicates a particular hypothesis about which effect provoked each observed transition, and how often.

Proposition 2. *Let O be a multiset of observations and $D = \{(p_i, e_i) \mid i \in I\}$ be an effect distribution with $\forall o \in O, \exists i \in I, e_i \in o$. Then O is ϵ-fair to D if and only if the following inequality holds:*

$$\min \left(\sum_{e \in D} |p_D(e) - \sum_{o \in O} c_{e,o}| \right) \leq \epsilon$$

where the minimum is taken over all combinations of nonnegative values for $c_{e,o}$'s (for all effects e, observed intervals o) which satisfy the three conditions:

$$\forall o \in O, e \in D \qquad if\ e \notin o\ then\ c_{e,o} = 0$$

$$\forall o \in O \qquad \sum_{e \in D} c_{e,o} = f_o$$

$$\forall o \in O \qquad \exists e\ such\ that\ p_e \neq 0\ and\ c_{e,o} \neq 0$$

Proposition 3 below shows that the minimum is indeed well-defined (as the objective value of a linear program).

Example 3 (continued). The multiset O from Example 2 is 0-fair to $D_3 = \{(x_1, .4), (x_1 x_2, .3), (x_2, .3)\}$, as witnessed by contributions $c_{x_1,o_1} = .4$, $c_{x_1 x_2,o_1} = .1$, $c_{x_1 x_2,o_2} = .2$, $c_{x_2,o_2} = .3$. Now for $D = \{(x_1 x_2, .9), (\overline{x}_1, .1)\}$ we get 0.2-fairness with $c_{x_1 x_2,o_1} = c_{x_1 x_2,o_2} = .5$, and for $D = \{x_1, .45), (x_2, .55)\}$ we get 0.1-fairness. Finally, O is *not* fair to $D = \{(x_1, 1)\}$, since no effect explains o_2.

Linearity. Fairness provides an effective way to compute most likely distributions. To see this, consider the following linear program[1] for given O and D:

Variables:	$c_{e,o}$	$(\forall o \in O, e \in o)$
Minimize:	$\sum_{e \in 3^X} \|p_e - \sum_{o \in O} c_{e,o}\|$	
Subject to:	$c_{e,o} \geq 0$	$(\forall o \in O, e \in o)$
	$\sum_{e \in o} c_{e,o} = f_o$	$(\forall o \in O)$

This program implements Proposition 2, but ignoring the last constraint $\forall o \in O, \exists e \in 3^X, p_e \neq 0$ and $c_{e,o} \neq 0$ (this would be a *strict* inequality constraint). Still, it can be shown that it correctly computes how fair O is to D.

Proposition 3. *Assume for all $o \in O$ there is an $e \in o$ with $p_e > 0$. Then the optimal value of the program above is the minimal ϵ such that O is ϵ-fair to D.*

We note that an arbitrary optimal solution (i.e., combination of $c_{e,o}$'s) of the program is not necessarily a witness of ϵ-fairness, in the sense that it may let some observation $o \in O$ be unexplained by any effect e with nonzero probability. Nevertheless, from such a solution a correct witness can always be straightforwardly retrieved (with the same, optimal value).

---

[1] The objective is indeed linear, since it consists in *minimizing* the absolute values.

Example 4 (continued). Let again $O = \{[x_1, x_1x_2] \times 5, [x_2, x_1x_2] \times 5\}$, and let $D = \{(x_1, 0), (x_2, .25), (x_1x_2, .25), (\overline{x}_1, .5)\}$. Observe that the fairness of O to D cannot be better than 1 (due to a .5-excess on effect $\overline{x}_1$). Hence the contributions $c_{x_1,o_1} = .5$, $c_{x_2,o_2} = .25$, $c_{x_1x_2,o_2} = .25$ are optimal. Nevertheless, they do not respect the constraint that there is an effect $e \in o_1$ with $p_e > 0$ and $c_{e,o_1} > 0$. Still, an optimal solution respecting this constraint can be easily retrieved by "redirecting" the contribution of x_1 to o_1 to the effect x_1x_2.

Clearly, this program is not practical, since it has exponentially many variables. We will see in Section 5 how to reduce this number in practice.

4   Variance of Sets of Observations

We now turn to the more general question of how likely it is that *several* multisets of observations $O_1, \ldots, O_q$ are all induced by the same effect distribution D. Naturally, we want to measure this by the radius of a ball centered at D and containing all O_i's. The smaller this radius, the more likely are all O_i's induced by D, and the smaller over all distributions D, the less the "variance" of O_i's. From the analysis in Section 3 it follows that fairness is the correct (unbiased) measure for defining the radius of such balls. The center of a ball containing all O_i's and with minimal radius is known as the *Chebyshev center* of the O_i's.

We have however taken care in our definition of fairness that any observed interval is accounted for by at least one effect with nonzero probability. As an unfortunate and rather surprising consequence, the center does not always exist.

Example 5. Let $O_1 = \{[x_1], [x_2] \times 2, [x_3] \times 2\}$, $O_2 = \{[x_2] \times 4, [x_4]\}$, and $O_3 = \{[x_3] \times 4, [x_4]\}$. It can be shown that O_2 and O_3 cannot be both 0.8-fair to any D, while for any $\epsilon > 0$ there is a D to which all of O_1, O_2, O_3 are $(0.8 + \epsilon)$-fair. Hence the center of O_1, O_2, O_3 does not exist (it is the argmin of an open set).

Still, we define $O_1, \ldots, O_q$ to be ϵ-*variant* (together) if there is an effect distribution D such that for all $i = 1, \ldots, q$, O_i is ϵ-fair to D. Clearly, variance inherits from fairness the property of being a correct, unbiased measure (of homogeneity), and the property of underestimating the real (hidden) variance.

It is important to note that when restricted to two multisets of observations, variance does not define a distance. Intuitively, this is simply because the witness D_{12} for the variance of O_1, O_2 needs not be the same as the witness D_{23} for O_2, O_3. In fact, the triangle inequality can even be made "very false".

Example 6. Consider the multisets $O_1 = \{[x_1, x_1x_2]\}$, $O_2 = \{[x_2, x_1x_2]\}$, and $O_3 = \{[x_2, \overline{x}_1x_2]\}$. Then O_1, O_2 are 0-variant (using $D = \{(x_1x_2, 1)\}$), and so are O_2, O_3 (using $D = \{(x_2, 1)\}$). Nevertheless, because the intervals in O_1 and O_3 are disjoint, for any distribution D the fairness of either O_1 or O_3 (or both) to D is at least 1, hence O_1, O_3 are not ϵ-variant for any $\epsilon < 1$.

The good news is that variance inherits the linear programming approach from fairness. Indeed, consider the following linear program (built on top of the one for fairness) for computing the variance of $O_1, \ldots, O_q$, as well as a witness distribution $D = \{(p_e, e) \mid e \in 3^X\}$:

Variables:	p_e	$(\forall e \in 3^X)$
	$c^i_{e,o}$	$(\forall i = 1, \ldots, q, o \in O_i, e \in o)$

Minimize: $\max_{i=1,\ldots,q} \sum_{e \in 3^X} |p_e - \sum_{o \in O_i} c^i_{e,o}|$

Subject to:	$p_e \geq 0$	$(\forall e \in 3^X)$
	$c^i_{e,o} \geq 0$	$(\forall i = 1, \ldots, q, o \in O_i, e \in o)$
	$\sum_{e \in 3^X} p_e = 1$	
	$\sum_{e \in o} c^i_{e,o} = f_o$	$(\forall i = 1, \ldots, q, o \in O_i)$

As Example 5 shows, the program may compute a (nonattained) infimum value, but this is still sufficient for our application in Section 6.

Proposition 4. *The optimal value σ of the above linear program satisfies $\sigma \leq \epsilon$ if and only if $O_1, \ldots, O_q$ are $(\epsilon + \alpha)$-variant for all $\alpha > 0$.*

5   Restriction to Intersections of Intervals

The linear programming formulations of fairness and variance which we gave involve an exponential number of variables (enumerating all 3^n effects). We now show how to restrict this. The restriction will not give a polynomial bound on the number of variables in general, but it proves useful in practice.

The basic idea is to identify to only one (arbitrary) effect all the effects in some intersection of intervals $o_1 \cap \cdots \cap o_q$ (one per O_i).

Definition 2. *Let $O_1, \ldots, O_q$ be observations. A maximal intersecting family (MIF) for $O_1, \ldots, O_q$ is any maximal multiset of intervals $M = \{o_{i_1}, \ldots, o_{i_r}\}$ satisfying (for all $j, j' \neq j$): (1) $i_j \neq i_{j'}$, (2) $o_{i_j} \in O_{i_j}$, and (3) $o_{i_1} \cap \cdots \cap o_{i_r} \neq \emptyset$.*

A set of effects $E \subseteq 3^X$ is said to be sufficient *for $O_1, \ldots, O_q$ if for all MIFs $M = \{o_{i_1}, \ldots, o_{i_r}\}$, there is at least one effect $e \in E$ with $e \in o_{i_1} \cap \cdots \cap o_{i_r}$.*

Example 7. Let $o_1 = [x_1, x_1 x_2]$, $o'_1 = [x_2, x_1 x_2]$, $o_2 = [\emptyset, x_1 x_2]$, $o'_2 = [\overline{x}_1 \overline{x}_2]\}$, and $O_1 = \{o_1, o'_1\}$, $O_2 = \{o_2, o'_2\}$. The MIFs for O_1, O_2 are $\{o_1, o_2\}$ (with intersection $[x_1, x_1 x_2]$), $\{o'_1, o_2\}$ (intersection $[x_2, x_1 x_2]$), and $\{o'_2\}$ (intersection $[\overline{x}_1 \overline{x}_2]$). Hence, for example, $\{x_1, x_2, \overline{x}_1 \overline{x}_2\}$ and $\{x_1 x_2, \overline{x}_1 \overline{x}_2\}$ are sufficient sets of effects.

The proof of the following proposition (omitted) is simple though tedious.

Proposition 5. *Let $O_1, \ldots, O_q$ be multisets of observations, and E be a sufficient set of effects for $O_1, \ldots, O_q$. The program for computing variance has the same optimal value when effect variables p_e's range over $e \in E$ (instead of 3^X).*

Given multisets O_1, O_2, a sufficient set of effects can be easily computed by (1) intersecting all intervals in O_1 with those in O_2, yielding a set of intervals O_{12}, (2) retaining the $\subset$-minimal intervals in $(O_1 \cup O_2 \cup O_{12}) \setminus \{\emptyset\}$, and (3) choosing one effect in each remaining interval. For a greater number of intervals, Steps (1)–(2) are applied recursively first. Such computation is very efficient in practice, and indeed generates a small number of effects as compared to 3^n.

6   Application: Learning Conditions of Actions

We now show how the tools of Sections 3–5 can be applied to learning *conditional* actions. Precisely, we consider a hidden action a, consisting of formulas $c_1, \ldots, c_\ell$, each associated with an effect distribution D_i. Conditions c_i are assumed to partition the state space, so that when a is taken in s, an effect is sampled from D_i, where c_i is the unique condition satisfied by s. Such actions are known as *Probabilistic STRIPS Operators* (PSO [8]).

When $c_1, \ldots, c_\ell$ are unknown but a has independent effects on the variables, SLF-Rmax [12] learns a DBN representation with KWIK guarantees [9], provided a bound k on the number of variables used in $c_1 \cup \cdots \cup c_\ell$ is known *a priori*.

We also assume such a known bound k, and build on SLF-Rmax for addressing PSO's. Note that using synchronic arcs, SLF-RMAX could in principle deal with nonindependent (correlated) effects, but the bound k would then depend also on an *a priori* bound on the amount of correlation (maximum size of an effect). The approach we propose does not suffer from this dependence, and hence is more sample-efficient. The price we pay is that our approach is only *heuristic*.

Basic Principles. SLF-Rmax gathers statistics (observations) about the action behavior in specific families of states (using Rmax-like exploration). Precisely, for each set C of k variables and each term c on C, a multiset of observations O^c is gathered *in states satisfying* c. For instance, for $k = 2$ and $c = x_1\overline{x}_3$, O^c will contain transitions observed only in states s satisfying x_1 and $\overline{x}_3$. For technical reasons (see below), statistics are also gathered on terms over $2k$ variables.

The full approach works by first learning the conditions $c_1, \ldots, c_\ell$ of a, then learning the distribution D_i for each of them using maximum likelihood. As concerns conditions, define the following property, respective to a given measure of homogeneity μ for multisets of observations (think of our notion of variance).

Definition 3. *A term c over k variables is said to be an ϵ-likely condition for action a (wrt μ) if $\mu(O^{c \wedge c^1}, \ldots, O^{c \wedge c^K}) \leq \epsilon$ holds, where $c^1, \ldots, c^K$ enumerate all $\binom{n-k}{k}2^k$ terms over k variables disjoint from c.*

Example 8. Let $n = 3$ variables, $k = 1$ variable in the (hidden) conditions of a, and assume the following transitions (s, s') have been observed 10 times each:

$$(000, 000), (110, 111), (011, 000), (101, 101), (111, 111)$$

We have for instance $O^{x_1\overline{x}_3} = \{[x_3, x_1x_2x_3] \times 10\}$ and $O^{x_1x_3} = \{[\emptyset, x_1\overline{x}_2x_3] \times 10, [\emptyset, x_1x_2x_3] \times 10\}$. The term x_1 is an ϵ-likely condition because $O^{x_1x_2}$, $O^{x_1\overline{x}_2}$, $O^{x_1x_3}$, $O^{x_1\overline{x}_3}$ are homogeneous, e.g., they are all fair to $D = \{(x_3, 1)\}$. Contrastingly, $\overline{x}_3$ is not ϵ-likely because $O^{\overline{x}_2\overline{x}_3} = \{[\emptyset, \overline{x}_1\overline{x}_2\overline{x}_3] \times 10\}$ and $O^{x_2\overline{x}_3} = \{[x_3, x_1x_2x_3] \times 10\}$ do not have even one likely effect in common.

Given enough observations for each $2k$-term c, it turns out that with high probability, retaining ϵ-likely conditions (for small ϵ, as derived from Equation 1) is correct. The argument is essentially as follows [12]:

1. any correct condition $c = c_i$ is ϵ-likely; indeed, all $O^{c \wedge c^j}$'s are homogeneous because by definition they are all induced by D_i,
2. for any ϵ-likely c, by homogeneity O^c is close in particular to $O^{c \wedge c_i}$ for the *real* condition c_i, and since $O^{c \wedge c_i}$ is induced by D_i, by the triangle inequality the most likely explanations of O^c are close to the correct D_i.

Handling Ambiguity. With ambiguous effects, we propose to use variance as the measure μ in Definition 3. Because variance underestimates the radius of the L1-ball centered at the *real* distribution (Section 4), we get a complete approach.

Proposition 6. *With high confidence (as given by Equation 1), any real condition c_i of a is an ϵ-likely condition of a with respect to variance.*

However, because the soundness of SLF-Rmax (Observation 2 above) relies on the triangle inequality, our approach is not *sound* in general, that is, a condition c may be ϵ-likely while inducing an incorrect distribution of effects. Still, choosing the candidate condition c with the smallest ϵ (that is, with the smallest L1-ball centered at it) gives a promising heuristic. In particular, the worst cases where two most likely explanations are very distant from one another are not likely to be frequent in practice. We give some preliminary experimental results below.

Also observe that we could get a *sound but incomplete* approach by maximizing instead of minimizing the radius over all D's in the definition of variance. This however requires to solve a mixed integer instead of a linear program, and anyway did not prove interesting in our early experiments.

Importantly, our approach inherits the sample complexity of SLF-Rmax. A straightforward application of Equation 1 and an analysis parallel to the one of SLF-Rmax [12] shows that we need a similar number of samples: essentially, for each candidate $2k$-term $c \wedge c^j$ (there are $\binom{n}{2k} 2^{2k}$ of them) we need the (polynomial) number of observations given by Equation 1 for $O^{c \wedge c^j}$ to be fair to the target distribution. Observe that $1/\epsilon$ can be used as a default estimate of the number of effects in the target distribution (ℓ in Equation 1). Overall, the sample complexity is polynomial for a fixed bound k, just as for SLF-Rmax.

Preliminary Experiments. We ran some proof-of-concept experiments on two 512-state problems ($n = 9$ variables). These only aim at demonstrating that though heuristic, our approach can work well in practice, in the sense that it learns a good model with the expected sample complexity. Hence we depict here the reward cumulated over time in an RL setting, averaged over 10 runs. For both problems on which we report here, learning can be detected to have converged at this point where the cumulated reward starts to increase continuously.

We first tested the Builder domain [3] (slightly modified by removing the "wait" action, forcing the agent to act indefinitely). The number of variables in the conditions is $k = 2$ for one action, 5 for another, and 1 for all others. Actions modify up to 4 variables together and have up to 3 different effects.

Figure 1 (left) shows the behaviour of our approach ("PSO-Rmax"). The behaviour of (unfactored) Rmax [2] is shown as a baseline; recall that Rmax

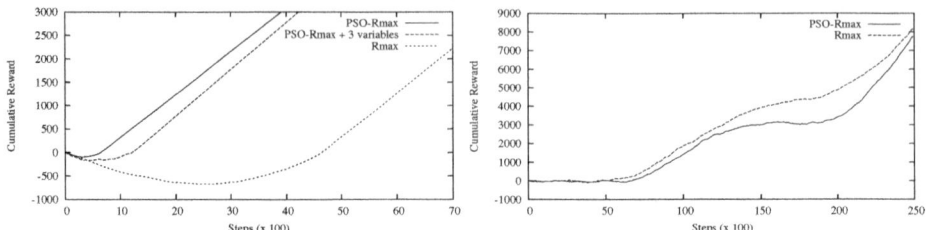

Fig. 1. cumulative rewards for Builder (left) and Stocktrading (right)

provably converges to optimal behaviour in a number of steps polynomial in the number of states. Both algorithms were run with $m = 10$ in Equation 1 (in the corresponding equation for Rmax), as appeared to be most efficient. We have not run SLF-Rmax here because the effects of actions on variables are not independent (we leave extensions of SLF-Rmax to this setting for future work).

Each "step" on the x-axis corresponds to 100 observed transitions. As can be seen, our algorithm efficiently exploits the structure and is very fast to converge to a near-optimal policy (by exploring an average of 57 states only). In no run have we observed behaviours keeping on deviating from the optimal one, as could happen in theory since our algorithm is not guaranteed to be sound and might learn an incorrect model. Inportantly, we also tested the behaviour of our approach with 3 dummy variables added along with actions to flip them, thus artificially extending the domain to 4096 states. As Figure 1 shows, this had almost no consequence on the convergence of our approach.

We also ran some experiments on a 3×2 Stocktrading domain [12]. Here the effects are independent, so that DBNs are a compact representation ($k = 2$ variables in conditions for SLF-Rmax) while PSOs are not ($k = 6$ and $k = 7$). Experiments confirm that SLF-Rmax converges much faster than our approach. Despite this, as Figure 1 (right) shows, our algorithm behaves similarily as $R_{\max}$, which could be expected since considering all $2k$-terms over 9 variables with $k = 6$ or 7 amounts to exploiting no structure. Nevertheless, again our method always converged to optimal behaviour despite its heuristic nature.

7   Conclusion

We have studied a maximum likelihood, unbiased approach to learning actions from transitions in which effects are ambiguous. We have shown that the maximally likely actions could be computed by a linear program with exponentially many variables, but that this number could be greatly reduced in pratice. Moreover, if a known, polynomial-size set of candidate effects is learnt or known in advance, then this can be exploited directly in the program.

We have proposed an application of our study to RL, for hidden actions with correlated effects, extending SLF-Rmax [12]. Our approach is complete but not sound in general, but some proof-of-concept experiments suggest that it may work in pratice. Nevertheless, this application is mainly illustrative, and our projects include using variance as a measure of information in techniques based on induction of decision trees [4], which prove to be more powerful in practice.

Another interesting perspective is to study information criteria able to bias the search for likely actions (e.g., MDL), and to integrate them efficiently in our linear programming approach. Finally, an important perspective in an RL context is to design exploration strategies able to help the agent disambiguate or learn the set of effects before trying to learn their relative probabilities.

Acknowledgements. This work is supported by the French National Research Agency under grant LARDONS (ANR-2010-BLAN-0215).

References

1. Boutilier, C., Dean, T., Hanks, S.: Decision-theoretic planning: Structural assumptions and computational leverage. J. Artificial Intelligence Research 11, 1–94 (1999)
2. Brafman, R.I., Tennenholtz, M.: R-max: A general polynomial time algorithm for near-optimal reinforcement learning. J. Machine Learning Research 3, 213–231 (2002)
3. Dearden, R., Boutilier, C.: Abstraction and approximate decision-theoretic planning. Artificial Intelligence 89, 219–283 (1997)
4. Degris, T., Sigaud, O., Wuillemin, P.-H.: Learning the structure of factored Markov Decision Processes in reinforcement learning problems. In: Proc. International Conference on Machine Learning (ICML 2006), pp. 257–264. ACM (2006)
5. Džeroski, S., De Raedt, L., Driessens, K.: Relational reinforcement learning. Machine Learning 43, 7–52 (2001)
6. Kearns, M., Koller, D.: Efficient reinforcement learning in factored MDPs. In: Proc. 16th International Joint Conference on Artificial Intelligence (IJCAI 1999), pp. 740–474. Morgan Kaufmann (1999)
7. Kearns, M., Singh, S.: Near-optimal reinforcement learning in polynomial time. Machine Learning 49(2-3), 209–232 (2002)
8. Kushmerick, N., Hanks, S., Weld, D.S.: An algorithm for probabilistic planning. Artificial Intelligence 76, 239–286 (1995)
9. Li, L., Littman, M.L., Walsh, T.J., Strehl, A.L.: Knows what it knows: a framework for self-aware learning. Machine Learning 82, 399–443 (2011)
10. Pasula, H.M., Zettlemoyer, L.S., Kaelbling, L.P.: Learning symbolic models of stochastic domains. J. Artificial Intelligence Research 29, 309–352 (2007)
11. Rodrigues, C., Gérard, P., Rouveirol, C., Soldano, H.: Incremental learning of relational action rules. In: Proc. 9th International Conference on Machine Learning and Applications (ICMLA 2010), pp. 451–458. IEEE Computer Society (2010)
12. Strehl, A.L., Diuk, C., Littman, M.L.: Efficient structure learning in factored-state MDPs. In: Proc. 22nd AAAI Conference on Artificial Intelligence (AAAI 2007), pp. 645–650. AAAI Press (2007)
13. Walsh, T.J., Szita, I., Diuk, C., Littman, M.L.: Exploring compact reinforcement-learning representations with linear regression. In: Proc. 25th Conference on Uncertainty in Artificial Intelligence, UAI 2009 (2009)
14. Weissman, T., Ordentlich, E., Seroussi, G., Verdu, S., Weinberger, M.J.: Inequalities for the l_1 deviation of the empirical distribution. Tech. Rep. HPL-2003-97, Hewlett-Packard Company (2003)

Feature Reinforcement Learning in Practice

Phuong Nguyen[1,2], Peter Sunehag[1], and Marcus Hutter[1,2,3]

[1] Australian National University
[2] NICTA
[3] ETHZ
nmphuong@cecs.anu.edu.au, first.last@anu.edu.au

Abstract. Following a recent surge in using history-based methods for resolving perceptual aliasing in reinforcement learning, we introduce an algorithm based on the feature reinforcement learning framework called ΦMDP [13]. To create a practical algorithm we devise a stochastic search procedure for a class of context trees based on parallel tempering and a specialized proposal distribution. We provide the first empirical evaluation for ΦMDP. Our proposed algorithm achieves superior performance to the classical U-tree algorithm [20] and the recent active-LZ algorithm [6], and is competitive with MC-AIXI-CTW [29] that maintains a bayesian mixture over all context trees up to a chosen depth. We are encouraged by our ability to compete with this sophisticated method using an algorithm that simply picks one single model, and uses Q-learning on the corresponding MDP. Our ΦMDP algorithm is simpler and consumes less time and memory. These results show promise for our future work on attacking more complex and larger problems.

1 Introduction

Reinforcement Learning (RL) [28] aims to learn how to succeed in a task through trial and error. This active research area is well developed for environments that are Markov Decision Processes (MDPs); however, real world environments are often partially observable and non-Markovian. The recently introduced Feature Markov Decision Process (ΦMDP) framework [13] attempts to reduce actual RL tasks to MDPs for the purpose of attacking the general RL problem where the environment's model as well as the set of states are unknown. In [27], Sunehag and Hutter take a step further in the theoretical investigation of Feature Reinforcement Learning by proving consistency results. In this article, we develop an actual Feature Reinforcement Learning algorithm and empirically analyze its performance in a number of environments.

One of the most useful classes of maps (Φs) that can be used to summarize histories as states of an MDP, is the class of context trees. Our stochastic search procedure, the principal component of our ΦMDP algorithm GSΦA, works on a subset of all context trees, called Markov trees. Markov trees have previously been studied in [23] but under names like FSMX sources or FSM closed tree sources. The stochastic search procedure employed for our empirical investigation utilizes a parallel tempering methodology [7,11] together with a specialized proposal distribution. In the experimental section, the performance of the ΦMDP algorithm where stochastic search is conducted over the space of context-tree maps is shown and compared with three other related context tree-based methods.

S. Sanner and M. Hutter (Eds.): EWRL 2011, LNCS 7188, pp. 66–77, 2012.

Our ΦMDP algorithm is briefly summarized as follows. First, perform a certain number of random actions, then use this history to find a high-quality map by minimizing a cost function that evaluates the quality of each map. The quality here refers to the ability to predict rewards using the created states. We perform a search procedure for uncovering high-quality maps followed by executing Q-learning on the MDP whose states are induced by the detected optimal map. The current history is then updated with the additional experiences obtained from the interactions with the environment through Q-Learning. After that, we may repeat the procedure but without the random actions. The repetition refines the current "optimal" map, as longer histories provide more useful information for map evaluation. The ultimate optimal policy of the algorithm is retrieved from the action values Q on the resulting MDP induced from the final optimal map.

Contributions. Our contributions are: extending the original ΦMDP cost function presented in [13] to allow for more discriminative learning and more efficient minimization (through stochastic search) of the cost; identifying the Markov action-observation context trees as an important class of feature maps for ΦMDP; proposing the GSΦA algorithm where several chosen learning and search procedures are logically combined; providing the first empirical analysis of the ΦMDP model; and designing a specialized proposal distribution for stochastic search over the space of Markov trees, which is of critical importance for finding the best possible ΦMDP agent.

Related Work. Our algorithm is a history-based method. This means that we are utilizing memory that in principle can be long, but in most of this article and in the related works is near term. Given a history h_t of observations, actions and rewards we define states $s_t = \Phi(h_t)$ based on some map Φ. The main class of maps that we will consider are based on context trees. The classical algorithm of this sort is U-tree [20], which uses a local criterion based on a statistical test for splitting nodes in a context tree; while ΦMDP employs a global cost function. This has the advantage that ΦMDP can be used in conjunction with any optimization methods to find the optimal model.

The advent of Universal AI [12] has caused a (re)surge of interest in history based methods: The first approximation and implementation of AIXI for repeated matrix games [22]; the active-LZ algorithm [6] which generalizes the widely used Lempel-Ziv compression scheme to the reinforcement learning setting and assumes n-Markov models of environments; and MC-AIXI-CTW [29] which uses a Bayesian mixture of context trees and incorporates both the Context Tree Weighting algorithm [32] as well as UCT Monte Carlo planning [15]. These can all be viewed as attempts at resolving perceptual aliasing problems with the help of short-term memory. This has turned out to be a more tractable approach than Baum-Welch methods for learning a Partially Observable Markov Decision Process (POMDP) [4] or Predictive State Representations [25]. The history based methods attempt to directly learn the environment states, thereby avoiding the POMDP-learning problem [14], [18] which is extremely hard to solve. Model minimization [8] is a line of works that also seek for a minimal representation of the state space, but focus on solving Markovian problems while ΦMDP and other aforementioned history-based methods target non-Markovian ones. It is also worthy to note that there are various other attempts to find compact representations of MDP state spaces [16]; most of which, unlike our approach, address the planning problem where the MDP model is given.

Paper Organization. The paper is organized as follows. Section 2 recaps Markov Decision Processes (AVI solver, Q-learning, and exploration). Section 3 describes the Feature Reinforcement Learning approach (ΦMDP). Section 4 defines the class of (action-)observation context trees used as Φ-maps. Section 5 explains our parallel tempering Monte Carlo tree search algorithm. These are the components from which the ΦMDP algorithm (GSΦA) is built. In Section 6 we put all of the components into our ΦMDP algorithm and also discuss our specialized search proposal distribution. Section 7 presents experimental results on four domains. Finally Section 8 summarizes the main results of the paper.

2   Markov Decision Processes (MDP)

An environment is a process which at any time, given action $a_t \in \mathcal{A}$ produces an observation $o_t \in \mathcal{O}$ and a corresponding reward $r_t \in \mathbb{R}$. When the process is a Markov Decision Process [28]; o_t represents the environment state, and hence is denoted by s_t instead. Formally, a finite MDP is denoted by a quadruple $\langle \mathcal{S}, \mathcal{A}, \mathcal{T}, \mathcal{R} \rangle$ in which $\mathcal{S}$ is a finite set of states; $\mathcal{A}$ is a finite set of actions; $\mathcal{T} = (T_{ss'}^a : s, s' \in \mathcal{S}, a \in \mathcal{A})$ is a collection of transition probabilities of the next state $s_{t+1} = s'$ given the current state $s_t = s$ and action $a_t = a$; and $R = (R_{ss'}^a : s, s' \in \mathcal{S}, a \in \mathcal{A})$ is a reward function $R_{ss'}^a = \mathbf{E}[r_{t+1}|s_t = s, a_t = a, s_{t+1} = s']$. The return at time step t is the total discounted reward $R_t = r_{t+1} + \gamma r_{t+2} + \gamma^2 r_{t+3} + \ldots$, where γ is the geometric discount factor ($0 \leq \gamma < 1$).

Similarly, the action value in state s following policy π is defined as $Q^\pi(s, a) = \mathbf{E}_\pi[R_t|s_t = s, a_t = a] = \mathbf{E}_\pi[\sum_{k=0}^\infty \gamma^k r_{t+k+1}|s_t = s, a_t = a]$. For a known MDP, a useful way to find an estimate of the optimal action values Q^* is to employ the Action-Value Iteration (AVI) algorithm, which is based on the optimal action-value Bellman equation [28], and iterates the update $Q(s, a) \leftarrow \sum_{s'} T_{ss'}^a [R_{ss'}^a + \gamma \max_{a'} Q(s', a')]$.

If the MDP model is unknown, an effective estimation technique is provided by Q-learning, which incrementally updates estimates Q_t through the equation

$$Q(s_t, a_t) \leftarrow Q(s_t, a_t) + \alpha_t(s_t, a_t) err_t$$

where the feedback error $err_t = r_{t+1} + \gamma \max_a Q(s_{t+1}, a) - Q(s_t, a_t)$, and $\alpha_t(s_t, a_t)$ is the learning rate at time t. Under the assumption of sufficient visits of all state-action pairs, Q-Learning converges if and only if some conditions on the learning rates are met [2], [28]. In practice a small constant value of the learning rates ($\alpha(s_t, a_t) = \eta$) is, however, often adequate to get a good estimate of Q^*. Q-Learning is off-policy; it directly approximates Q^* regardless of what actions are actually taken. This approach is particularly beneficial when handling the exploration-exploitation tradeoff in RL.

It is well known that learning by taking greedy actions retrieved from the current estimate $\widehat{Q}$ of Q^* to explore the state-action space generally leads to suboptimal behavior. The simplest remedy for this inefficiency is to employ the ϵ-greedy scheme, where with probability $\epsilon > 0$ we take a random action, and with probability $1 - \epsilon$ the greedy action is selected. This method is simple, but has shown to fail to properly resolve the exploration-exploitation tradeoff. A more systematic strategy for exploring unseen states, instead of just taking random actions, is to use optimistic initial values [28,3]. To apply this idea to Q-Learning, we simply initialize $Q(s, a)$ with large values. Suppose $R_{\max}$ is the maximal reward, Q initializations of at least $\frac{R_{\max}}{1-\gamma}$ are optimistic as $Q(s, a) \leq \frac{R_{\max}}{1-\gamma}$.

3 Feature Reinforcement Learning

Problem Description. An RL agent aims to find the optimal policy π for taking action a_t given the history $h_t = o_1 r_1 a_1 \ldots o_{t-1} r_{t-1} a_{t-1} o_t r_t$ at time t in order to maximize the long-term reward signal. If the problem satisfies an MDP; as can be seen above, efficient solutions are available. We aim to attack the most challenging RL problem where the environment's states and model are both unknown.

ΦMDP Framework. In [13], Hutter proposes a history-based method, a general statistical and information theoretic framework called ΦMDP. This approach offers a critical preliminary reduction step to facilitate the agent's ultimate search for the optimal policy. The general ΦMDP framework endeavors to extract relevant features for reward prediction from the past history h_t by using a feature map $\Phi \colon \mathcal{H} \to \mathcal{S}$, where $\mathcal{H}$ is the set of all finite histories. More specifically, we want the states $s_t = \Phi(h_t)$ and the resulting tuple $\langle \mathcal{S}, \mathcal{A}, \mathcal{T}, \mathcal{R} \rangle$ to satisfy the Markov property of an MDP. As aforementioned, one of the most useful classes of Φs is the class of context trees, where each tree maps a history to a single state represented by the tree itself. A more general class of Φ is Probabilistic-Deterministic Finite Automata (PDFA) [30], which map histories to the MDP states where the next state can be determined from the current state and the next observation [19]. The primary purpose of ΦMDP is to find a map Φ so that rewards of the MDP induced from the map can be predicted well. This enables us to use MDP solvers, like AVI and Q-learning, on the induced MDP to find a good policy. The reduction quality of each Φ is dictated by the capability of predicting rewards of the resulting MDP induced from that Φ. A suitable cost function that measures the utility of Φs for this purpose is essential, and the optimal Φ is the one that minimizes this cost function.

Cost Function. The cost used in this paper is an extended version of the original cost introduced in [13]. The cost is motivated by the minimum encoding length principles MDL [10] and MML [31]. We define a cost that measures the reward predictability of each Φ, or more specifically of the resulting MDP induced from that Φ. Based on this, our cost includes the description length of rewards; however, rewards depend on states as well, so the description length of states must be also added to the cost. In other words, the cost comprises coding of the rewards and resulting states, and is defined as follows:

$$\mathbf{Cost}_\alpha(\Phi|h_n) := \alpha \mathbf{CL}(s_{1:n}|a_{1:n}) + (1 - \alpha)\mathbf{CL}(r_{1:n}|s_{1:n}, a_{1:n})$$

where $s_{1:n} = s_1, ..., s_n$ and $a_{1:n} = a_1, ..., a_n$ and $s_t = \Phi(h_t)$ and $h_t = o r a_{1:t-1} r_t$ and $0 \leq \alpha \leq 1$. For coding we use a two-part code [31,10], hence the code length (CL) is $\mathbf{CL}(x) = \mathbf{CL}(x|\theta) + \mathbf{CL}(\theta)$ where x denotes the data sampled from the model specified by parameters θ. We employ the optimal Markov codes [5] for describing MDP data

$\mathbf{CL}(x|\theta) = \log(1/Pr_\theta(x))$, while the parameters θ based on frequency estimates of $\mathcal{T}$ and $\mathcal{R}$ are uniformly encoded to precision $1/\sqrt{\ell(x)}$ where $\ell(x)$ is the sequence length of x [10,13]: $\mathbf{CL}(\theta) = \frac{m-1}{2}\log\ell(x)$, where m is the number of parameters. The optimal Φ is found via the optimization problem $\Phi^{optimal} = \mathrm{argmin}_\Phi \mathbf{Cost}_\alpha(\Phi|h_n)$.

As we primarily want to find a Φ that has the best reward predictability, the introduction of α is primarily to stress on reward coding, making costs for high-quality Φs much lower with very small α values. In other words, α amplifies the differences among high-quality Φs and bad ones; and this accelerates our stochastic search process described below.

We furthermore replace $\mathbf{CL}(x)$ with $\mathbf{CL}_\beta(x) = \mathbf{CL}(x|\theta) + \beta\mathbf{CL}(\theta)$ in $\mathbf{Cost}_\alpha$ to define $\mathbf{Cost}_{\alpha,\beta}$ for the purpose of being able to faster select the right model given limited data. The motivation to introduce β is the following:

For stationary environments the cost function is analytically of the form $C_1(\Phi) \times n + C_2(\Phi) \times \beta \times \log(n)$. The asymptotically optimal Φ is the one which minimizes $C_1(\Phi)$. The second term $C_2(\Phi)\beta\log(n)$ prevents overfitting for finite n. However, in practice, this complexity penalty is often too large. So in order to obtain the optimal Φ with limited data, a small value of β will help. We assert that with a very large number of samples n, α and β can be ignored in the above cost function ($\alpha = 0.5$ and $\beta = 1$ are used as the cost in [13]). The choice of small α and β helps us more quickly to overcome the model penalty and find the optimal map. This strategy is a quite common practice in statistics, and even in the Minimum Description Length (MDL) community [10]. For instance, AIC [1] uses a very small $\beta = 2/\log n$.

The interested reader is referred to [13] or our technical report [21] for more detailed analytical formulas; and [27] for further motivation and consistency proofs of the ΦMDP model.

4   Context Trees

The class of maps that we will base our algorithm on is a class of context trees.

Observation Context Tree (OCT). OCT is a class of maps Φ used to extract relevant information from histories that include only past observations, not actions and rewards. The presentation of OCT is mainly to facilitate the definitions of the below Action-Observation Context Tree.

Definition. Given an $|\mathcal{O}|$-ary alphabet $\mathcal{O} = \{o^1, o^2, \ldots, o^{|\mathcal{O}|}\}$, an OCT constructed from the alphabet $\mathcal{O}$ is defined as a $|\mathcal{O}|$-ary tree in which edges coming from any internal node are labeled by letters in $\mathcal{O}$ from left to right in the order given.

Given an OCT $\mathcal{T}$ constructed from the alphabet $\mathcal{O}$, the state suffix set, or briefly state set $\mathcal{S} = \{s^1, s^2, \ldots, s^m\} \subseteq \mathcal{O}^*$ induced from $\mathcal{T}$ is defined as the set of all possible strings of edge labels forming along a path from a leaf node to the root node of $\mathcal{T}$. $\mathcal{T}$ is called a Markov tree if it has the so-called Markov property for its associated state set, that is, for every $s^i \in \mathcal{S}$ and $o^k \in \mathcal{O}$, $s^i o^k$ has a unique suffix $s^j \in \mathcal{S}$. The state set of a Markov OCT is called Markov state set. OCTs that do not have the Markov property are identified as non-Markov OCTs. Non-Markov state sets are similarly defined.

Example. Figure 1(a)(A) and 1(a)(B) respectively represent two binary OCTs of depths two and three; also Figures 1(b)(A) and 1(b)(B) illustrate two ternary OCTs of depths two and three.

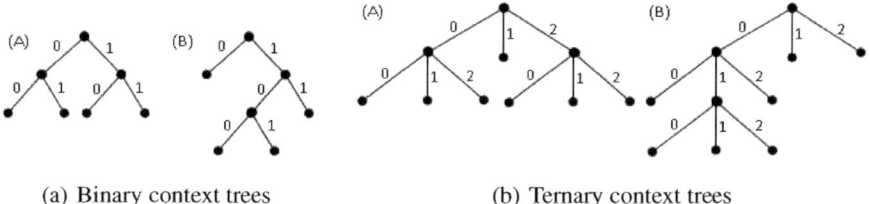

(a) Binary context trees   (b) Ternary context trees

Fig. 1. Context Trees

As can be seen from Figure 1, trees 1(a)(A) and 1(b)(A) are Markov; on the other hand, trees 1(a)(B) and 1(b)(B) are non-Markov. The state set of tree 1(a)(A) is $\mathcal{S}^{(a)(A)} = \{00, 01, 01, 11\}$; and furthermore with any further observation $o \in \mathcal{O}$ and $s \in \mathcal{S}^{(a)(A)}$, there exists a unique $s' \in \mathcal{S}$ which is a suffix of so. Hence, tree 1(a)(A) is Markov. Table 1(a) represents the deterministic relation between s, o and s'.

Table 1. Markov and Non-Markov properties

(a) Markov property of $\mathcal{S}^{(a)(A)}$								(b) Non-markov property of $\mathcal{S}^{(a)(B)}$							
s	00	01	10	11	00	01	10	11	s	0 001 101 11				0	001 101 11	
o		0				1			o		0				1	
s'	00	10	00	10	01	11	01	11	s'	0	0	0	0	101 or 001	11	11 11

However, there is no such relation in tree 1(a)(B), or state set $\mathcal{S}^{(a)(B)} = \{0, 001, 101, 11\}$; for $s = 0$ and $o = 1$, it is ambiguous whether $s' = 101$ or 001. Table 1(b) clarifies the non-Markov property of tree 1(a)(B). Similar arguments can be applied for trees 1(b)(A) and 1(b)(B) to identify their (non) Markov property. It is also worthy to specify how an OCT can be used as a map. We illustrate the mapping using again the OCTs in Figure 1. Given two histories $h_5 = 11101$ and $h'_6 = 211210$, then $\Phi^{(a)(A)}(h_5) = 01$, $\Phi^{(a)(B)}(h_5) = 101$, $\Phi^{(b)(A)}(h'_6) = 10$, and $\Phi^{(b)(B)}(h'_6) = 210$.

Action-Observation Context Tree (AOCT). AOCTs extend the OCTs presented above to the generic RL problem where relevant histories contain both actions and observations.

Definition. Given two alphabets, $\mathcal{O} = \{o^1, o^2, \ldots, o^{|\mathcal{O}|}\}$ named observation set, and $\mathcal{A} = \{a^1, a^2, \ldots, a^{|\mathcal{A}|}\}$ named action set, an AOCT constructed from the two alphabets is defined as a tree where any internal node at even depths has branching factor $|\mathcal{O}|$, and edges coming from such nodes are labeled by letters in $\mathcal{O}$ from left to right in the order given; and similarly any internal node at odd depths has branching factor $|\mathcal{A}|$, and edges coming from these nodes are labeled by letters in $\mathcal{A}$ also from left to right in the specified order.

The definitions of Markov and non-Markov AOCTs are similar to those of OCTs except that a next observation is now replaced by the next action and observation.

5   Stochastic Search

While we have defined the cost criterion for evaluating maps, the problem of finding the optimal map remains. When the Φ space is huge, such as the context-tree map space where the number of Φs grows doubly exponentially with the tree depth, exhaustive search is unable to deal with domains where the optimal Φ is non-trivial. Stochastic search is a powerful tool for solving optimization problems where the landscape of the objective function is complex and it appears impossible to analytically or numerically find the exact or even approximate global optimal solution. A typical stochastic search algorithm starts with a predefined or arbitrary configuration (initial argument of the objective function or state of a system), and from this generates a sequence of configurations based on some predefined probabilistic criterion; the configuration with the best objective value will be retained. There are a wide range of stochastic search methods proposed in the literature [24]; the most popular among these are simulated-annealing-type algorithms [17,26]. An essential element of a simulated-annealing (SA) algorithm is a Markov Chain Monte Carlo (MCMC) sampling scheme where a proposed new configuration $\tilde{y}$ is drawn from a proposal distribution $q(\tilde{y}|y)$, and we then change from configuration y to $\tilde{y}$ with probability $\min\{1, \frac{\pi_T(y)q(y|\tilde{y})}{\pi_T(\tilde{y})q(\tilde{y}|y)}\}$ where π_T is a target distribution.

The traditional SA uses an MCMC scheme with some temperature-decreasing strategy. Although shown to be able to find the global optimum asymptotically [9], it can work badly in practice as we usually do not know which temperature cooling scheme is appropriate for the problem under consideration. Fortunately in the ΦMDP cost function we know typical cost differences between two Φs ($C\beta \times \log(n)$), so the range of appropriate temperatures can be significantly reduced. The search process may be improved if we run a number of SA procedures with various different temperatures. Parallel Tempering (PT) [7,11], an interesting variant of the traditional SA, significantly improves this stochastic search process by smartly offering a swapping step, letting the search procedure use low temperatures for exploitation and high ones for exploration.

Parallel Tempering. PT performs stochastic search over the product space $\mathcal{X}_1 \times \ldots \times \mathcal{X}_I(\mathcal{X}_i = \mathcal{X} \; \forall 1 \leq i \leq I)$, where $\mathcal{X}$ is the objective function's domain, and I is the parallel factor. Fixed temperatures T_i ($i = 1, \ldots, I$, and $0 < T_1 < T_2 < \ldots < T_I$) are chosen for spaces $\mathcal{X}_i$ ($i = 1, \ldots, I$). Temperatures T_i ($i = 1, \ldots, I$) are selected based on the formula $(\frac{1}{T_i} - \frac{1}{T_{i+1}})|\Delta H| \approx -\log p_a$.

where ΔH is the "typical" difference between function values of two successive configurations; and p_a is the lower bound for the swapping acceptance rate. The main steps of each PT loop are as follows:

- $(x_1^{(t)}, \ldots, x_I^{(t)})$ is the current sampling; draw $u \sim \text{Uniform}[0,1]$

- If $u \leq \alpha_0$, update every $x_i^{(t)}$ to $x_i^{(t+1)}$ via some Markov Chain Monte Carlo (MCMC) scheme like Metropolis-Hasting (Parallel step).
- If $u > \alpha_0$, randomly choose a neighbor pair, say i and $i + 1$, and accept the swap of $x_i^{(t)}$ and $x_{i+1}^{(t)}$ with probability $\min\{1, \frac{\pi_{T_i}(x_{i+1}^{(t)})\pi_{T_{i+1}}(x_i^{(t)})}{\pi_{T_i}(x_i^{(t)})\pi_{T_{i+1}}(x_{i+1}^{(t)})}\}$ (Swapping step).

6 The ΦMDP Algorithm

We now describe how the generic ΦMDP algorithm works. It first takes a number of random actions (5000 in all our experiments). Then it defines the cost function $Cost_{\alpha,\beta}$ based on this history. Stochastic search is used to find a map Φ with low cost. Based on the optimal Φ the history is transformed into a sequence of states, actions and rewards. We use optimistic frequency estimates from this history to estimate probability parameters for state transitions and rewards. More precisely, we use $\frac{R_{\max}+r_1+...+r_m}{m+1}$ instead of the average $\frac{r_1+...+r_m}{m}$ to estimate expected reward, where $r_1, ..., r_m$ are the rewards that have been observed for a certain state-action pair, and $R_{\max}$ is the highest possible reward. The statistics are used to estimate Q values using AVI. After this, the agent starts to interact with the environment again using Q-learning initialized with the values that resulted from the performed AVI. The switch from AVI to Q-Learning is effective since we can incrementally update state-action values immediately after each interaction with the environment; while AVI requires updating the whole MDP dynamics and running a number of value iterations. The first set of random actions might not be sufficient to characterize what the best maps Φ look like, so it might be beneficial to add the new history gathered by the Q-Learning interactions with the environment to the old history, and then repeat the process but without the initial sampling.

In our experiments in Section 7, PT is employed to search over the Φ space of Markov AOCTs.

Proposal Distribution for Stochastic Search over the Markov-AOCT Space. The principal optional component of the above high-level algorithm, GSΦA, is a stochastic search procedure of which some algorithms have been presented in Section 5. In these algorithms, an essential technical detail is the proposal distribution q. It is natural to generate the next tree (the next proposal or configuration) from the current tree by splitting or merging nodes. This can, however, cause us to leave the set of Markov trees and we have to perform a sequence of operations to get back into the class. If we propose a merge, all of the operations are merges and if we split all of the operations are splits. This leads us to taking larger steps than if we worked with the whole class of trees and we are more quickly reaching useful configurations. We refer the interested reader to our technical report for full details on the proposal distribution [21].

7 Experiments

Experimental Setup. In this section we present our empirical studies of the ΦMDP algorithm GSΦA described in Section 6. For all of our experiments, stochastic search (PT) is applied in the Φ space of Markov AOCTs.

For a variety of tested domains, our algorithm produces consistent results using the same set of parameters. These parameters are shown in Table 2, and are **not** fine tuned. For full descriptions of our test domains, see our technical report [21], or [29]. We compare our results with the competing algorithms presented in [29].

The results of ΦMDP and the three competitors in the four environments, Cheese Maze, Kuhn Poker, 4×4 grid, and Tiger, are shown in Figure 3. In each of the plots, various time points are chosen to assess and compare the quality of the policies learned by the four approaches. In order to evaluate how good a learned policy is, at each point, the learning process of each agent, and the exploration of the three competitors are temporarily switched off. The selected statistic to compare the quality of learning is the averaged reward over 5000 actions using the current policy. For stability, the statistic is averaged over 10 runs.

As shown in more detail below, ΦMDP is superior to U-tree and active-LZ, and is comparable to MC-AIXI-CTW in short-term memory domains. Overall conclusions are clear, and we, therefore, omit error bars.

Table 2. Parameter setting for the GSΦA algorithm

Parameter	Component	Value
α	$Cost_{\alpha,\beta}$	0.1
β	$Cost_{\alpha,\beta}$	0.1
$initialSample-$ $-Number$	GSΦA	5000
$agentLearning-$ $-Loops$	GSΦA	1
Iterations	PT	100
I	PT	10
$T_i, \ i \leq I$	PT	$T_i = \beta \times$ $i \times \log(n)$
α_0	PT	0.7
γ	AVI, Q-Learning	0.999999
η	Q-Learning	0.01

Environments and Results. We provide results from four test domains, and only discuss the cheese maze more closely. Please refer to our technical report for more details [21].

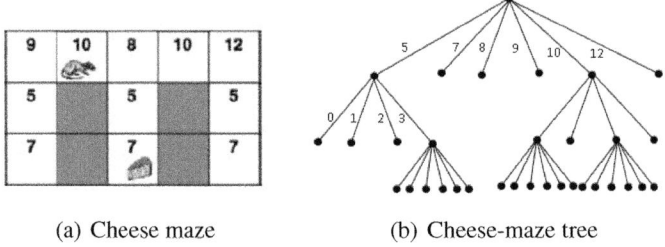

(a) Cheese maze   (b) Cheese-maze tree

Fig. 2. Cheese-maze domain and tree

Cheese Maze. As can be seen from Figure 2(a), some observations themselves alone are insufficient for the (shortsighted) mouse to locate itself unambiguously in the maze. Hence, the mouse must learn to resolve these ambiguities of observations in the maze to be able to find the optimal policy. Our algorithm found a context tree consisting of 43 states that contains the tree as shown in Figure 2(b). The states in this tree resolve the most important ambiguities of the raw observations and an optimal policy can be found.

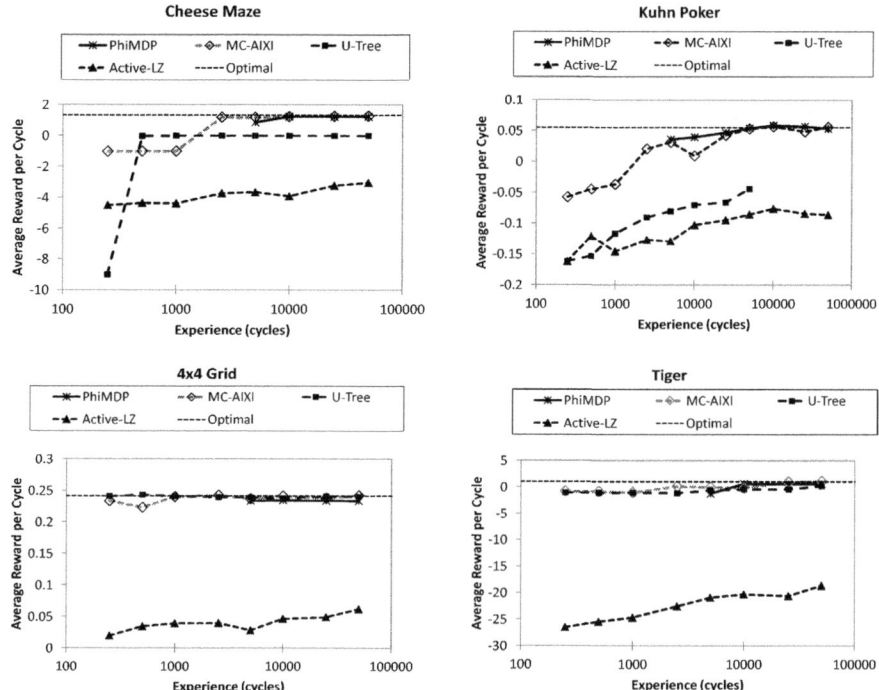

Fig. 3. Cheese maze and Kuhn poker 4×4 Grid and Tiger

8 Conclusions

Based on the Feature Reinforcement Learning framework [13] we defined actual practical reinforcement learning agents that perform very well empirically. We evaluated a reasonably simple instantiation of our algorithm that first takes 5000 random actions followed by finding a map through a search procedure and then performs Q-learning on the MDP defined by the map's state set.

We performed an evaluation on four test domains used to evaluate MC-AIXI-CTW in [29]. Those domains are all suitably attacked with context tree methods. We defined a ΦMDP agent for a class of maps based on context trees, and compared it to three other context tree-based methods. Key to the success of our ΦMDP agent was the development of a suitable stochastic search method for the class of Markov AOCTs. We combined parallel tempering with a specialized proposal distribution that results in an effective stochastic search procedure. The ΦMDP agent outperforms both the classical U-tree algorithm [20] and the recent Active-LZ algorithm [6], and is competitive with the newest state of the art method MC-AIXI-CTW [29]. The main reason that ΦMDP outperforms U-tree is that ΦMDP uses a global criterion (enabling the use of powerful global optimizers) whereas U-tree uses a local split-merge criterion. ΦMDP also performs significantly better than Active-LZ. Active-LZ learns slowly as it overestimates the environment model (assuming n-Markov or complete context-tree environment models); and this leads to unreliable value-function estimates.

Below are some detailed advantages of ΦMDP over MC-AIXI-CTW:

- ΦMDP is more efficient than MC-AIXI-CTW in both computation and memory usage. ΦMDP only needs an initial number of samples and then it finds the optimal map and uses AVI to find MDP parameters. After this it only needs a Q-learning update for each iteration. On the other hand, MC-AIXI-CTW requires model updating, planning and value-reverting at every single cycle which together are orders of magnitude more expensive than Q-learning. In the experiments ΦMDP finished in minutes while MC-AIXI-CTW needed hours.
- ΦMDP learns a single state representation and can use many classical RL algorithms, e.g. Q-Learning, for MDP learning and planning.
- Another key benefit is that ΦMDP represents a more discriminative approach than MC-AIXI-CTW since it aims primarily for the ability to predict future rewards and not to fully model the observation sequence. If the observation sequence is very complex, this becomes essential.

On the other hand, to be fair, it should be noted that compared to ΦMDP, MC-AIXI-CTW is more principled. The results presented in this paper are encouraging since they show that we can achieve comparable results to the more sophisticated MC-AIXI-CTW algorithm on problems where only short-term memory is needed. We plan to utilize the aforementioned advantages of the ΦMDP framework, like flexibility in environment modeling and computational efficiency, to attack more complex and larger problems.

Acknowledgement. This work was supported by ARC grant DP0988049 and by NICTA. We also thank Joel Veness and Daniel Visentin for their assistance with the experimental comparison.

References

1. Akaike, H.: A new look at the statistical model identification. IEEE Transactions on Automatic Control 19, 716–723 (1974)
2. Bertsekas, D.P., Tsitsiklis, J.N.: Neuro-Dynamic Programming. Anthena Scientific, Belmont (1996)
3. Brafman, R.I., Tennenholz, M.: R-max -a general polynomial time algorithm for near-optimal reinforcement learning. Journal of Machine Learing Research 3, 213–231 (2002)
4. Chrisman, L.: Reinforcement learning with perceptual aliasing: The perceptual distinctions approach. In: AAAI, pp. 183–188 (1992)
5. Cover, T.M., Thomas, J.A.: Elements of Information Theory. John Willey and Sons (1991)
6. Farias, V., Moallemi, C., Van Roy, B., Weissman, T.: Universal reinforcement learning. IEEE Transactions on Information Theory 56(5), 2441–2454 (2010)
7. Geyer, C.J.: Markov chain Monte Calro maximum likelihood. In: Computing Science and Statistics: the 23rd Symposium on the Interface, pp. 156–163. Interface Foundation, Fairfax (1991)
8. Givan, R., Dean, T., Greig, M.: Equivalence notions and model minimization in Markov decision process. Artificial Intelligence 147, 163–223 (2003)
9. Granville, V., Křivánek, M., Rasson, J.P.: Simulated annealing: A proof of convergence. IEEE Transactions on Pattern Analysis and Machine Intelligence 16(6), 652–656 (1994)
10. Grünwald, P.D.: The Minimum Description Length Principle. The MIT Press (2007)

11. Hukushima, K., Nemoto, K.: Exchange Monte Carlo method and application to spin glass simulations. Journal of the Physical Socieity of Japan 65(4), 1604–1608 (1996)
12. Hutter, M.: Universal Articial Intelligence: Sequential Decisions based on Algorithmic Probability. Springer, Berlin (2005)
13. Hutter, M.: Feature reinforcement learning: Part I. Unstructured MDPs. Journal of General Artificial Intelligence (2009)
14. Kaelbling, L.P., Littman, M.L., Cassandra, A.R.: Planning and acting in paritally observable stochastic domains. Artifical Intelligence 101, 99–134 (1998)
15. Kocsis, L., Szepesvári, C.: Bandit Based Monte-Carlo Planning. In: Fürnkranz, J., Scheffer, T., Spiliopoulou, M. (eds.) ECML 2006. LNCS (LNAI), vol. 4212, pp. 282–293. Springer, Heidelberg (2006)
16. Li, L., Walsh, T.J., Littmans, M.L.: Towards a unified theory of state abstraction for MDPs. In: Proceedings of the 9th International Symposium on Artificial Intelligence and Mathematics (2006)
17. Liu, J.S.: Monte Carlo Strategies in Scientific Computing. Springer, Heidelberg (2001)
18. Madani, O., Handks, S., Condon: On the undecidability of probabilistic planning and related stochastic optimization problems. Artifical Intelligence 147, 5–34 (2003)
19. Mahmud, M.M.H.: Constructing states for reinforcement learning. In: Fürnkranz, J., Joachims, T. (eds.) Proceedings of the 27th International Conference on Machine Learning (ICML 2010), Haifa, Israel, pp. 727–734 (June 2010), http://www.icml2010.org/papers/593.pdf
20. McCallum, A.K.: Reinforcement Learning with Selective Perception and Hidden State. Ph.D. thesis, Department of Computer Science, University of Rochester (1996)
21. Nguyen, P., Sunehag, P., Hutter, M.: Feature refinrocement learning in practice. Tech. rep., Australian National University (2011)
22. Poland, J., Hutter, M.: Universal learning of repeated matrix games. In: Proc. 15th Annual Machine Learning Conf. of Belgium and The Netherlands (Benelearn 2006), pp. 7–14. Ghent (2006), http://arxiv.org/abs/cs.LG/0508073
23. Rissanen, J.: A universal data compression system. IEEE Transactions on Information Theory 29(5), 656–663 (1983)
24. Schneider, J., Kirkpatrick, S.: Stochastic Optimization, 1st edn. Springer, Heidelberg (2006)
25. Singh, S.P., James, M.R., Rudary, M.R.: Predictive state representations: A new theory for modeling dynamical systems. In: Proceedings of the 20th Conference in Uncertainty in Artificial Intelligence, Banff, Canada, pp. 512–518 (2004)
26. Suman, B., Kumar, P.: A survey of simulated annealing as a tool for single and multiobjecctive optimization. Journal of the Operational Research Society 57, 1143–1160 (2006)
27. Sunehag, P., Hutter, M.: Consistency of Feature Markov Processes. In: Hutter, M., Stephan, F., Vovk, V., Zeugmann, T. (eds.) ALT 2010. LNCS(LNAI), vol. 6331, pp. 360–374. Springer, Heidelberg (2010)
28. Sutton, R., Barto, A.: Reinforcement Learning. The MIT Press (1998)
29. Veness, J., Ng, K.S., Hutter, M., Uther, W., Silver, D.: A Monte-Carlo AIXI approximation. Journal of Artifiicial Intelligence Research 40(1), 95–142 (2011)
30. Vidal, E., Thollard, F., Higuera, C.D.L., Casacuberta, F., Carrasco, R.C.: Probabilitic finite-state machines. IEEE Transactions on Pattern Analysis and Machine Intelligence 27(7), 1013–1025 (2005)
31. Wallace, C.S.: Statistical and Inductive Inference by Minimum Message Length. Springer, Berlin (2005)
32. Wilems, F.M.J., Shtarkov, Y.M., Tjalkens, T.J.: The context tree weighting method: Basic properties. IEEE Transactions on Information Theory 41, 653–664 (1995)

Reinforcement Learning with a Bilinear Q Function

Charles Elkan

Department of Computer Science and Engineering
University of California, San Diego
La Jolla, CA 92093-0404
elkan@ucsd.edu

Abstract. Many reinforcement learning methods are based on a function $Q(s, a)$ whose value is the discounted total reward expected after performing the action a in the state s. This paper explores the implications of representing the Q function as $Q(s, a) = s^T W a$, where W is a matrix that is learned. In this representation, both s and a are real-valued vectors that may have high dimension. We show that action selection can be done using standard linear programming, and that W can be learned using standard linear regression in the algorithm known as fitted Q iteration. Experimentally, the resulting method learns to solve the mountain car task in a sample-efficient way. The same method is also applicable to an inventory management task where the state space and the action space are continuous and high-dimensional.

1 Introduction

In reinforcement learning (RL), an agent must learn what actions are optimal for varying states of the environment. Let s be a state, and let a be an action. Many approaches to RL are centered around the idea of a Q function. This is a function $Q(s, a)$ whose value is the total reward achieved by starting in state s, performing action a, and then performing the optimal action, whatever that might be, in each subsequent state.

In general, a state can have multiple aspects and an action can have multiple sub-actions, so s and a are vectors. In general also, the entries in state and action vectors can be continuous or discrete. For domains where s and a are both real-valued vectors, this paper investigates the advantages of representing $Q(s, a)$ as a linear function of both s and a. Specifically, this representation, which is called bilinear, is $Q(s, a) = s^T W a$, where W is a matrix. The goal of reinforcement learning is then to induce appropriate values for the entries of W. We show that learning W can be reduced to standard linear regression. When the bilinear approach is combined with the batch RL method known as fitted Q iteration [Murphy, 2005] [Ernst et al., 2005], experiments show that the resulting algorithm learns to solve the well-known mountain car task with very few training episodes, and that the algorithm can be applied to a challenging

S. Sanner and M. Hutter (Eds.): EWRL 2011, LNCS 7188, pp. 78–88, 2012.

problem where the state space has 33 dimensions and the action space has 82 dimensions.

The approach suggested in this paper is surprisingly simple. We hope that readers see this as an advantage, not as a lack of sophistication; although simple, the approach is novel. The paper is organized as follows. First, Section 2 explains the consequences of writing $Q(s,a)$ as $s^T W a$. Next, Section 3 discusses using fitted Q iteration to learn W. Then, Section 4 explains how learning W within fitted Q iteration can be reduced to linear regression. Finally, Sections 5 and 6 describe initial experimental results, and Section 7 is a brief conclusion. Related work is discussed throughout the paper, rather than in a separate section.

2   The Bilinear Representation of the Q Function

As mentioned, we propose to represent Q functions as bilinear, that is, to write $Q(s,a) = s^T W a$ where s and a are real-valued column vectors. Given a Q function, the recommended action a^* for a state s is $a^* = \text{argmax}_a\ Q(s,a)$. With the bilinear representation, the recommended action is

$$a^* = \text{argmax}_a\ Q(s,a) = \text{argmax}_a\ x \cdot a$$

where $x = s^T W$. This maximization is a linear programming (LP) task. Hence, it is tractable in practice, even for high-dimensional action and state vectors.

When actions are real-valued, in general they must be subject to constraints in order for optimal actions to be well-defined and realistic. The maximization $\text{argmax}_a\ x \cdot a$ remains a tractable linear programming task as long as the constraints on a are linear. As a special case, linear constraints may be lower and upper bounds for components (sub-actions) of a. Constraints may also involve multiple components of a. For example, there may be a budget limit on the total cost of some or all sub-actions, where each sub-action has a different cost proportional to its magnitude.[1]

When following the policy defined directly by a Q function, at each time step the action vector a is determined by an optimization problem that is a function only of the present state s. The key that can allow the approach to take into account future states, indirectly, is that the Q function can be learned. A learned Q function can emphasize aspects of the present state and action vector that are predictive of long-term reward. With the bilinear approach, the Q function is learnable in a straightforward way. The bilinear approach is limited, however, because the optimal Q function that represents long-term reward perfectly may be not bilinear. Of course, other parametric representations for Q functions are subject to the same criticism.

---

[1] Given constraints that combine two or more components of the action vector, the optimal value for each action component is in general not simply its lower or upper bound. Even when optimal action values are always lower or upper bounds, the LP approach is far from trivial in high-dimensional action spaces: it finds the optimal value for every action component in polynomial time, whereas naive search might need $O(2^d)$ time where d is the dimension of the action space.

The solution to an LP problem is typically a vertex in the feasible region. Therefore, the maximization operation selects a so-called "bang-bang" action vector, each component of which is an extreme value. In many domains, bang-bang actions are optimal given simplified reward functions, but actions that change smoothly over time are desirable. In at least some domains, the criteria to be minimized or maximized to achieve desirably smooth solutions are known, for example the "minimum jerk" principle [Viviani and Flash, 1995]. These criteria can be included as additional penalties in the maximization problem to be solved to find the optimal action. If the criteria are nonlinear, the maximization problem will be harder to solve, but it may still be convex, and it can still be possible to learn a bilinear Q function representing how long-term reward depends on the current state and action vectors.

Related Work. Other approaches to reinforcement learning with continuous action spaces include [Lazaric et al., 2007], [Melo and Lopes, 2008], and [Pazis and Lagoudakis, 2009]. Linear programming has been used before in connection with reinforcement learning, in [De Farias and Van Roy, 2003] and in [Pazis and Parr, 2011] recently, among other papers. However, previous work does not use a bilinear Q function.

The bilinear representation of Q functions contains an interaction term for every component of the action vector and of the state vector. Hence, it can represent the consequences of each action component as depending on each aspect of the state. Some previously proposed representations for Q functions are less satisfactory. The simplest representation is tabular: a separate value is stored for each combination of a state value s and an action value a. This representation is usable only when both the state space and the action space are discrete and of low cardinality. A linear representation is a weighted combination of fixed basis functions [Lagoudakis and Parr, 2003]. In this representation $Q(s, a) = w \cdot [\phi_1(s, a), \cdots, \phi_p(s, a)]$ where ϕ_1 to ϕ_p are fixed real-valued functions and w is a weight vector of length p. The bilinear approach is a special case of this representation where each basis function is the product of one state component and one action component; see Equation (1) below. An intuitive drawback of previously used basis function representations is that they ignore the distinction between states and actions: essentially, s and a are concatenated as inputs to the basis functions ϕ_i.

When the action space is continuous, how to perform the maximization operation $\text{argmax}_a Q(s, a)$ efficiently is a crucial issue; see the discussion in [Pazis and Parr, 2011] where the problem is called action selection. With the bilinear representation, finding the optimal action exactly in a multidimensional continuous space of candidate actions is tractable, and fast in practice. No similarly general approach is known when other basis function representations are used, or when neural networks [Riedmiller, 2005] or ensembles of decision trees [Ernst et al., 2005] are used. Note however that in the context of computational neuroscience, a deep approach to avoiding the maximization has been proposed [Todorov, 2009].

3   Fitted Q Iteration

In what is called the batch setting for RL, an optimal policy or Q function must be learned from historical data [Neumann, 2008]. A single training example is a quadruple $\langle s, a, r, s' \rangle$ where s is a state, a is the action actually taken in this state, r is the immediate reward that was obtained, and s' is the state that was observed to come next. The method named fitted Q iteration is the following algorithm:

> Define $Q_0(s, a) = 0$ for all s and a.
> For horizon $h = 0, 1, 2, \ldots$
> > For each example $\langle s, a, r, s' \rangle$
> > > let the label $v = r + \gamma \max_b Q_h(s', b)$
> > Train Q_{h+1} with labeled tuples $\langle s, a, v \rangle$

The next section shows how to train Q_{h+1} efficiently using the bilinear representation. The variable h is called the horizon, because it counts how many steps of lookahead are implicit in the learned Q function. In some applications, in particular medical clinical trials, the maximum horizon is a small fixed integer [Murphy, 2005]. In these cases the discount factor γ can be set to one.

The Q iteration algorithm is deceptively simple. It was proposed essentially in the form above in parallel by [Murphy, 2005] and [Ernst et al., 2005], and it can be traced back to [Gordon, 1995b, Gordon, 1995a]. However, learning a Q function directly from multiple examples of the form $\langle s, a, v \rangle$ is also part of the Q-RRL algorithm of [Džeroski et al., 2001], and there is a history of related research in papers by economists [Judd and Solnick, 1994] [Stachurski, 2008].

Some of the advantages of Q iteration can be seen by considering the update rule of standard Q learning, which is

$$Q(s, a) := (1 - \alpha)Q(s, a) + \alpha[r + \gamma \max_b Q(s', b)].$$

This rule has two drawbacks. First, no general method is known for choosing a good learning rate α. Second, if $Q(s, a)$ is approximated by a continuous function, then when its value is updated for one $\langle s, a \rangle$ pair, its values for different state-action pairs are changed in unpredictable ways. In contrast, in the Q iteration algorithm all $Q(s, a)$ values are fitted simultaneously, so the underlying supervised learning algorithm (linear regression in our case) can make sure that all of them are fitted reasonably well, and no learning rate is needed.

Another major advantage of the fitted Q iteration algorithm is that historical training data can be collected in alternative ways. There is no need to collect trajectories starting at specific states. Indeed, there is no need to collect consecutive trajectories of the form $\langle s_1, a_1, r_1, s_2, a_2, r_2, s_3, \ldots \rangle$. And, there is no need to know the policy π that was followed to generate historical episodes of the form $\langle s, a = \pi(s), r, s' \rangle$. Q iteration is an off-policy method, meaning that data collected while following one or more non-optimal policies can be used to learn a better policy.

Off-policy methods for RL that do not do active exploration face the problem of sample selection bias. The probability distribution of training examples $\langle s, a \rangle$ is different from the probability distribution of optimal examples $\langle s, \pi^*(s) \rangle$. A reason why Q iteration is correct as an off-policy method is that it is discriminative as opposed to generative. Q iteration does not model the distribution of state-action pairs $\langle s, a \rangle$. Instead, it uses a discriminative method to learn only how the values to be predicted depend on s and a.

A different type of bias faced by fitted Q iteration in general is excessive optimism implicit in the maximization operation [Chakraborty et al., 2008]. In the experiments below, this bias does not cause major problems. In future work we will explore how the double Q learning idea [van Hasselt, 2010] can be combined with fitted Q iteration to reduce the optimism bias.

4   Learning the Matrix W

A function of the form $s^T W a$ is called bilinear because it is linear in its first argument s and also linear in its second argument a. Consider a training set of examples of the form $\langle s, a, v \rangle$ where v is a target value. It is not immediately obvious how to learn a matrix W such that $s^T W a = v$ approximately. However, we show that the training task can be reduced to standard linear regression. Let the vectors s and a be of length m and n respectively, so that $W \in \mathbb{R}^{m \times n}$. The key is to notice that

$$s^T W a = \sum_{i=1}^{m} \sum_{j=1}^{n} (W \circ sa^T)_{ij} = vec(W) \cdot vec(sa^T). \qquad (1)$$

In this equation, sa^T is the matrix that is the outer product of the vectors s and a, and $\circ$ denotes the elementwise product of two matrices, which is sometimes called the Hadamard product. The notation $vec(A)$ means the matrix A converted into a vector by concatenating its columns.

Equation (1) leads to faster training than alternative approaches. In particular, for any square or rectangular matrices A and B with compatible dimensions, $trace(AB) = trace(BA)$. This identity implies the cyclic property of traces, namely $trace(ABC) = trace(CBA)$. A special case of the cyclic property is $s^T W a = trace(s^T W a) = trace(W a s^T)$ which resembles Equation (1). However, computing the product $W(as^T)$ has cubic time complexity while applying Equation (1) has time complexity only $O(mn)$.

Equation (1) is simple, but as far as we know its implications have not been explored before now. Based on it, each training triple $\langle s, a, v \rangle$ can be converted into the pair $\langle vec(sa^T), v \rangle$ and $vec(W)$ can be learned by standard linear regression. The vectors $vec(sa^T)$ are potentially large, since they have length mn, so the exact linear regression computation may be expensive. Stochastic gradient descent is faster than an exact solution, especially if there are many training examples and the matrices sa^T are sparse.

Each entry of the matrix sa^T represents the interaction of a feature of the state vector and a feature of the action vector. In some domains, background knowledge may tell us that many interaction terms have no predictive value. These terms can simply be omitted from the vectors $vec(sa^T)$, making them shorter and making the linear regression training faster. The corresponding entries of the vectorized W matrix are set to zero.

In a reinforcement learning domain, some states may have high value regardless of what action is selected. The basic bilinear model cannot represent this directly; specifically, it does not allow some components of s to lead to high values $Q(s, a)$ regardless of a. We add a constant additional component to each a vector to obtain this extra expressiveness. The additional constant component is a pseudo-action that always has magnitude 1. For similar reasons, we also add a constant component to each s vector.

If the s and a vectors are short, then the matrix W may not be sufficiently expressive. In this case, it is possible to expand the s vector and/or the a vector before forming the outer product sa^T. For example, s itself may be re-represented as ss^T. This re-representation allows the bilinear model to be a quadratic function of the state. Achieving nonlinearity by expanding training examples explicitly, rather than by using a nonlinear kernel, is in line with current trends in machine learning [Chang et al., 2010]. It may sometimes be appropriate to expand just one of s and a, but not both.

To prevent overfitting, regularized linear regression can be used to learn $vec(W)$. Another approach to prevent overfitting is to require W to have a simplified structure. This can be achieved by defining $W = AB^T$ where A and B are rectangular matrices, and learning A and B instead of W. Training W directly by linear regression is a convex optimization problem, with or without many forms of regularization, and with or without some entries of W constrained to be zero. However, finding the optimal representation AB^T is in general not a convex problem, and has multiple local minima.

5 Mountain Car Experiments

The mountain car task is perhaps the best-known test case in research on reinforcement learning [Sutton and Barto, 1998]. The state space has two dimensions, the position x and velocity v of the car. The action space has one dimension, acceleration a. In most previous work acceleration is assumed to be a discrete action. Here, it is allowed to be any real value between -0.001 and $+0.001$, but as explained above, linear programming leads to it always being either -0.001 or $+0.001$. In order to allow the bilinear model W to be sufficiently expressive, we use a six-dimensional expanded state space $\langle x, v, x^2, xv, v^2, x^3 \rangle$, and a two-dimensional action vector $\langle 1, a \rangle$ whose first component is a constant pseudo-action. The matrix W then has 12 trainable parameters.

Table 1 shows that fitted Q iteration with a bilinear Q function learns to control the car well with just 400 training tuples. This sample efficiency (that is, speed of learning) is orders of magnitude better than what is achievable with

Table 1. Average length (number of steps) of a testing episode as a function of training set size, in the mountain car domain

tuples	length
100	285.6
200	317.8
400	88.1
800	88.6
1600	87.4
3200	87.2
6400	87.3
12800	85.9

Notes: The size of a training set is the number of $\langle s, a, r, s' \rangle$ tuples used for bilinear Q iteration. For each tuple independently, s and a are chosen from a uniform distribution over the legal state and action spaces. The discount factor is 0.9; performance is not sensitive to its exact value. Following previous work, the starting state for each testing episode is the bottom of the valley with zero velocity. A testing episode terminates when the goal state is reached, or after 500 steps.

variants of Q learning [Smart and Kaelbling, 2000]. The most sample-efficient previously published method appears to be fitted Q iteration with a neural network [Riedmiller, 2005]; the bilinear method requires several times fewer training examples. We conjecture that the bilinear method is sample-efficient because it needs to learn values for fewer parameters. The bilinear method also requires less computation, because both learning and the argmax operation are linear.

Some details of the experiment are important to mention. The precise scenario described in [Sutton and Barto, 1998] is used. Many subsequent papers, including [Riedmiller, 2005], have made changes to the scenario, which makes experimental results not directly comparable. Table 1 is based on testing from a fixed state; we can also consider test episodes starting in arbitrary states. About 1/4 of training sets of size 400 yield Q functions that are successful (reach the goal) from every possible initial state. The other 3/4 of training sets lead to Q functions that are successful for the majority of initial states.

6   Inventory Management Experiments

In many potential applications of reinforcement learning, the state space, and possibly also the action space, is high-dimensional. For example, in many business domains the state is essentially the current characteristics of a customer, who is represented by a high-dimensional real-valued vector [Simester et al., 2006]. Most existing RL methods cannot handle applications of this nature, as discussed by [Dietterich, 2009]. Research in this area is multidisciplinary, with successful current methods arising from both the operations research community [Powell, 2007] and the machine learning community [Hannah and Dunson, 2011].

In inventory management applications, at each time step the agent sees a demand vector and/or a supply vector for a number of products. The agent must decide how to satisfy each demand in order to maximize long-term benefit, subject to various rules about the substitutability and perishability of products. Managing the stocks of a blood bank is an important application of this type. In this section we describe initial results based on the formulation described in

Table 2. Short-term reward functions for blood bank management

	[Yu, 2007]	new
supply exact blood type	50	0
substitute O- blood	60	0
substitute other type	45	0
fail to meet demand	0	-60
discard blood	-20	0

[Yu, 2007] and Chapter 12 of [Powell, 2007]. The results are preliminary because some implementation issues are not yet resolved.

The blood bank stores blood of eight types: AB+, AB-, A+, A-, B+, B-, O+, and O-. Each period, the manager sees a certain level of demand for each type, and a certain level of supply. Some types of blood can be substituted for some others; 27 of the 64 conceivable substitutions are allowed. Blood that is not used gets older, and beyond a certain age must be discarded. With three discrete ages, the inventory of the blood bank is a vector in $\mathbb{R}^{24}$. The state of the system is this vector concatenated with the 8-dimensional demand vector and a unit constant component. The action vector has $3 \cdot 27 + 1 = 82$ dimensions since there are 3 ages, 27 possible allocations, and a unit constant pseudo-action. Each component of the action vector is a quantity of blood of one type and age, used to meet demand for blood of the same or another type. The LP solved to determine the action at each time step has learned coefficients in its objective function, but fixed constraints such as that the total quantity supplied from each of the 24 stocks must be not be more than the current stock amount.

The blood bank scenario is an abstraction of a real-world situation where the objectives to be maximized, both short-term and long-term, are subject to debate. Previous research has used measures of long-term success that are intuitively reasonable, but not mathematically consistent with the immediate reward functions used. Table 2 shows two different immediate reward functions. Each entry is the benefit accrued by meeting one unit of demand with supply of a certain nature. The middle column is the short-term reward function used in previous work, while the right column is an alternative that is more consistent with the long-term evaluation measure used previously (described below).

For training and for testing, a trajectory is a series of periods. In each period, the agent sees a demand vector drawn randomly from a specific distribution. The agent supplies blood according to its policy and sees its immediate reward as a consequence. Note that the immediate reward is defined by a function given in Table 2, but the agent does not know this function. All training and testing is in the standard reinforcement learning scenario, where the agent sees only random realizations of the MDP.

Then, blood remaining in stock is aged by one period, blood that is too old is discarded, and fresh blood arrives according to a different probability distribution. Training is based on 1000 trajectories, each 10 periods long, in

Table 3. Success of alternative learned policies

	average unmet A+ demand	frequency of severe unmet A+ demand
greedy policy	7.3%	46%
policy of [Yu, 2007]	7.9%	30%
bilinear method (i)	18.4%	29%
bilinear method (ii)	7.55%	12%

Notes: Column headings are explained in the text. Current and previous percentages are not directly comparable, because of differing numbers of periods and other differences.

which the agent follows a greedy policy.[2] Testing is based on trajectories of the same length where supply and demand vectors are drawn from the same distributions, but the agent follows a learned policy.

According to [Yu, 2007], the percentage of unsatisfied demand is the best long-term measure of success for a blood bank. If a small fraction of demand is not met, that is acceptable because all high-priority demands can still be met. However, if more than 10% of demand for any type is not met in a given period, then some patients may suffer seriously. Therefore, we measure both the average percentage of unmet demand and the frequency of unmet demand over 10%. Again according to [Yu, 2007], the A+ blood type is a good indicator of success, because the discrepancy between supply and demand is greatest for it: on average, 34.00% of demand but only 27.94% of supply is for A+ blood.

Table 3 shows the performance reported by [Yu, 2007], and the preliminary performance of the bilinear method using the two reward functions of Table 2. Given the original immediate reward function, (i), fitted Q iteration with a bi-linear Q function satisfies on average less of the demand for A+ blood. However, the original reward function does not include an explicit penalty for failing to meet demand. The implementation of [Yu, 2007] is tuned in a domain-specific way to satisfy more demand for A+ blood even though doing so does not accrue explicit immediate reward. The last column of Table 2 shows a different im-mediate reward function that does penalize failures to meet demand. With this reward function, the bilinear approach learns a policy that reduces the frequency of severe failures.

7   Discussion

Reinforcement learning has been an important research area for several decades, but it is still a major challenge to use training examples efficiently, to be

---

[2] The greedy policy is the policy that supplies blood to maximize one-step reward, at each time step. Many domains are like the blood bank domain in that the optimal one-step policy, also called myopic or greedy, can be formulated directly as a lin-ear programming problem with known coefficients. With bilinear fitted Q iteration, coefficients are learned that formulate a long-term policy as a problem in the same class. The linear representation is presumably not sufficiently expressive to represent the optimal long-term policy, but it is expressive enough to represent a policy that is better than the greedy one.

computationally tractable, and to handle action and state spaces that are high-dimensional and continuous. Fitted Q iteration with a bilinear Q function can meet these criteria, and initial experimental results confirm its usefulness. The constraints on what actions are legal in each state are accommodated in a straightforward way by the linear programming algorithm that computes the optimal action vector at each time step. We intend to explore this approach further in the blood bank domain and in other large-scale reinforcement learning scenarios.

Acknowledgments. The author is grateful to Vivek Ramavajjala for the initial implementation of the bilinear Q iteration method, used for the experiments described above. Thanks are also due to anonymous referees and others for beneficial comments that led to definite improvements in the paper.

References

[Chakraborty et al., 2008] Chakraborty, B., Strecher, V., Murphy, S.: Bias correction and confidence intervals for fitted Q-iteration. In: NIPS Workshop on Model Uncertainty and Risk in Reinforcement Learning (2008)

[Chang et al., 2010] Chang, Y.W., Hsieh, C.J., Chang, K.W., Ringgaard, M., Lin, C.J.: Training and testing low-degree polynomial data mappings via linear SVM. Journal of Machine Learning Research 11, 1471–1490 (2010)

[De Farias and Van Roy, 2003] De Farias, D.P., Van Roy, B.: The linear programming approach to approximate dynamic programming. Operations Research 51(6), 850–865 (2003)

[Dietterich, 2009] Dietterich, T.G.: Machine Learning and Ecosystem Informatics: Challenges and Opportunities. In: Zhou, Z.-H., Washio, T. (eds.) ACML 2009. LNCS, vol. 5828, pp. 1–5. Springer, Heidelberg (2009)

[Džeroski et al., 2001] Džeroski, S., De Raedt, L., Driessens, K.: Relational reinforcement learning. Machine Learning 43(1), 7–52 (2001)

[Ernst et al., 2005] Ernst, D., Geurts, P., Wehenkel, L.: Tree-based batch mode reinforcement learning. Journal of Machine Learning Research 6(1), 503–556 (2005)

[Gordon, 1995a] Gordon, G.J.: Stable fitted reinforcement learning. In: Advances in Neural Information Processing Systems (NIPS), pp. 1052–1058 (1995a)

[Gordon, 1995b] Gordon, G.J.: Stable function approximation in dynamic programming. In: Proceedings of the International Conference on Machine Learning (ICML), pp. 261–268 (1995b)

[Hannah and Dunson, 2011] Hannah, L.A., Dunson, D.B.: Approximate dynamic programming for storage problems. In: Proceedings of the International Conference on Machine Learning, ICML (2011)

[Judd and Solnick, 1994] Judd, K.L., Solnick, A.J.: Numerical dynamic programming with shape-preserving splines. Unpublished paper from the Hoover Institution (1994), http://bucky.stanford.edu/papers/dpshape.pdf

[Lagoudakis and Parr, 2003] Lagoudakis, M.G., Parr, R.: Least-squares policy iteration. Journal of Machine Learning Research 4, 1107–1149 (2003)

[Lazaric et al., 2007] Lazaric, A., Restelli, M., Bonarini, A.: Reinforcement learning in continuous action spaces through sequential Monte Carlo methods. In: Advances in Neural Information Processing Systems 20 (NIPS). MIT Press (2007)

[Melo and Lopes, 2008] Melo, F.S., Lopes, M.: Fitted Natural Actor-Critic: A New Algorithm for Continuous State-Action MDPs. In: Daelemans, W., Goethals, B., Morik, K. (eds.) ECML PKDD 2008, Part II. LNCS (LNAI), vol. 5212, pp. 66–81. Springer, Heidelberg (2008)

[Murphy, 2005] Murphy, S.A.: A generalization error for Q-learning. Journal of Machine Learning Research 6, 1073–1097 (2005)

[Neumann, 2008] Neumann, G.: Batch-mode reinforcement learning for continuous state spaces: A survey. ÖGAI Journal 27(1), 15–23 (2008)

[Pazis and Lagoudakis, 2009] Pazis, J., Lagoudakis, M.G.: Binary action search for learning continuous-action control policies. In: Proceedings of the 26th Annual International Conference on Machine Learning (ICML), pp. 100–107 (2009)

[Pazis and Parr, 2011] Pazis, J., Parr, R.: Generalized value functions for large action sets. In: Proceedings of the International Conference on Machine Learning, ICML (2011)

[Powell, 2007] Powell, W.B.: Approximate Dynamic Programming. John Wiley & Sons, Inc. (2007)

[Riedmiller, 2005] Riedmiller, M.: Neural Fitted Q Iteration - First Experiences with a Data Efficient Neural Reinforcement Learning Method. In: Gama, J., Camacho, R., Brazdil, P.B., Jorge, A.M., Torgo, L. (eds.) ECML 2005. LNCS (LNAI), vol. 3720, pp. 317–328. Springer, Heidelberg (2005)

[Simester et al., 2006] Simester, D.I., Sun, P., Tsitsiklis, J.N.: Dynamic catalog mailing policies. Management Science 52(5), 683–696 (2006)

[Smart and Kaelbling, 2000] Smart, W.D., Kaelbling, L.P.: Practical reinforcement learning in continuous spaces. In: Proceedings of the 17th International Conference on Machine Learning (ICML), pp. 903–910 (2000)

[Stachurski, 2008] Stachurski, J.: Continuous state dynamic programming via nonexpansive approximation. Computational Economics 31(2), 141–160 (2008)

[Sutton and Barto, 1998] Sutton, R.S., Barto, A.G.: Reinforcement learning: An introduction. MIT Press (1998)

[Todorov, 2009] Todorov, E.: Efficient computation of optimal actions. In: Proceedings of the National Academy of Sciences 106(28), 11478–11483 (2009)

[van Hasselt, 2010] van Hasselt, H.P.: Double Q-learning. In: Advances in Neural Information Processing Systems (NIPS), vol. 23 (2010)

[Viviani and Flash, 1995] Viviani, P., Flash, T.: Minimum-jerk, two-thirds power law, and isochrony: converging approaches to movement planning. Journal of Experimental Psychology 21, 32–53 (1995)

[Yu, 2007] Yu, V.: Approximate dynamic programming for blood inventory management. Honors thesis, Princeton University (2007)

ℓ_1-Penalized Projected Bellman Residual

Matthieu Geist[1] and Bruno Scherrer[2]

[1] Supélec, IMS Research Group, Metz France
[2] INRIA, MAIA Project-Team, Nancy France

Abstract. We consider the task of feature selection for value function approximation in reinforcement learning. A promising approach consists in combining the Least-Squares Temporal Difference (LSTD) algorithm with ℓ_1-regularization, which has proven to be effective in the supervised learning community. This has been done recently whit the LARS-TD algorithm, which replaces the projection operator of LSTD with an ℓ_1-penalized projection and solves the corresponding fixed-point problem. However, this approach is not guaranteed to be correct in the general off-policy setting. We take a different route by adding an ℓ_1-penalty term to the projected Bellman residual, which requires weaker assumptions while offering a comparable performance. However, this comes at the cost of a higher computational complexity if only a part of the regularization path is computed. Nevertheless, our approach ends up to a supervised learning problem, which let envision easy extensions to other penalties.

1 Introduction

A core problem of reinforcement learning (RL) [19] is to assess the quality of some control policy (for example within a policy iteration context), quantified by an associated value function. In the less constrained setting (large state space, unknown transition model), there is a need for estimating this function from sampled trajectories. Often, a parametric representation of the value function is adopted, and many algorithms have been proposed to learn the underlying parameters [2,21]. This implies to choose *a priori* the underlying architecture, such as basis functions for a parametric representation or the neural topology for a multi-layered perceptron. This problem-dependent task is more difficult in RL that in the more classical supervised setting because the value function is never directly observed, but defined as the fixed-point of an associated Bellman operator.

A general direction to alleviate this problem is the study of non-parametric approaches for value function approximation. This implies many different methods, such as feature construction [10,15] or Kernel-based approaches [7,22]. Another approach consists in defining beforehand a (very) large number of features and then choosing automatically those which are relevant for the problem at hand. This is generally known as feature selection. In the supervised learning setting, this general idea is notably instantiated by ℓ_1-regularization [24,5], which has been recently extended to value function approximation using different approaches [13,12,11,16].

S. Sanner and M. Hutter (Eds.): EWRL 2011, LNCS 7188, pp. 89–101, 2012.

In this paper, we propose an alternative ℓ_1-regularization of the Least-Squares Temporal Difference (LSTD) algorithm [4]. One searches for an approximation of the value function V (being a fixed-point of the Bellman operator T) belonging to some (linear) hypothesis space $\mathcal{H}$, onto which one projects any function using the related projection operator Π. LSTD provides $\hat{V} \in \mathcal{H}$, the fixed-point of the composed operator ΠT. The sole generalization of LSTD to ℓ_1-regularization has been proposed in [12] ([11] solves the same problem, [13] regularizes a - biased- Bellman Residual and [16] considers linear programming). They add an ℓ_1-penalty term to the projection operator and solve the consequent fixed-point problem, the corresponding algorithm being called LARS-TD in reference to the homotopy path algorithm LARS (Least Angle Regression) [6] which inspired it. However, their approach does not correspond to any convex optimization problem and is improper if some conditions are not met.

In this paper, we propose to take a different route to combine LSTD with ℓ_1-regularization. Instead of searching for a fixed-point of the Bellman operator combined with the ℓ_1-regularized projection operator, we add an ℓ_1 penalty term to the minimization of a projected Bellman residual, introducing the ℓ_1-PBR (Projected Bellman Residual) algorithm. Compared to [12], the proposed approach corresponds to a convex optimization problem. Consequently, it is correct under much weaker assumptions, at the cost of a generally higher computational cost. Section 2 reviews some useful preliminaries, notably the LSTD and the LARS-TD algorithms. Section 3 presents the proposed approach and discusses some of its properties, in light of the state of the art. Section 4 illustrates our claims and intuitions on simple problems and Section 5 opens perspectives.

2   Preliminaries

A Markovian decision process (MDP) is a tuple $\{S, A, P, R, \gamma\}$ where S is the finite[1] state space, A the finite action space, $P : s, a \in S \times A \to p(.|s,a) \in \mathcal{P}(S)$ the family of Markovian transition probabilities, $R : s \in S \to r = R(s) \in \mathbb{R}$ the bounded reward function and γ the discount factor weighting long-term rewards. According to these definitions, the system stochastically steps from state to state conditionally on the actions the agent performs. Let i be the discrete time step. To each transition (s_i, a_i, s_i') is associated an immediate reward r_i. The action selection process is driven by a policy $\pi : s \in S \to \pi(s) \in A$. The quality of a policy is quantified by the value function V^π, defined as the expected discounted cumulative reward starting in a state s and then following the policy π: $V^\pi(s) = E[\sum_{i=0}^\infty \gamma^i r_i | s_0 = s, \pi]$. Thanks to the Markovian property, the value function of a policy π is the unique fixed-point of the Bellman operator

$$T^\pi : V \in \mathbb{R}^S \to T^\pi V \in \mathbb{R}^S : T^\pi V(s) = E_{s'|s,\pi(s)}[R(s) + \gamma V(s')]. \qquad (1)$$

Let $P^\pi = (p(s'|s, \pi(s)))_{1 \le s, s' \le |S|}$ be the associated transition matrix, the value function is therefore the solution of the linear system $V^\pi = R + \gamma P^\pi V^\pi$.

---

[1] The finite state space assumption is made for simplicity, but this work can easily be extended to continuous state spaces.

In the general setting addressed here, two problems arise. First, the model (that is R and P^π) is unknown, and one should estimate the value function from sampled transitions. Second, the state space is too large to allow an exact representation and one has to rely on some approximation scheme. Here, we search for a value function $\hat{V}^\pi$ being a linear combination of p basis functions $\phi_i(s)$ chosen beforehand, the parameter vector being noted θ:

$$\hat{V}^\pi(s) = \sum_{i=1}^{p} \theta_i \phi_i(s) = \theta^T \phi(s), \theta \in \mathbb{R}^p, \phi(s) = \left(\phi_1(s) \ldots \phi_p(s) \right)^T. \tag{2}$$

Let us note $\Phi \in \mathbb{R}^{|S| \times p}$ the feature matrix whose rows contain the feature vectors $\phi(s)^T$ for any state $s \in S$. This defines an hypothesis space $\mathcal{H} = \{\Phi\theta | \theta \in \mathbb{R}^p\}$ into which we should search for a good approximation $\hat{V}^\pi$ of V^π.

2.1   LSTD

The LSTD algorithm [4] minimizes the distance between the value function $\hat{V}$ and the back-projection onto $\mathcal{H}$ of its image under the Bellman operator (this image having no reason to belong to $\mathcal{H}$):

$$\hat{V}^\pi = \operatorname*{argmin}_{V \in \mathcal{H}} \| V - \Pi T^\pi V \|_D^2, \quad \Pi T^\pi V = \operatorname*{argmin}_{h \in \mathcal{H}} \| T^\pi V - h \|_D^2, \tag{3}$$

with $D \in \mathbb{R}^{|S| \times |S|}$ being a diagonal matrix whose components are some state distribution. With a linear parameterization, $\hat{V}^\pi$ is actually the fixed-point of the composed ΠT^π operator: $\hat{V}^\pi = \Pi T^\pi \hat{V}^\pi$.

However, the model is unknown (and hence T^π), so LSTD actually solves a samples-based fixed-point problem. Assume that we have a set of n transitions $\{(s_i, a_i, r_i, s_i')\}_{1 \le i \le n}$, not necessarily sampled along one trajectory. Let us introduce the sampled based feature and reward matrices:

$$\tilde{\Phi} = \begin{pmatrix} \phi(s_1)^T \\ \vdots \\ \phi(s_n)^T \end{pmatrix} \in \mathbb{R}^{n \times p}, \quad \tilde{\Phi}' = \begin{pmatrix} \phi(s_1')^T \\ \vdots \\ \phi(s_n')^T \end{pmatrix} \in \mathbb{R}^{n \times p}, \quad \tilde{R} = \begin{pmatrix} r_1 \\ \vdots \\ r_n \end{pmatrix} \in \mathbb{R}^n. \tag{4}$$

The LSTD estimate θ^* is thus given by the following nested optimization problems, the first equation depicting the projection and the second the minimization:

$$\begin{cases} \omega_\theta = \operatorname{argmin}_{\omega \in \mathbb{R}^p} \| \tilde{R} + \gamma \tilde{\Phi}' \theta - \tilde{\Phi}\omega \|^2 \\ \theta^* = \operatorname{argmin}_{\theta \in \mathbb{R}^p} \| \tilde{\Phi}\omega_\theta - \tilde{\Phi}\theta \|^2 \end{cases}. \tag{5}$$

The parameterization being linear, this can be easily solved:

$$\theta^* = \tilde{A}^{-1}\tilde{b}, \quad \tilde{A} = \tilde{\Phi}^T \Delta\tilde{\Phi}, \quad \Delta\tilde{\Phi} = \tilde{\Phi} - \gamma\tilde{\Phi}', \quad \tilde{b} = \tilde{\Phi}^T \tilde{R}. \tag{6}$$

Asymptotically, $\Phi\theta^*$ converges to the fixed-point of ΠT^π.

2.2   LARS-TD

In supervised learning, ℓ_1-regularization [24,5] consists in adding a penalty on the minimized objective function, this penalty being proportional to the ℓ_1-norm of the parameter vector, $\|\theta\|_1 = \sum_{j=1}^{p} |\theta_j|$. As ℓ_2-regularization, this prevents overfitting, but the use of the ℓ_1-norm also produces sparse solutions (components of θ being exactly set to zero). Therefore, adding such a penalty is often understood as performing feature selection.

In order to combine LSTD with ℓ_1-regularization, It has been proposed to add an ℓ_1-penalty term to the projection equation [12]. This corresponds to the following optimization problem:

$$\begin{cases} \omega_\theta = \mathrm{argmin}_{\omega \in \mathbb{R}^p} \|\tilde{R} + \gamma \tilde{\Phi}'\theta - \tilde{\Phi}\omega\|^2 + \lambda\|\omega\|_1 \\ \theta^* = \mathrm{argmin}_{\theta \in \mathbb{R}^p} \|\tilde{\Phi}\omega_\theta - \tilde{\Phi}\theta\|^2 \end{cases}, \tag{7}$$

where λ is the regularization parameter. Equivalently, this can be seen as solving the following fixed-point problem:

$$\theta^* = \underset{\theta \in \mathbb{R}^p}{\mathrm{argmin}} \|\tilde{R} + \gamma \tilde{\Phi}'\theta^* - \tilde{\Phi}\theta\|^2 + \lambda\|\theta\|_1. \tag{8}$$

This optimization problem cannot be formulated as a convex one [12]. However, based on subdifferential calculus, equivalent optimality conditions can be derived, which can be used to provide a LARS-like homotopy path algorithm. Actually, under some conditions, the estimate θ^* as a function of λ is piecewise linear. In the supervised setting, LARS [6] is an algorithm which compute efficiently the whole regularization path, that is the solutions $\theta^*(\lambda)$ for any $\lambda \geq 0$, by identifying the breaking points of the regularization path. In [12], a similar algorithm solving optimization problem (7) is proposed. A finite sample analysis of LARS-TD has been provided recently [8], in the on-policy case.

For LARS-TD to be correct (that is admitting a continuous and unique regularization path [11]), it is sufficient for $\tilde{A}$ to be a P-matrix[2] [12]. In the on-policy case (that is the state distribution is the MDP stationary distribution induced by the policy π), given enough samples, $\tilde{A}$ is positive definite and hence a P-matrix. However, if the state distribution is different from the stationary distribution, which is typically the case in an off-policy setting, no such guarantee can be given. This is a potential weakness of this approach, as policy evaluation often occurs in some off-policy policy iteration context.

3 ℓ_1-penalized Projected Bellman Residual

Starting from the same classical formulation of LSTD in Equation (5), we take a different route to add regularization, to be compared to the LARS-TD approach depicted in Equation (7):

---

[2] A square matrix is a P-matrix if all its principle minors are strictly positive. This is a strict superset of the class of (non-symmetric) definite positive matrices.

$$\begin{cases} \omega_\theta = \operatorname{argmin}_{\omega \in \mathbb{R}^p} \| \tilde{R} + \gamma \tilde{\Phi}' \theta - \tilde{\Phi} \omega \|^2 \\ \theta^* = \operatorname{argmin}_{\theta \in \mathbb{R}^p} \| \tilde{\Phi} \omega_\theta - \tilde{\Phi} \theta \|^2 + \lambda \| \theta \|_1 \end{cases} . \tag{9}$$

Instead of adding the ℓ_1-penalty term to the projection equation, as in LARS-TD, we propose to add it to the minimization equation.

In order to investigate the conceptual difference between both approaches, we will now consider their asymptotic behavior. Let Π_λ be the ℓ_1-penalized projection operator: $\Pi_\lambda V = \operatorname{argmin}_{h \in \mathcal{H}} \| V - h \|_D^2 + \lambda \| h \|_1$. The LARS-TD algorithm searches for a fixed-point of the composed operator $\Pi_\lambda T^\pi$, which appears clearly from its analysis [8]: $\hat{V} = \Pi_\lambda T^\pi \hat{V}$. On the other hand, the approach proposed in this paper adds an ℓ_1-penalty term to the minimization of the (classical) projection of the Bellman residual:

$$\hat{V} = \operatorname*{argmin}_{\Phi \theta \in \mathcal{H}} \| \Pi (\Phi \theta - T^\pi (\Phi \theta)) \|_D^2 + \lambda \| \theta \|_1 \tag{10}$$

Because of this, we name it ℓ_1-PBR (Projected Bellman Residual). Both approaches make sense, both with their own pros and cons. Before discussing this and studying the properties of ℓ_1-PBR, we provide a practical algorithm.

3.1   Practical Algorithm

The proposed ℓ_1-PBR turns out to be much simpler to solve than LARS-TD. We assume that the matrix $\tilde{\Phi}^T \tilde{\Phi} \in \mathbb{R}^{p \times p}$ is invertible[3]. The projection equation can be solved analytically:

$$\Phi \omega_\theta = \hat{\Pi} (\tilde{R} + \gamma \tilde{\Phi}' \theta), \quad \hat{\Pi} = \tilde{\Phi} (\tilde{\Phi}^T \tilde{\Phi})^{-1} \tilde{\Phi}^T \tag{11}$$

Therefore, optimization problem (9) can be written in the following equivalent form:

$$\theta^* = \operatorname*{argmin}_{\theta \in \mathbb{R}^p} \| \tilde{y} - \tilde{\Psi} \theta \|^2 + \lambda \| \theta \|_1, \quad \tilde{y} = \hat{\Pi} \tilde{R}, \quad \tilde{\Psi} = \tilde{\Phi} - \gamma \hat{\Pi} \tilde{\Phi}' \tag{12}$$

The interesting thing here is that $\tilde{y}$ and $\tilde{\Psi}$ being defined, we obtain a purely supervised learning problem. First, this allows solving it by applying directly the LARS algorithm, or any other approache such as LCP. Second, this would still hold for any other penalization term. Therefore, extensions of the proposed ℓ_1-PBR to adaptive lasso [25] or elastic-net [26] (among others) are straightforward.

The pseudo-code of ℓ_1-PBR is provided in Alg. 1 ($\tilde{\Psi}_\mathcal{A}$ denotes the columns of $\tilde{\Psi}$ corresponding to the indices in the current active set $\mathcal{A}$, and similarly for a vector). As it is a direct use of LARS, we refer the reader to [6] for

---

[3] If this is not the case, it is sufficient to add an arbitrary small amount of ℓ_2-regularization to the projection equation.

Algorithm 1. ℓ_1-PBR

Initialization;
Compute $\hat{\Pi} = \tilde{\Phi}(\tilde{\Phi}^T \tilde{\Phi})^{-1} \tilde{\Phi}^T$, $\tilde{y} = \hat{\Pi} \tilde{R}$, $\tilde{\Psi} = \tilde{\Phi} - \gamma \hat{\Pi} \tilde{\Phi}'$;
Set $\theta = \mathbf{0}$ and initialize the correlation vector $c = \tilde{\Psi}^T \tilde{R}$;
Let $\{\bar{\lambda}, i\} = \max_j(|c_j|)$ and initialize the active set $\mathcal{A} = \{i\}$;

while $\bar{\lambda} > \lambda$ **do**

$\quad$ Compute update direction: $\varDelta\theta_{\mathcal{A}} = (\tilde{\Psi}_{\mathcal{A}}^T \tilde{\Psi}_{\mathcal{A}})^{-1} \mathrm{sgn}(c_{\mathcal{A}})$;

$\quad$ Find step size to add element: $\{\delta_1, i_1\} = \min^+_{j \notin \mathcal{A}}(\frac{c_j - \bar{\lambda}}{d_j - 1}, \frac{c_j + \bar{\lambda}}{d_j + 1})$ with $d = \tilde{\Psi}^T \tilde{\Psi}_{\mathcal{A}} \varDelta\theta_{\mathcal{A}}$;

$\quad$ Find step size to remove element: $\{\delta_2, i_2\} = \min^+_{j \in \mathcal{A}}(-\frac{\theta_j}{\varDelta\theta_j})$;

$\quad$ Compute $\delta = \min(\delta_1, \delta_2, \bar{\lambda} - \lambda)$. Update $\theta \leftarrow \theta_{\mathcal{A}} - \delta_{\mathcal{A}} \varDelta\theta_{\mathcal{A}}$, $\bar{\lambda} \leftarrow \bar{\lambda} - \delta$ and $c \leftarrow c - \delta d$;

$\quad$ Add i_1 to ($\delta_1 < \delta_2$) or remove i_2 from ($\delta_2 < \delta_1$) the active set $\mathcal{A}$;

---

full details and provide only the general idea. It can be shown (see Sec. 3.2) that the regularization path is piecewise linear. Otherwise speaking, there exists $\{\lambda_0 = 0, \dots, \lambda_k\}$ such that for any $\lambda \in]\lambda_i, \lambda_{i+1}[$, $\nabla_\lambda \theta^*$ is a constant vector. Let us call this regularization values the breaking points. As for λ large enough, the trivial solution is $\theta^* = 0$, $\lambda_k < \infty$. The LARS algorithm starts by identifying the breaking point λ_k at which $\theta^* = 0$. Then, it sequentially discovers other breaking points until $\lambda_0 = 0$ or until a specified regularization factor is reached (or possibly until a specified number of features have been added to the active set). For each interval $]\lambda_i, \lambda_{i+1}[$, it computes the constant vector $\nabla_\lambda \theta^* = \varDelta\theta$, which allows inferring the solution for any point of the path. These breaking points correspond to activation or deactivation of a basis function (that is a parameter becomes nonzero or zero), this aspect being inherent to the fact that the ℓ_1-norm is not differentiable at zero. Computing the candidate breaking points in the algorithm corresponds to detecting when one of the equivalent optimality conditions is violated.

Compared to LARS-TD, the disadvantage of ℓ_1-PBR is its higher time and memory complexities. Both algorithms share the same complexities per iteration of the LARS-like homotopy path algorithm. However, ℓ_1-PBR additionally requires projecting the reward and some features onto the hypothesis space $\mathcal{H}$ (that is computing $\tilde{y}$ and $\tilde{\Psi}$ in Equation (12)). This adds the complexity of a full least-squares. Computing the full regularization path with LARS-TD presents also the same complexity as a full least-squares; in this case, both approaches requires the same order of computations and memory. However, if only a part of the regularization path is computed, the complexity of LARS-TD decreases to solving a least-squares with as many parameters as there are active features for the smallest value of λ, whereas the complexity of ℓ_1-PBR keeps the same order.

3.2   Correctness of ℓ_1-PBR

The LARS-TD algorithm requires the matrix $\tilde{A} = \tilde{\Phi}^T \Delta \tilde{\Phi}$ to be a P-matrix in order to find a solution. The next straightforward property shows that ℓ_1-PBR requires much weaker conditions.

Theorem 1. *If $\tilde{A} = \tilde{\Phi}^T \Delta \tilde{\Phi}$ and $\tilde{M} = \tilde{\Phi}^T \tilde{\Phi}$ are invertible, then the ℓ_1-PBR algorithm finds a unique solution for any $\lambda \geq 0$, and the associated regularization path is piecewise linear.*

Proof. As Equation (12) defines a supervised optimization problem, checking that $\tilde{\Psi}^T \tilde{\Psi}$ is symmetric positive definite is sufficient for the result (this matrix being symmetric positive by construction). The matrix $\tilde{M}$ being invertible, the empirical projection operator $\hat{\Pi}$ is well defined. Moreover, $\hat{\Pi}$ being a projection, $\hat{\Pi}\tilde{\Phi} = \tilde{\Phi}$, $\hat{\Pi}^T = \hat{\Pi}$ and $\hat{\Pi}^2 = \hat{\Pi}$. Therefore, we have that $\tilde{\Psi}^T \tilde{\Psi} = \tilde{A}^T \tilde{M}^{-1} \tilde{A}$. The considered optimization problem is then strictly convex, hence the existence and uniqueness of its solution. Piecewise linearity of the regularization path is a straightforward consequence of Prop. 1 of [17].

These conditions are much weaker than the ones of LARS-TD. Moreover, even if they are not satisfied (for example if there are more basis functions than samples, that is $p > n$), one can add a small ℓ_2-penalty term to each equation of (9). This would correspond to replacing the projection by an ℓ_2-penalized projection and the ℓ_1-penalty term by an elastic net one. In [12], it is argued that the matrix $\tilde{A}$ can be ensured to be a P-matrix by adding an ℓ_2-penalty term. This should be true only for a high enough associated regularization parameter, whereas any strictly positive parameter is sufficient in our case. Moreover, for LARS-TD, a badly chosen regularization parameter can lead to instabilities (if an eigenvalue of $\tilde{A}$ is too close to zero).

3.3   Discussion

Using an ℓ_1-penalty term for value function approximation has been considered before in [16] in an approximate linear programming context or in [13] where it is used to minimize a Bellman residual[4]. Our work is closer to LARS-TD [12], briefly presented in Section 2.2, and both approaches are compared next. In [11], it is proposed to solve the same fixed-point optimization problem (7) using a Linear Complementary Problem (LCP) approach instead of a LARS-like algorithm. This has several advantages. Notably, it allows using warm starts (initializing the algorithm with starting points from similar problems), which is useful in a policy iteration context. Notice that this LCP approach can be easily adapted to the proposed ℓ_1-PBR algorithm. Recall also that ℓ_1-PBR can be easily adapted to many penalty terms, as it ends up to a supervised learning problem (see Equation (12)). This is less clear for other approaches.

---

[4] As noted in [11], they claim to adapt LSTD while actually regularizing a Bellman residual minimization, which is well known to produce biased estimates [1].

As explained in Section 2.2, LARS-TD requires $\tilde{A}$ to be a P-matrix in order to be correct (the LCP approach requires the same assumption [11]). This condition is satisfied in the on-policy case, given enough transitions. However, there is no such result in the more general off-policy case, which is of particular interest in a policy iteration context. An advantage of ℓ_1-PBR is that it relies on much weaker conditions, as shown in Proposition 1. Therefore, the proposed approach can be used safely in an off-policy context, which is a clear advantage over LARS-TD. Regarding this point, an interesting analogy for the difference between LARS-TD and ℓ_1-BRM is the difference between the classical TD algorithm [19] and the recent TDC (TD with gradient Correction) [20].

TD is an online stochastic gradient descent algorithm which aims at solving the fixed-point problem $\hat{V} = \Pi T^\pi \hat{V}$ using a bootstrapping approach. One of its weaknesses is that it can be unstable in an off-policy setting. The TDC algorithm has been introduced in [20] in order to alleviate this problem. TDC is also an online stochastic gradient descent algorithm, but it minimizes the projected Bellman residual $\|\hat{V} - \Pi T^\pi \hat{V}\|^2$. When both approaches converge, they do so to the same solution. However, contrary to TD, TDC is provably convergent in an off-policy context, the required conditions being similar to those of Proposition 1 (A and M should not be singular). This is exactly the difference between LARS-TD and ℓ_1-PBR: LARS-TD penalizes the projection defining the fixed-point problem of interest, whereas ℓ_1-PBR penalizes the projected Bellman residual.

However, this weaker usability conditions have a counterpart: ℓ_1-PBR has generally higher time and memory complexities than LARS-TD, as explained in Section 3.1. Nevertheless, off-policy learning usually suggests batch learning, so this increased cost might not be such a problem. Also, if ℓ_1-PBR is used in a policy iteration context, the computation of the projection can be factorized over iterations, as it does not depend on transiting states.

The proposed algorithm can also be linked to an (unbiased) ℓ_1-penalized Bellman residual minimization (BRM). Let us consider the asymptotic form of the optimization problem solved by ℓ_1-PBR, depicted in Equation (10). Using the Pythagorean theorem, it can be rewritten as follows:

$$\theta^* = \underset{\theta \in \mathbb{R}^p}{\operatorname{argmin}} \|\Phi\theta - T^\pi \Phi\theta\|_D^2 - \|\Pi T^\pi \Phi\theta - T^\pi \Phi\theta\|_D^2 + \lambda\|\theta\|_1 \qquad (13)$$

Assume that the hypothesis space is rich enough to represent $T^\pi V$, for any $V \in \mathcal{H}$. Then, the term $\|\Pi T^\pi \Phi\theta - T^\pi \Phi\theta\|_D^2$ vanishes and ℓ_1-PBR ends up to add an ℓ_1-penalty term to the Bellman Residual Minimization (BRM) cost function. Surely, in practice this term will not vanish, because of the finite number of samples and of a not rich enough hypothesis space. Nevertheless, we expect it to be small given a large enough $\mathcal{H}$, so ℓ_1-PBR should behave similarly to some hypothetic ℓ_1-BRM algorithm, unbiased because computed using the transition model. BRM is known to be more stable and more predictable than LSTD [14,18]. However, it generally leads to a biased estimate, unless a double sampling approach is used or the model is known [1], a problem we do not have. We illustrate this intuition in the next section.

4   Illustration

Two simple problems are considered here. The first one is a two-state MDP which shows that ℓ_1-PBR finds solutions when LARS-TD does not and illustrates the improved stability of the proposed approach. The second problem is the Boyan chain [3]. It is used to illustrate our intuition about the relation between ℓ_1-PBR and ℓ_1-BRM, depicted in Section 3.3, and to compare prediction abilities of LARS-TD and of our approach.

4.1   The Two-State MDP

The first problem is a simple two-state MDP [2,12,18]. The transition matrix is $P = \begin{pmatrix} 0 & 1 \\ 0 & 1 \end{pmatrix}$ and the reward vector $R = \begin{pmatrix} 0 & -1 \end{pmatrix}^T$. The optimal value function is therefore $v^* = \frac{-1}{1-\gamma} \begin{pmatrix} \gamma & 1 \end{pmatrix}^T$. Let us consider the one-feature linear approximation $\Phi = \begin{pmatrix} 1 & 2 \end{pmatrix}^T$ with uniform distribution $D = \begin{pmatrix} \frac{1}{2} & 0 \\ 0 & \frac{1}{2} \end{pmatrix}$. Let γ be the discount factor. Consequently, we have that $A = \Phi^T D(\Phi - \gamma P \Phi) = \frac{5}{2}(1 - \frac{6}{5}\gamma)$ and $b = \Phi^T DR = -1$. The value $\gamma = \frac{5}{6}$ is singular. Below it, A is a P-matrix, but above it is not the case (obviously, in this case $A < 0$).

The solutions to problems (7) and (9), respectively noted $\theta_\lambda^{\text{lars}}$ and $\theta_\lambda^{\text{pbr}}$, can be easily computed analytically in this case. If $\gamma < \frac{5}{6}$, both approaches have an unique regularization path. For LARS-TD, we have $\theta_\lambda^{\text{lars}} = 0$ if $\lambda > 1$ and $\theta_\lambda^{\text{lars}} = -\frac{2}{5(1-\frac{6}{5}\gamma)}(1 - \lambda)$ else. For ℓ_1-PBR, we have $\theta_\lambda^{\text{pbr}} = 0$ if $\lambda > |1 - \frac{6}{5}\gamma|$ and $\theta_\lambda^{\text{pbr}} = -\frac{2}{5(1-\frac{6}{5}\gamma)}(1 - \frac{\lambda \operatorname{sgn}(\theta_\lambda^{\text{pbr}})}{1-\frac{6}{5}\gamma})$ else. If $\gamma > \frac{5}{6}$, the ℓ_1-PBR solution still holds, but LARS-TD no longer admits a unique solution, A being not a P-matrix. The solutions of LARS-TD are the following: $\theta_\lambda^{\text{lars}} = 0$ if $\lambda > 1$, $\theta_\lambda^{\text{lars}} = -\frac{2}{5(1-\frac{6}{5}\gamma)}(1 - \lambda)$ if $\lambda > 1$, and $\theta_\lambda^{\text{lars}} = -\frac{2}{5(1-\frac{6}{5}\gamma)}(1 + \lambda)$ for any $\lambda \geq 0$.

Figure 1 shows the regularization paths of LARS-TD and ℓ_1-PBR for $\gamma = 0.9$, in the just depicted off-policy case (left panel) as well as in the on-policy case (right panel). For $\lambda = 0$, both approaches coincide, as they provide the LSTD solution. In the off-policy case, LARS-TD has up to three solutions, and the regularization path is not continuous, which was already noticed in [12]. ℓ_1-PBR

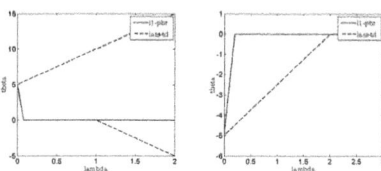

Fig. 1. Two-state MDP, regularization paths (left panel: off-policy; right panel: on-policy)

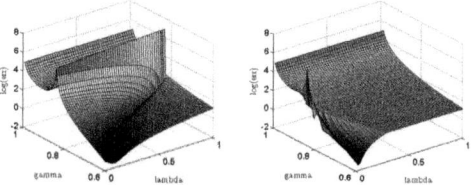

Fig. 2. Two-state MDP, error surface (left: LARS-TD; right: ℓ_1-PBR)

has not this problem, this illustrates Proposition 1. In the on-policy case, both approaches work, providing different regularization paths.

Figure 2 shows the error (defined here as $\|v^* - \Phi\theta_\lambda\|_D$) as a function of the discount factor γ and of the regularization factor λ, in the off-policy case. We restrict ourselves to $\lambda \in [0,1]$, such that LARS-TD has a unique solution for any value of λ. The left panel show the error surface of LARS-TD and the right panel the one of ℓ_1-PBR. For γ small enough, A is a P-matrix and the error is usually slightly lower for LARS-TD than for ℓ_1-PBR. However, when γ is close to the singular value, LARS-TD presents a high error for any value of λ whereas ℓ_1-PBR is more stable (high errors only occurs for small values of λ, close to the singular discount factor). Consequently, on this (somehow pathological) simple example, LARS-TD may have a slightly better prediction ability, but at the cost of a larger zone of instabilities.

4.2   The Boyan Chain

The Boyan chain is a 13-state Markov chain where state s^0 is an absorbing state, s^1 transits to s^0 with probability 1 and a reward of -2, and s^i transits to either s^{i-1} or s^{i-2}, $2 \leq i \leq 12$, each with probability 0.5 and reward -3. The feature vectors $\phi(s)$ for states s^{12}, s^8, s^4 and s^0 are respectively $[1,0,0,0]^T$, $[0,1,0,0]^T$, $[0,0,1,0]^T$ and $[0,0,0,1]^T$, and they are obtained by linear interpolation for other states. The optimal value function is exactly linear in these features, and the corresponding optimal parameter vector is $\theta^* = [-24, -16, -8, 0]^T$. In addition to these 4 relevant features, we added 9 irrelevant features, containing Gaussian random noise for each state (adding more than 9 features would prevent computing the whole regularization path: if $p > |S|$, A and M are necessarily singular).

First, Figure 4 illustrates the regularization paths for ℓ_1-PBR (left panel), LARS-TD (middle panel) and ℓ_1-BRM (Bellman Residual Minimization, right panel). This last algorithm minimizes the classical (unbiased) BRM cost function penalized with an ℓ_1-norm. More formally, the considered optimization problem is:

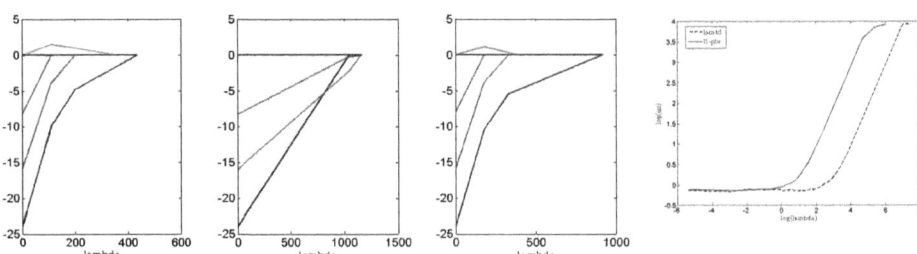

Fig. 3. Boyan chain, regularization paths (left: ℓ_1-PBR; middle: LARS-TD; right: ℓ_1-BRM)   **Fig. 4.** Boyan chain, error curves

$$\theta^* = \underset{\theta \in \mathbb{R}^p}{\mathrm{argmin}} \, \|\tilde{R} - (\tilde{\Phi} - \gamma \tilde{\Xi})\theta\|^2 + \lambda \|\theta\|_1 \text{ with } \tilde{\Xi} = \begin{bmatrix} E_{s'|s_1}[\phi(s')] & \dots & E_{s'|s_n}[\phi(s')] \end{bmatrix}^T.$$

(14)

This can be easily solved using the LARS algorithm, by treating it as a supervised learning approach with observations $\tilde{R}$ and predictors $\tilde{\Phi} - \gamma \tilde{\Xi}$.

The regularization paths have been computed using samples collected from 50 trajectories. One can see on Figure 4 that ℓ_1-PBR and LARS-TD have quite different regularization paths, whereas those of ℓ_1-PBR and ℓ_1-BRM are really close. Irrelevant features have small weights along the whole regularization path for all approaches (most of them cannot even be seen on the figure), and all algorithms converge to the LSTD solution. This tends to confirm the intuition discussed in Section 3.3: with a rich enough hypothesis space, ℓ_1-PBR is close to unbiased ℓ_1-BRM (which is not practical in general, as it requires knowing the transition model for computing the $\tilde{\Xi}$ features).

As regularization paths are quite different for LARS-TD and ℓ_1-PBR, it is interesting to compare their prediction abilities. Figure 4 shows the prediction error (more formally $\|v^* - \Phi\theta\|$) as a function of the regularization parameter for both algorithms (notice the logarithmic scale for both axes). This figure is an average of 1000 independent learning runs using samples generated from 50 trajectories. Error curves are similar, whereas not for the same range of regularization values. Therefore, both approaches offer similar performance on this example.

5   Conclusion

In this paper, we have proposed and alternative to LARS-TD, which searches for the fixed-point of the ℓ_1-projection composed with the Bellman operator. Instead, we add an ℓ_1-penalty term to the minimization of the projected Bellman residual. Notice that the same algorithm has been proposed in parallel and independently in [9], which provides a complementary point of view. Our approach is somehow reminiscent of how TDC [20] has been introduced in order to alleviate the inherent drawback of the classical TD. The proposed approach is correct under weaker conditions and can therefore be used safely in an off-policy setting, contrary to LARS-TD (even if this seems not to be a problem according to the few experiments published [12,11]). Preliminary experiments suggest that both approaches offer comparable performance. As it ends up to a supervised learning problem, ℓ_1-PBR can also be easily extended to other penalty terms. However, this comes at the cost of a higher computational cost. Even if not described in the paper, extension of ℓ_1-PBR to the state-action value function approximation is straightforward. In the future, we plan to perform a deeper theoretical study of the proposed approach (the analysis of [7] in the case of ℓ_2-penalized LSTD can be a lead) and to apply it to control problems (notably Tetris [23] should be an interesting application, as features are quite interpretable).

References

1. Antos, A., Szepesvári, C., Munos, R.: Learning near-optimal policies with Bellman-residual minimization based fitted policy iteration and a single sample path. Machine Learning 71(1), 89–129 (2008)
2. Bertsekas, D.P., Tsitsiklis, J.N.: Neuro-Dynamic Programming. Athena Scientific
3. Boyan, J.A.: Technical Update: Least-Squares Temporal Difference Learning. Machine Learning 49(2-3), 233–246 (1999)
4. Bradtke, S.J., Barto, A.G.: Linear Least-Squares algorithms for temporal difference learning. Machine Learning 22(1-3), 33–57 (1996)
5. Chen, S.S., Donoho, D.L., Saunders, M.A.: Atomic Decomposition by Basis Pursuit. SIAM Journal on Scientific Computing 20, 33–61 (1999)
6. Efron, B., Hastie, T., Johnstone, I., Tibshirani, R.: Least Angle Regression. Annals of Statistics 32(2), 407–499 (2004)
7. Farahmand, A., Ghavamzadeh, M., Szepesvári, C., Mannor, S.: Regularized policy iteration. In: 22nd Annual Conference on Neural Information Processing Systems (NIPS 21), Vancouver, Canada (2008)
8. Ghavamzadeh, M., Lazaric, A., Munos, R., Hoffman, M.: Finite-Sample Analysis of Lasso-TD. In: International Conference on Machine Learning (2011)
9. Hoffman, M.W., Lazaric, A., Ghavamzadeh, M., Munos, R.: Regularized least squares temporal difference learning with nested ℓ_2 and ℓ_1 penalization. In: European Workshop on Reinforcement Learning (2011)
10. Johns, J., Mahadevan, S.: Constructing basis functions from directed graphs for value function approximation. In: Proceedings of the 24th International Conference on Machine Learning, ICML 2007, pp. 385–392. ACM, New York (2007)
11. Johns, J., Painter-Wakefield, C., Parr, R.: Linear Complementarity for Regularized Policy Evaluation and Improvement. In: Lafferty, J., Williams, C.K.I., Shawe-Taylor, J., Zemel, R.S., Culotta, A. (eds.) NIPS 23, pp. 1009–1017 (2010)
12. Kolter, J.Z., Ng, A.Y.: Regularization and Feature Selection in Least-Squares Temporal Difference Learning. In: Proceedings of the 26th International Conference on Machine Learning (ICML 2009), Montreal, Canada (2009)
13. Loth, M., Davy, M., Preux, P.: Sparse Temporal Difference Learning using LASSO. In: IEEE International Symposium on Approximate Dynamic Programming and Reinforcement Learning, Hawaï, USA (2007)
14. Munos, R.: Error bounds for approximate policy iteration. In: International Conference on Machine Learning (2003)
15. Parr, R., Li, L., Taylor, G., Painter-Wakefield, C., Littman, M.L.: An analysis of linear models, linear value-function approximation, and feature selection for reinforcement learning. In: Proceedings of the 25th International Conference on Machine Learning, ICML 2008, pp. 752–759. ACM, New York (2008)
16. Petrik, M., Taylor, G., Parr, R., Zilberstein, S.: Feature Selection Using Regularization in Approximate Linear Programs for Markov Decision Processes. In: Proceedings of ICML (2010)
17. Rosset, S., Zhu, J.: Piecewise linear regularized solution paths. The Annals of Statistics 35(3), 1012–1030 (2007)
18. Scherrer, B.: Should one compute the Temporal Difference fix point or minimize the Bellman Residual? The unified oblique projection view. In: 27th International Conference on Machine Learning - ICML 2010, Haïfa, Israël (2010)
19. Sutton, R.S., Barto, A.G.: Reinforcement Learning: An Introduction (Adaptive Computation and Machine Learning). The MIT Press (1998)

20. Sutton, R.S., Maei, H.R., Precup, D., Bhatnagar, S., Silver, D., Szepesvári, C., Wiewiora, E.: Fast gradient-descent methods for temporal-difference learning with linear function approximation. In: Proceedings of ICML, pp. 993–1000. ACM, New York (2009)
21. Szepesvári, C.: Algorithms for Reinforcement Learning. Morgan and Kaufmann (2010)
22. Taylor, G., Parr, R.: Kernelized value function approximation for reinforcement learning. In: Proceedings of the 26th Annual International Conference on Machine Learning, ICML 2009, pp. 1017–1024. ACM, New York (2009)
23. Thiery, C., Scherrer, B.: Building Controllers for Tetris. International Computer Games Association Journal 32, 3–11 (2009)
24. Tibshirani, R.: Regression Shrinkage and Selection via the Lasso. Journal of the Royal Statistical Society. Series B (Methodological) 58(1), 267–288 (1996)
25. Zou, H.: The adaptive lasso and its oracle properties. Journal of the American Statistical Association 101(476), 1418–1429 (2006)
26. Zou, H., Zhang, H.H.: On the adaptive elastic-net with a diverging number of parameters. The Annals of Statistics 37(4), 1733–1751 (2009)

Regularized Least Squares Temporal Difference Learning with Nested ℓ_2 and ℓ_1 Penalization

Matthew W. Hoffman[1], Alessandro Lazaric[2], Mohammad Ghavamzadeh[2],
and Rémi Munos[2]

[1] University of British Columbia, Computer Science, Vancouver, Canada
[2] INRIA Lille - Nord Europe, Team SequeL, France

Abstract. The construction of a suitable set of features to approximate value functions is a central problem in reinforcement learning (RL). A popular approach to this problem is to use high-dimensional feature spaces together with least-squares temporal difference learning (LSTD). Although this combination allows for very accurate approximations, it often exhibits poor prediction performance because of overfitting when the number of samples is small compared to the number of features in the approximation space. In the linear regression setting, regularization is commonly used to overcome this problem. In this paper, we review some regularized approaches to policy evaluation and we introduce a novel scheme (L_{21}) which uses ℓ_2 regularization in the projection operator and an ℓ_1 penalty in the fixed-point step. We show that such formulation reduces to a standard Lasso problem. As a result, any off-the-shelf solver can be used to compute its solution and standardization techniques can be applied to the data. We report experimental results showing that L_{21} is effective in avoiding overfitting and that it compares favorably to existing ℓ_1 regularized methods.

1 Introduction

In the setting of reinforcement learning (RL), least-squares temporal difference learning (LSTD) [2] is a very popular mechanism for approximating the value function V^π of a given policy π. More precisely, this approximation is accomplished using a linear function space $\mathcal{F}$ spanned by a set of k features $\{\phi_i\}_{i=1}^k$. Here, the choice of feature set greatly determines the accuracy of the value function estimate. In practice, however, there may be no good, *a priori* method of selecting these features. One solution to this problem is to use high-dimensional feature spaces in the hopes that a good set of features lies somewhere in this basis. This introduces other problems, though, as the number of features can outnumber the number of samples $n \leq k$, leading to overfitting and poor prediction. In the linear regression setting, regularization is commonly used to overcome this problem. The two most common regularization approaches involve using ℓ_1 or ℓ_2 penalized least-squares, known as *Lasso* or *ridge regression* respectively (see e.g., [7]). The approach of Lasso is of particular interest due to its *feature selection* property, wherein the geometric form of the ℓ_1 penalty tends to encourage

S. Sanner and M. Hutter (Eds.): EWRL 2011, LNCS 7188, pp. 102–114, 2012.

sparse solutions. This property is especially beneficial in high-dimensional problems, as they allow solutions to be expressed as a linear combination of a small number of features. The application of the Lasso to the problem of value function approximation in high dimensions would thus seem to be a perfect match. However, the RL setting differs greatly from regression in that the objective is not to recover a target function given its noisy observations, but is instead to approximate the fixed-point of the Bellman operator given sample trajectories. The addition of this fixed-point creates difficulties when attempting to extend Lasso results to the RL setting. Despite these difficulties, one particular form of ℓ_1 penalization has been previously studied, both empirically [11,10] and theoretically [9], with interesting results.

In this paper, we consider the two-level nested formulation of LSTD used in [1,5], where the first level optimization is defined by the projection of a Bellman image onto the linear space $\mathcal{F}$, and the second level accounts for the fixed point part of the algorithm. This formulation allows us to define a wide range of algorithms depending on the specific implementation of the projection operator and the fixed point step. In particular, we will discuss a number of regularization methods for LSTD, two based on ℓ_2 penalties (one of which is introduced in [5]), one based on an ℓ_1 penalized projection [11], and finally we will introduce a novel approach which solves the problem via two nested optimization problems including an ℓ_2 and an ℓ_1 penalty (L_{21}). Unlike previous methods using ℓ_1 regularization in LSTD, this new approach introduces the ℓ_1 penalty in the fixed point step and this allows us to cast the problem as a standard Lasso problem. As a result, we do not need to develop any specific method to solve the corresponding optimization problem (as in [11], where a specific implementation of LARS has been defined) and we can apply general-purpose solvers to compute its solution. This additional flexibility also allows us to perform an explicit standardization step on the data similar to what is done in regression. We show in the experiments that all the methods using an ℓ_1 regularization successfully take advantage of the sparsity of V^π and avoid overfitting. Furthermore, we show that, similar to regression, standardization may improve the prediction accuracy, thus allowing L_{21} to achieve a better performance than LARSTD [11].

2   Preliminaries

We consider the standard RL framework [14] wherein a learning agent interacts with a stochastic environment by following a policy π. This interaction is modeled as a Markov Decision Process (MDP) given as a tuple $(\mathcal{X}, \mathcal{A}, P, r, \gamma)$, where $\mathcal{X}$ is a set of states; $\mathcal{A}$ a set of actions; the transition kernel P is such that for all $x \in \mathcal{X}$ and $a \in \mathcal{A}$, $P(\cdot|x, a)$ is a distribution over $\mathcal{X}$; $r : \mathcal{X} \to \mathbb{R}$ is a reward function and $\gamma \in [0, 1]$ a discount factor. Given a deterministic policy $\pi : \mathcal{X} \to \mathcal{A}$, we denote by P^π the transition operator with kernel $P^\pi(\cdot|x) = P(\cdot|x, \pi(x))$.

The value function we are interested in learning maps x to its long-term expected value $V^\pi(x) = \mathbb{E}\left[\sum_{t=0}^\infty \gamma^t r(x_t)|x_0 = x, \pi\right]$. We can also define this quantity as the unique fixed-point of the Bellman operator $V^\pi = T^\pi V^\pi$, where the operator T^π is defined as

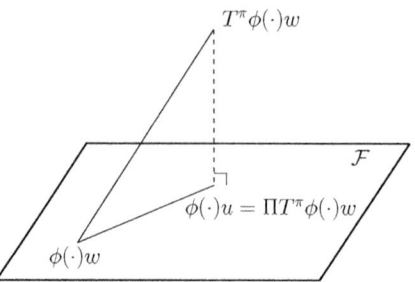

Fig. 1. A graphical illustration of the LSTD problem. Here we see the Bellman operator which takes us out of the space $\mathcal{F}$ and the orthogonal projection back onto this space.

$$(T^\pi V)(x) = r(x) + \gamma \int_{\mathcal{X}} P(dx'|x, \pi(x))V(x')$$

or more concisely as $T^\pi V = r + \gamma P^\pi V$. When both r and P^π are known, this quantity can be solved for analytically, however we will consider the situation that these quantities are either unknown, too large to evaluate, or both.

Rather than directly computing the value function, we will instead seek to approximate V^π with a space $\mathcal{F}$ consisting of linear combinations of k features $\phi : \mathcal{X} \to \mathbb{R}^k$ and weights $w \in \mathbb{R}^k$, i.e. $V^\pi(x) \approx \phi(x)^T w$. The result of the Bellman operator will not necessarily lie in the span of the basis ϕ. Instead, we will adopt the approach of LSTD and approximate the resulting vector with the closest vector that *does* lie in $\mathcal{F}$. We do so by introducing a projection operator Π such that $\Pi V(x) = \phi(x)^T u^*$ where the corresponding weight is the solution to the least-squares problem: $u^* = \arg\min_{u \in \mathbb{R}^k} \|\phi^T u - V\|_\nu^2$, and where $\|f\|_\nu^2 = \int_{\mathcal{X}} f(x)^2 \nu(dx)$ is the $\ell_2(\nu)$-norm of f w.r.t. the distribution ν. By combining the Bellman and projection operators, we can write the LSTD fixed-point as $\widehat{V}^\pi = \Pi T^\pi \widehat{V}^\pi$ which, for $\widehat{V}^\pi = \phi^T w$, can be written as

$$w = u^* = \arg\min_{u \in \mathbb{R}^k} \|\phi^T u - (r + \gamma P^\pi \phi^T w)\|_\nu^2. \tag{1}$$

Alternatively, we can write the value function as the solution $\widehat{V}^\pi = \phi^T w^*$ to the following nested optimization problem [1]:

$$u^* = \arg\min_{u \in \mathbb{R}^k} \|\phi^T u - (r + \gamma P^\pi \phi^T w)\|_\nu^2$$
$$w^* = \arg\min_{w \in \mathbb{R}^k} \|\phi^T w - \phi^T u^*\|_\nu^2. \tag{2}$$

We will refer to the first problem as the *projection* step and the second as the *fixed-point* step. We should also note that the fixed-point step is not the same as the projection of $\phi^T u^*$ since u^* itself depends on w. Further, this interpretation gives us a better picture of what is being optimized by LSTD, namely that we are finding the value function $\widehat{V}^\pi$ which minimizes the distance between itself and its projected Bellman image $\Pi T^\pi \widehat{V}^\pi$ (see Figure 1).

Next, we assume an n-length trajectory consisting of transitions (x_i, a_i, r_i, x_i') sampled from the MDP of interest, and we define the sample matrices

$$\Phi = \begin{bmatrix} \phi(x_1)^T \\ \vdots \\ \phi(x_n)^T \end{bmatrix}, \quad \Phi' = \begin{bmatrix} \phi(x_1')^T \\ \vdots \\ \phi(x_n')^T \end{bmatrix}, \quad R = \begin{bmatrix} r(x_1) \\ \vdots \\ r(x_n) \end{bmatrix}.$$

We can then write an empirical version of (1) and solve for the fixed point by setting $w = u^*$,

$$u^* = \arg\min_{u \in \mathbb{R}^k} \|\Phi u - (R + \gamma \Phi' w)\|_2^2 = (\Phi^T \Phi)^{-1} \Phi^T (R + \gamma \Phi' w), \tag{3}$$

$$w = (\Phi^T (\Phi - \gamma \Phi'))^{-1} \Phi^T R = A^{-1} b, \tag{4}$$

where we have defined $A = \Phi^T (\Phi - \gamma \Phi')$ and $b = \Phi^T R$. While this solution provides an unbiased estimate of the value function, it can perform quite poorly when the number of samples is small in relation to the number of features. This type of scenario often results in *overfitting*, i.e. where we have more free parameters than observations, resulting in an overly complex model that is able to fit the noise of the system rather than the underlying system itself. In the next section, we examine various regularization methods designed to avoid overfitting.

3   Regularized LSTD

In this section we describe four different regularization methods which apply different penalty terms to the *projection* or to the *fixed-point* step introduced in (2). In the first two such schemes we do not penalize the fixed-point step, and we thus leave this step implicit. The final two schemes, however, rely on penalizing both sub-problems. Ultimately we describe a method which uses a mixture of ℓ_2 and ℓ_1 penalties, but that can be expressed as a standard Lasso problem.

3.1 ℓ_2 Penalization (L_2)

The simplest form of regularization we can utilize involves adding an ℓ_2 penalty to the projection operator presented in (3), i.e.

$$u^* = \arg\min_{u \in \mathbb{R}^k} \|\Phi u - (R + \gamma \Phi' w)\|_2^2 + \beta \|u\|_2^2$$

$$= (\Phi^T \Phi + \beta I)^{-1} \Phi^T (R + \gamma \Phi' w). \tag{5}$$

Similar to (4), we solve for the fixed point $w = u^*$ and obtain:

$$w = (\Phi^T (\Phi - \gamma \Phi') + \beta I)^{-1} \Phi^T R = (A + \beta I)^{-1} b. \tag{6}$$

We also see here the standard LSTD components A and b, the only difference with the original formulation being the addition of β along the diagonal of A.

3.2 ℓ_1 Penalization (L_1)

We can also consider adopting an ℓ_1 penalty in the projection step, i.e.

$$u^* = \underset{u \in \mathbb{R}^k}{\arg\min} \|\Phi u - (R + \gamma \Phi' w)\|_2^2 + \beta \|u\|_1. \tag{7}$$

The difficulty with this approach lies in the fact that there is now no closed-form solution to the optimization problem, a fact which causes difficulties when attempting to solve for the fixed-point $w = u^*$. Even though the projection step is just one of ℓ_1 penalized least-squares, the use of the fixed-point results in the problem not being equivalent to the Lasso. In fact, a more specialized algorithm is required to solve this problem. For a full description of this approach, and a related algorithm to solve this problem (LARSTD), we refer the reader to [11].

3.3 ℓ_2 and ℓ_2 Penalization (L_{22})

Another approach we can take involves using the nested-optimization formulation of LSTD and applying regularization to both the projection and fixed-point steps. Such regularization was utilized in [5]. We can write this problem as

$$u^* = \underset{u \in \mathbb{R}^k}{\arg\min} \|\Phi u - (R + \gamma \Phi' w)\|_2^2 + \beta \|u\|_2^2$$
$$w^* = \underset{w \in \mathbb{R}^k}{\arg\min} \|\Phi w - \Phi u^*\|_2^2 + \beta' \|w\|_2^2. \tag{8}$$

Just as we did in (5) we can find a closed-form solution to u^* and can then simplify the residual term of the fixed-point subproblem as

$$\Phi w - \Phi u^* = \Phi w - \underbrace{\Phi(\Phi^T \Phi + \beta I)^{-1} \Phi^T}_{\Sigma}(R + \gamma \Phi' w). \tag{9}$$

Here the matrix Σ represents the empirical ℓ_2 penalized projection operator, or *hat matrix*, which projects n-vectors onto the space spanned by the features Φ. We can then solve for w^* in closed form as

$$w^* = \underset{w \in \mathbb{R}^k}{\arg\min} \| \overbrace{(\Phi - \gamma \Sigma \Phi')}^{X} w - \overbrace{\Sigma R}^{y} \|_2^2 + \beta' \|w\|_2^2 = (X^T X + \beta' I)^{-1} X^T y. \tag{10}$$

We can also, however, formulate this problem in terms of the standard LSTD matrices (as defined in Section 2) by noting that for $C = \Phi(\Phi^T \Phi + \beta I)^{-1}$ we can write $X = C(A + \beta I)$ and $y = Cb$.

3.4 ℓ_2 and ℓ_1 Penalization (L_{21})

Finally, we can also consider the same nested optimization problem as in in the previous scheme, but with an ℓ_1 penalty used in the fixed-point operator, i.e.

$$u^* = \underset{u \in \mathbb{R}^k}{\arg\min} \|\Phi u - (R + \gamma \Phi' w)\|_2^2 + \beta \|u\|_2^2$$
$$w^* = \underset{w \in \mathbb{R}^k}{\arg\min} \|\Phi w - \Phi u^*\|_2^2 + \beta' \|w\|_1 \tag{11}$$

Here we can again use the simplification from (10) to write the solution as

$$w^* = \underset{w \in \mathbb{R}^k}{\arg\min} \| \underbrace{(\varPhi - \gamma \varSigma \varPhi')}_{X} w - \underbrace{\varSigma R}_{y} \|_2^2 + \beta' \|w\|_1. \tag{12}$$

As a result we have now transformed L_{21} into a standard Lasso problem, in terms of X and y, to which we can apply any off-the-shelf solution method.

4 Standardizing the Data

In standard applications of the Lasso, it is often assumed that the feature matrix has columns that are standardized (i.e. that they are centered and zero-mean) and that the response vector is centered. Although we will briefly discuss the reasons for this, a more comprehensive treatment is given in e.g. [7, Section 3.4]. The centering is assumed because we generally want to estimate an unpenalized bias term w_0 so as to avoid making the problem dependent on the responses' origin. In doing so, the bias is given by the mean response and the remaining weights w can then be estimated using no bias term and centering the features and responses. The scaling of the features is perhaps more important and we can first note that the Lasso estimate is not invariant to this scaling. The scaling essentially evens the playing field for deciding which features are important, but it can also greatly impact the convergence speed of solution methods [7]. In this section we will now describe how to incorporate these assumptions into the L_{21} scheme introduced in the previous section.

We will now explicitly introduce a bias term w_0 into the value function approximation with $V^\pi(x) \approx \phi(x)^T w + w_0$. We can then rewrite the nested optimization problem as

$$(u^*, u_0^*) = \underset{u, u_0}{\arg\min} \|\varPhi u + u_0 - (R + \gamma(\varPhi' w + w_0))\|_2^2 + \beta_2 \|u\|_2^2 \tag{13}$$

$$(w^*, w_0^*) = \underset{w, w_0}{\arg\min} \|\varPhi w + w_0 - (\varPhi u^* + u_0^*)\|_2^2 + \beta_1 \|w\|_1. \tag{14}$$

Before solving these problems we will first introduce the following notation: $\overline{\varPhi} = \text{mean}(\varPhi)$ is the row-vector of feature means, $\widetilde{\varPhi} = \varPhi - \mathbf{1}_n \overline{\varPhi}$ are the centered features where $\mathbf{1}_n$ is a column vector of n ones, and $\widehat{\varPhi} = \widetilde{\varPhi}\varOmega$ consists of the centered and rescaled features given a scaling matrix $\varOmega$ whose diagonal elements consist of the inverse standard deviations of the feature matrix $\varPhi$. Similar terms are introduced for centering both $\varPhi'$ and R.

For both the projection and fixed-point sub-problems we can solve for the bias and weight terms invididually. In both cases the bias is given by the mean response minus the mean of the features and the optimal weights. For the bias of the projection step this can be written as

$$u_0^* = (\overline{R} + \gamma \overline{\varPhi}' w + \gamma w_0) - \overline{\varPhi} u^*. \tag{15}$$

The bias now depends upon finding u^*, but we can solve for this by centering both the features and responses and solving the following problem:

$$\Omega^{-1} u^* = \underset{u \in \mathbb{R}^k}{\arg\min} \|\widehat{\Phi} u - (\widetilde{R} + \gamma \widetilde{\Phi}' w)\|_2^2 + \beta_2 \|u\|_2^2 \tag{16}$$

$$= (\widehat{\Phi}^T \widehat{\Phi} + \beta_2)^{-1} \widehat{\Phi}^T (\widetilde{R} + \gamma \widetilde{\Phi}' w). \tag{17}$$

Note, however, that since we are solving this minimization problem using a scaled feature matrix (i.e. $\widehat{\Phi}$) we must remember to rescale back into the original units, which accounts for the the inverse of Ω.

We can now write the projected value function as

$$\Phi u^* + u_0^* = (\Phi - \mathbf{1}_n \overline{\Phi}) u^* + (\overline{R} + \gamma \overline{\Phi}' w + \gamma w_0)$$

$$= \underbrace{\widehat{\Phi} (\widehat{\Phi}^T \widehat{\Phi} + \beta_2)^{-1} \widehat{\Phi}^T}_{\Sigma} (\widetilde{R} + \gamma \widetilde{\Phi}' w) + (\overline{R} + \gamma \overline{\Phi}' w + \gamma w_0).$$

Here we have combined terms using the definitions introduced earlier and we can see the Σ term is the ℓ_2 penalized projection matrix onto the *centered and rescaled* features. Plugging this projected value function into the fixed-point problem we arrive at

$$(w^*, w_0^*) = \underset{w, w_0}{\arg\min} \|\Phi w + w_0 - \Phi u^* - u_0\|_2^2 + \beta_1 \|w\|_1$$

$$= \underset{w, w_0}{\arg\min} \| \underbrace{(\Phi - \gamma \Sigma \widetilde{\Phi}' - \gamma \mathbf{1}_n \overline{\Phi}')}_{X} w + (1 - \gamma) w_0 - \underbrace{(\Sigma \widetilde{R} + \overline{R})}_{y} \|_2^2 + \beta_1 \|w\|_1.$$

We can see then that this is very closely related to the original formulation from (12), however here we are projecting according to the centered/rescaled features and X and y have additional components related to the mean next-state features and mean rewards respectively.

Now we truly have a standard Lasso problem. We can solve for the optimum w^* using any off-the-shelf Lasso solver using the scaled/centered matrix X and centered vector y (and again if we rescale X we must return the output to the original scaling). Given the optimal weights we can then solve for the bias term as $(1 - \gamma) w_0^* = \overline{y} - \overline{X} w^*$, where these terms again denote the mean response and the mean of the features respectively.

5   Discussion of the Different Regularization Schemes

In this work, we are particularly interested in the situation where the number of features is greater than the number of samples, $k > n$. Drawing on results from standard regression problems, we would expect the ℓ_2-based methods to perform poorly in this situation. In fact, we will see in the later experiments that just such a drop-off in performance occurs at the point when k overtakes n. We would, however, expect the ℓ_1-based methods, due to the feature selection properties of this penalty, to continue performing well so long as there is some small subset of relevant features that fit the value function well. In the setting of regression this behavior is well-established both empirically and theoretically [15,3]. The extension of this behavior to the RL setting, although expected, is not entirely straight-forward due to the fixed-point aspect of the learning process. In the rest of this section, we will discuss and contrast how the ℓ_1 regularization is implemented in the two penalized methods, L_1 and L_{21}.

The main difference between the two schemes, L_1 and L_{21}, lies in the choice of where to place the ℓ_1 penalty. The L_1 approach uses a penalty directly in the projection operator. This is a straight-forward modification of the projection, and we would expect that, if the result of applying the Bellman operator can be well-represented by a sparse subset of features, the projection will find this. In fact, there are some recent theoretical guarantees that, if the target value function V^π is sparse, the L_1 scheme will be able to take advantage of this fact [9]. More precisely, if $s \ll k$ is the number of features needed to represent V^π, the prediction error of L_1 is shown to directly scale with s instead of k [9]. This suggests that L_1 will perform well until the number of samples is bigger than the number of relevant features, $n \geq s$. However, we must note that when the projection step is combined with the fixed point, it is not entirely clear what is being optimized by the L_1 procedure. In fact, in [11] it is claimed that this approach does not correspond to any optimization problem in w. Further, from an algorithmic point of view, since the fixed-point is applied *after* the ℓ_1 optimization, this approach does not correspond to a Lasso problem, resulting in the need to use more specialized algorithms to solve for w.

Alternatively, the L_{21} approach places an ℓ_1 penalty in the fixed-point step. The main benefit of this approach is that it allows us to cast the regularized LSTD problem as a Lasso problem, to which we can apply general-purpose Lasso or ℓ_1 solvers. This also allows us to more straightforwardly apply results such as those of Section 4, something that is not easily done in the L_1 scheme. The application of standardization to the feature matrix has a great deal of impact on the results of Lasso in regression problems and we would expect it to have a great deal of impact for RL as well (in the later experiments we will see some evidence of this). We also found that the ability to standardize the features played a role in the previously mentioned flexibility of L_{21}. In fact, we found that without standardization of the features it was difficult to apply certain iterative methods such as those discussed in [13,6] due to slow convergence.

One potential downside to the L_{21} approach is the necessity of using an ℓ_2 penalty in the projection step. It is not entirely clear what effect this has on the resulting algorithm, although in our preliminary experiments we found that this penalty was primarily useful in computing the projection matrix Σ, and making sure that the required matrix was non-singular. Further, the necessity of computing Σ does make L_{21} somewhat more expensive than L_1, however as this matrix only depends on Φ, using this procedure inside of a policy iteration scheme would only require the matrix being computed once. Finally, while there does exist some theoretical evidence that the L_1 approach is able to take advantage of the sparsity of V^π, no such guarantees exist yet for L_{21}. Our experiments, however, seem to indicate that L_{21} is able to capitalize on such value functions, and may even perform better when the value function is "truly sparse".

6   Experimental Results

In comparing the performance of the regularization schemes introduced in Section 3, we will consider the *chain problem* introduced in [12]. In these experiments we will utilize a 20-state, 2-action MDP wherein states are connected in a chain and where upon taking action *left* from state x the system transitions to state $x - 1$ with probability p and $x + 1$ with probability $1 - p$. The same holds for action *right* but in reverse and at both ends of the chain a successful right (or left) action leaves the state unchanged. We will restrict ourselves to the problem of policy evaluation and will evaluate the performance of the regularization schemes as the relative number of features k (as compared to the number of samples n) is varied. In particular, we will consider $k = s + \bar{s}$ features consisting of s "relevant" features including a single constant feature and some number of radial-basis functions (RBFs) spread uniformly over the state space. The additional features consist of "irrelevant" or "noise" features implemented by extending the state-space of the chain model to include $\bar{s}$ additional dimensions where each additional dimension is independently distributed $x_t^{i+1} \sim \mathcal{N}(0, \sigma^2)$ for all time indices t. The value of each noise feature is then given by the corresponding state's value, i.e. $\phi_{s+i}(x_t) = x_t^{i+1}$ for $1 \leq i \leq \bar{s}$. These additional features can also be thought of as *random* features of the standard chain model.

We will first consider the case where the value function can be constructed as a sparse combination of features. In order to test this scenario, we start with a sparse value function and work backwards. We will consider a value function given as a linear combination of s relevant features, i.e. $V(x) = \sum_{i=1}^{s} \phi_i(x) w_i^*$ and we then define the reward of our model as $r(x, x') = V(x) - \gamma V(x')$. We have, in this case, artificially constructed our reward in order to enforce the desired sparse value function. In this experiment we used $s = 10$ and varied the number irrelevant features. In each of 20 runs we sample n samples on-policy and use them to approximate the value function. Here we used a policy π which takes action *left* on the first 10 states and action *right* on the next 10, although due to our construction of the reward function for this problem, the actual policy used does not affect the value function. The resulting approximation was then evaluated at 500 test points and the empirical error between this and the true value function is computed. The policy is kept fixed across all runs and within each run we perform cross-validation over the regularization parameter.

In Figure 2 we compare the L_2, L_1, and L_{21} variants described in Section 3; here we omit the L_{22} variant to avoid clutter[1]. L_{21} can be run using any Lasso solver. In our implementation we both tested LARS [4] and a number of gradient descent procedures [13,6]. In order to generate the value function we sampled our true weights w_i^* uniformly in the range $[-5, 5]$ for every run. From our

---

[1] Preliminary experimental results seemed to show that the L_{22} variant performed similarly to the L_2 scheme.

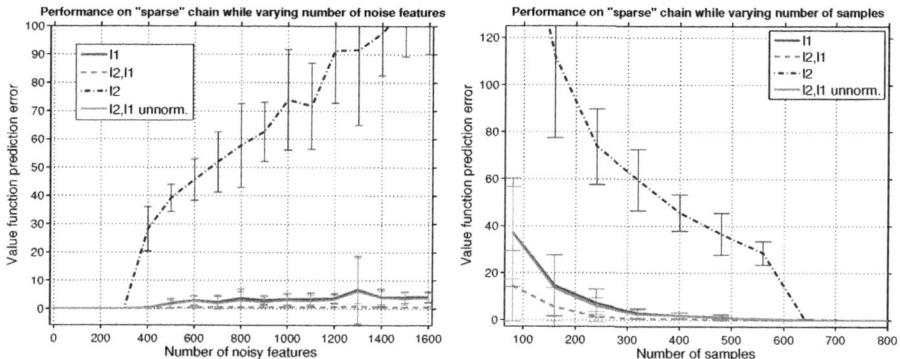

Fig. 2. Performance of policy evaluation on a sparse value function

experiments, however, we saw that the true values of w did not play a significant role in explaining the error of each method: this variance was more attributable to the data samples generated. The first of these plots shows the behavior using $n = 400$ samples (consisting of 20 episodes of length 20) while varying the number of noise features. We can first note that all algorithms perform quite well until the number of features k approaches the number of samples n, but after around 400 noise features the algorithms begin to differentiate. L_2 immediately starts overfitting the data and its prediction error grows almost linearly with the total number of features k. On the other hand, the ℓ_1 regularization approaches are expected to depend more directly on the *true* dimensionality s rather than k. In fact, we can see that both L_1 and L_{21} are not drastically affected by overfitting when k becomes greater than n. In the plot, we also include the performance of L_{21} without the rescaling described in Section 4 and here this performs about as well as the L_1 approach (using LARSTD). The rescaled version, however, is much better able to deal with the increase in number of noise features as can be seen by its very low prediction error, and in comparison to the other methods this is nearly constant. The second plot shows the results of $\bar{s} = 600$ noise features and varying the number of samples n, and we see similar differences in performance.

Next we consider the standard chain problem which uses a reward of 1 when the relevant state (x_1) is at either endpoint and 0 elsewhere. In the first of these experiments we use $s = 6$ relevant features and again vary the number irrelevant features. In each of 20 runs we again sample 400 data points (consisting of 20 episodes of length 20) on-policy and use these points to approximate the value function. Given this model we can compute the true value function and in order to generate the plots we evaluate the empirical error between this and the estimated value function at 500 test points. Again, the policy is kept fixed and within each run we perform cross-validation over the regularization parameter.

The results of Figure 3 closely echo the results of Figure 2. Here, however, we can see that there is a somewhat more significant difference between the unscaled version of L_{21} and the L_1 variant, although this performance is still significantly better than L_2 alone. We also see that the prediction error is no longer zero (as

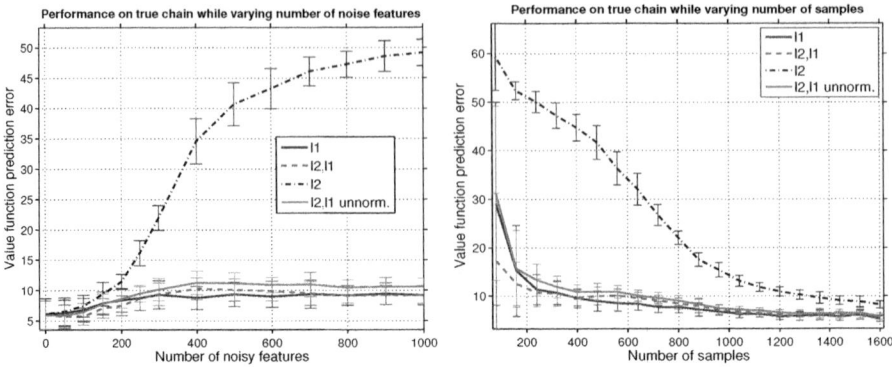

Fig. 3. Performance of policy evaluation on the chain model for a fixed policy

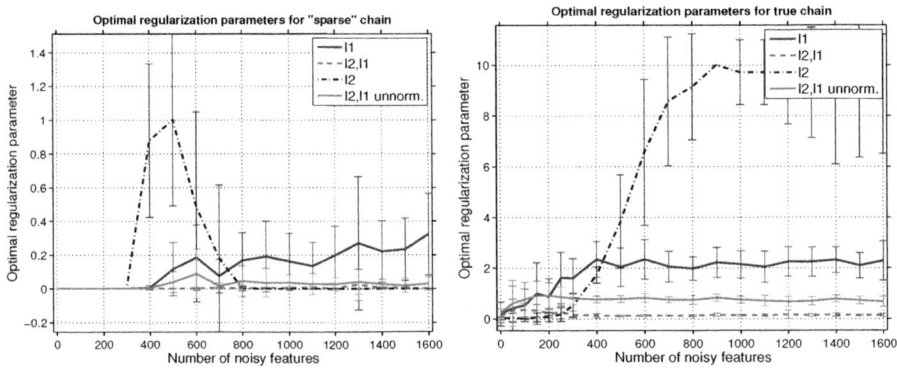

Fig. 4. Effect of the number of irrelevant features on the optimal penalty parameters chosen via cross-validation

would be expected due to the fact that now the value function can be exactly represented in the linear space). We can also see, however, that there is a more significant increase in this error as we approach the point where the number of features equals the number of samples. Again, the second plot shows the results of $\bar{s} = 600$ noise features while varying the number of samples n, and again we see similar differences in performance.

In Figure 4 we show the effect of the number of irrelevant features on the best penalty parameters chosen via cross-validation on both the sparse chain and the standard chain model. The ℓ_1 based methods use LARS (or the LARSTD modification) to find the regularization path and for the ℓ_2 based method we used a grid of 20 parameters logarithmically spaced between 10^{-6} and 10. In both figures we show results where the number of samples is fixed at $n = 400$ and the number of noisy features is increased. The main point to see here is the much greater variability in this parameter value for the L_2 method. The L_1 method then may also be more variable than the L_{21} method, but this level of increase could also be attributable to different scaling in this parameter. For these experiments we also found that the ℓ_2 parameter of L_{21} was primarily useful in ensuring the relevant matrix Σ was nonsingular, and as a result we were able to set this parameter to a quite low-value 10^{-6} and in these plots we only show the ℓ_1 penalty. Based on this it would seem that the tuning of the penalty parameter for L_{21} is at least *no harder* than that of L_1 due to the minimal effect of the second parameter Finally, we point out that the decline in the penalty parameter of L_2 for the sparse chain is attributable to the fact that after approximately 600–800 noise features all parameters were "equally bad".

Finally, we analyzed the effect of the number irrelevant features on the number of iterations when solving the problem using L_1/LARSTD and when solving the L_{21} problem using LARS [4]. Although these results are not entirely conclusive, it appears that the L_{21} approach may require fewer iterations for sparse models, but more for non-sparse models. As mentioned earlier, the additional flexibility of the L_{21} scheme allows us to apply more sophisticated optimizations schemes such as the coordinate descent approach of [6] or other iterative schemes [13]. We applied both of these procedures, code for which is available online and is quite fast, to our problem. However, the comparison of the complexity of these different algorithms is not simple and we leave it for future work.

7 Conclusion

In this paper we presented a number of regularization schemes for LSTD, culminating in a novel approach using nested ℓ_2 and ℓ_1 penalties. We then showed how to apply standardization techniques from the regression literature to this problem and discussed preliminary experiments comparing this approach to other schemes based on ℓ_2 and ℓ_1 penalized projection. The main benefit of this approach is the additional flexibility it provides over ℓ_1 projected LSTD approaches (such as LARSTD), and the ability to use standard Lasso solvers and techniques. This means that any new solver for Lasso could be immediately used in L_{21}, whereas this would need a rethinking of the whole algorithm for L_1. We should also note a similar approach, derived in parallel, of [8].

There are also a number of open areas of research with this approach. A more thorough experimental analysis is needed, including an extension of this method to the problem of policy iteration via LSPI. A theoretical analysis of this approach is also called for, in line with the previously noted analysis of the L_1 scheme presented in [9]. Finally, a better understanding of the ramifications of including the ℓ_2 penalty would be most beneficial.

References

1. Antos, A., Szepesvári, C., Munos, R.: Learning near-optimal policies with Bellman-residual minimization based fitted policy iteration and a single sample path. Machine Learning 71(1) (2008)
2. Bradtke, S., Barto, A.: Linear least-squares algorithms for temporal difference learning. Machine Learning 22, 33–57 (1996)

3. Bunea, F., Tsybakov, A., Wegkamp, M.: Sparsity oracle inequalities for the lasso. Electronic Journal of Statistics 1, 169–194 (2007)
4. Efron, B., Hastie, T., Johnstone, I., Tibshirani, R.: Least angle regression. Annals of Statistics 32(2) (2004)
5. Farahmand, A., Ghavamzadeh, M., Szepesvari, C., Mannor, S.: Regularized policy iteration. In: Advances in Neural Information Processing Systems 21 (2009)
6. Friedman, J., Hastie, T., Höfling, H., Tibshirani, R.: Pathwise coordinate optimization. The Annals of Applied Statistics 1(2), 302–332 (2007)
7. Friedman, J., Hastie, T., Tibshirani, R.: The elements of statistical learning. Springer, Heidelberg (2001)
8. Geist, M., Scherrer, B.: ℓ_1-penalized projected bellman residual. In: European Workshop on Reinforcement Learning (2011)
9. Ghavamzadeh, M., Lazaric, A., Munos, R., Hoffman, M.: Finite-sample analysis of Lasso-TD. In: Proceedings of the International Conference on Machine Learning (2011)
10. Johns, J., Painter-Wakefield, C., Parr, R.: Linear complementarity for regularized policy evaluation and improvement. In: Advances in Neural Information Processing Systems 23 (2010)
11. Kolter, J.Z., Ng, A.Y.: Regularization and feature selection in least-squares temporal difference learning. In: Proceedings of the International Conference on Machine Learning (2009)
12. Lagoudakis, M.G., Parr, R.: Least-squares policy iteration. Journal of Machine Learning Research 4 (2003)
13. Schmidt, M.: Graphical Model Structure Learning with l1-Regularization. Ph.D. thesis, University of British Columbia (2010)
14. Sutton, R., Barto, A.: Reinforcement Learning: An Introduction. MIT Press (1998)
15. Tibshirani, R.: Regression shrinkage and selection via the lasso. Journal of the Royal Statistical Society. Series B (Methodological) 58(1), 267–288 (1996)

Recursive Least-Squares Learning with Eligibility Traces

Bruno Scherrer[1] and Matthieu Geist[2]

[1] INRIA, MAIA Project-Team, Nancy, France
[2] Supélec, IMS Research Group, Metz, France

Abstract. In the framework of Markov Decision Processes, we consider the problem of learning a linear approximation of the value function of some fixed policy from one trajectory possibly generated by some other policy. We describe a systematic approach for adapting *on-policy* learning least squares algorithms of the literature (LSTD [5], LSPE [15], FPKF [7] and GPTD [8]/KTD [10]) to *off-policy* learning *with eligibility traces*. This leads to two known algorithms, LSTD(λ)/LSPE(λ) [21] and suggests new extensions of FPKF and GPTD/KTD. We describe their recursive implementation, discuss their convergence properties, and illustrate their behavior experimentally. Overall, our study suggests that the state-of-art LSTD(λ) [21] remains the best least-squares algorithm.

1 Introduction

We consider the problem of learning a linear approximation of the value function of some fixed policy in a Markov Decision Process (MDP) framework, in the most general situation where learning must be done from a single trajectory possibly generated by some other policy, *a.k.a.* *off-policy* learning. Given samples, well-known methods for estimating a value function are temporal difference (TD) learning and Monte Carlo [19]. TD learning with eligibility traces [19], known as TD(λ), provide a nice bridge between both approaches, and by controlling the bias/variance trade-off [12], their use can significantly speed up learning. When the value function is approximated through a linear architecture, the depth λ of the eligibility traces is also known to control the quality of approximation [20]. Overall, the use of these traces often plays an important practical role.

In the *on-policy* case (where the policy to evaluate is the same as the one that generated data), there has been a significant amount of research on linear Least-Squares (LS) approaches, which are more sample-efficient than TD/Monte-Carlo. Notable such works include LSTD(λ) [5], LSPE(λ) [15], FPKF [7] and GPTD [8]/ KTD [10]. Works on off-policy linear learning are sparser: [16] proposed a variation of TD(λ) that could combine off-policy learning with linear approximation and eligibility traces. Recently, [21] proposed and analysed off-policy versions of LSTD(λ) and LSPE(λ). The first motivation of this article is to argue that it is conceptually simple to extend *all* the LS algorithms we have just mentionned

S. Sanner and M. Hutter (Eds.): EWRL 2011, LNCS 7188, pp. 115–127, 2012.

so that they can be applied to the off-policy setting *and* use eligibility traces. If this allows to rederive the off-policy versions of LSTD(λ) and LSPE(λ) [21], it also leads to new candidate algorithms, for which we will derive recursive formulations. The second motivation of this work is to describe the subtle differences between these intimately-related algorithms on the analytical side, and to provide some comparative insights on their empirical behavior (a topic that has to our knowledge not been considered in the literature, even in the particular *on-policy* and *no-trace* situation).

The rest of the paper is organized as follows. Sec. 2 introduces the background of Markov Decision Processes and describes the state-of-the-art algorithms for on-policy learning with recursive LS methods. Sec. 3 shows how to adapt these methods so that they can both deal with the off-policy case and use eligibility traces. The resulting algorithms are formalized, the formula for their recursive implementation is derived, and we discuss their convergence properties. Sec. 4 illustrates empirically the behavior of these algorithms and Sec. 5 concludes.

2   Background and State-of-the-art On-policy Algorithms

A Markov Decision Process (MDP) is a tuple $\{S, A, P, R, \gamma\}$ in which S is a finite state space identified with $\{1, 2, \ldots, N\}$, A a finite action space, $P \in \mathcal{P}(S)^{S \times A}$ the set of transition probabilities, $R \in \mathbb{R}^{S \times A}$ the reward function and γ the discount factor. A mapping $\pi \in \mathcal{P}(A)^S$ is called a policy. For any policy π, let P^π be the corresponding stochastic transition matrix, and R^π the vector of mean reward when following π, *i.e.* of components $E_{a|\pi,s}[R(s,a)]$. The value $V^\pi(s)$ of state s for a policy π is the expected discounted cumulative reward starting in state s and then following the policy π: $V^\pi(s) = E_\pi[\sum_{i=0}^\infty \gamma^i r_i | s_0 = s]$ where E_π denotes the expectation induced by policy π. The value function satisfies the (linear) Bellman equation: $\forall s, V^\pi(s) = E_{s',a|s,\pi}[R(s,a) + \gamma V^\pi(s')]$. It can be rewritten as the fixed-point of the Bellman evaluation operator: $V^\pi = TV^\pi$ where for all V, $TV = R^\pi + \gamma P^\pi V$.

In this article, we are interested in learning an approximation of this value function V^π under some constraints. First, we assume our approximation to be linearly parameterized: $\hat{V}_\theta(s) = \theta^T \phi(s)$ with $\theta \in \mathbb{R}^p$ being the parameter vector and $\phi(s)$ the feature vector. Also, we want to estimate the value function V^π (or equivalently associated parameters) from a single finite trajectory generated using a possibly different behaviorial policy π_0. Let μ_0 be the stationary distribution of the stochastic matrix $P_0 = P^{\pi_0}$ of the *behavior policy* π_0 (we assume it exists and is unique). Let D_0 be the diagonal matrix of which the elements are $(\mu_0(i))_{1 \leq i \leq N}$. Let Φ be the matrix of feature vectors: $\Phi = [\phi(1) \ldots \phi(N)]^T$. The projection Π_0 onto the hypothesis space spanned by Φ with respect to the μ_0-quadratic norm, which will be central for the understanding of the algorithms, has the following closed-form: $\Pi_0 = \Phi(\Phi^T D_0 \Phi)^{-1} \Phi^T D_0$.

In the rest of this section, we review existing on-policy least-squares based temporal difference learning algorithms. In this case, the behavior and target policies are the same so we omit the subscript 0 for the policy (π) and the projection (Π). We assume that a trajectory $(s_1, a_1, r_1, s_2, \ldots, s_j, a_j, r_j, s_{j+1}, \ldots s_{i+1})$ sampled according to the policy π is available. Let us introduce the sampled Bellman operator $\hat{T}_j$, defined as: $\hat{T}_j : V \in \mathbb{R}^S \rightarrow \hat{T}_j V = r_j + \gamma V(s_{j+1}) \in \mathbb{R}$ so that $\hat{T}_j V$ is an unbiased estimate of $TV(s_j)$. If values were observable, estimating the projected parameter vector θ would reduce to project the value function onto the hypothesis space using the empirical projection operator. This would be the classical least-squares approach. Since values are not observed — only transitions (rewards and next states) are —, we will rely on *temporal differences* (terms of the form $\hat{T}_j V - V(s_j)$) to estimate the value function.

The Least-Squares Temporal Differences (LSTD) algorithm of [6] aims at finding the fixed point of the operator being the composition of the projection onto the hypothesis space and of the Bellman operator. Otherwise speaking, it searches for the fixed point $\hat{V}_\theta = \Pi T \hat{V}_\theta$, Π being the just introduced projection operator. Using the available trajectory, LSTD solves the following fixed-point problem: $\theta_i = \mathrm{argmin}_{\omega \in \mathbb{R}^p} \sum_{j=1}^{i} (\hat{T}_j \hat{V}_{\theta_i} - \hat{V}_\omega(s_j))^2$. The Least-Squares Policy Evaluation (LSPE) algorithm of [15] searches for the same fixed point, but in an iterative way instead of directly (informally, $\hat{V}_{\theta_i} \simeq \Pi T \hat{V}_{\theta_{i-1}}$). The corresponding optimization problem is: $\theta_i = \mathrm{argmin}_{\omega \in \mathbb{R}^p} \sum_{j=1}^{i} (\hat{T}_j \hat{V}_{\theta_{i-1}} - \hat{V}_\omega(s_j))^2$. The Fixed-Point Kalman Filter (FPKF) algorithm of [7] is a least-squares variation of the classical temporal difference learning algorithm [19]. Value function approximation is treated as a supervised learning problem, and unobserved values are bootstrapped: the unobserved value $V^\pi(s_j)$ is replaced by the estimate $\hat{T}_j \hat{V}_{\theta_{j-1}}$. This is equivalent to solving the following optimization problem: $\theta_i = \mathrm{argmin}_{\omega \in \mathbb{R}^p} \sum_{j=1}^{i} (\hat{T}_j \hat{V}_{\theta_{j-1}} - \hat{V}_\omega(s_j))^2$. Finally, the Bellman Residual Minimization (BRM) algorithm aims at minimizing the distance between the value function and its image through the Bellman operator, $\|V - TV\|^2$. When the sampled operator is used, this leads to biased estimates (*e.g.* see [1]). The corresponding optimization problem is: $\theta_i = \mathrm{argmin}_{\omega \in \mathbb{R}^p} \sum_{j=1}^{i} (\hat{T}_j \hat{V}_\omega - \hat{V}_\omega(s_j))^2$. This cost function has originally been proposed by [2] who minimized it using a stochastic gradient approach. It has been considered by [14] with a least-squares approach, however with a double sampling scheme to remove the bias. The parametric Gaussian Process Temporal Differences (GPTD) algorithm of [8] and the linear Kalman Temporal Differences (KTD) algorithm of [10] can be shown to minimize this cost using a least-squares approach (so with bias).

All these algorithms can be summarized as follows:

$$\theta_i = \underset{\omega \in \mathbb{R}^p}{\mathrm{argmin}} \sum_{j=1}^{i} \left(\hat{T}_j \hat{V}_\xi - \hat{V}_\omega(s_j) \right)^2. \tag{1}$$

One of the presented approach is obtained by instantiating $\xi =$ θ_i, θ_{i-1}, θ_{j-1} or ω and solving the corresponding optimization problem. If more algorithms can be summarized under this generic equation (see [11]), the current paper will restrict its focus on linear least-squares based approaches.

3   Extension to Eligibility Traces and Off-policy Learning

This section contains the core of our contribution: we are going to describe a systematic approach in order to adapt the previously mentionned algorithms so that they can deal with eligibility traces and off-policy learning. The actual formalization of the algorithms, along with the derivation of their recursive implementation, will then follow.

Let $0 \le \lambda \le 1$ be the eligibility factor. Using eligibility traces amounts to looking for the fixed point of the following variation of the Bellman operator [4]: $\forall V \in \mathbb{R}^S$, $T^\lambda V = (1 - \lambda) \sum_{i=0}^{\infty} \lambda^i T^{i+1} V$, that makes a geometric average with parameter λ of the powers of the original Bellman operator T. Clearly, any fixed point of T is a fixed point of T^λ and vice-versa. An equivalent *temporal difference* based definition of T^λ is (see *e.g.* [15]): $\forall s$,
$$T^\lambda V(s) = V(s) + E_\pi[\sum_{j=i}^{\infty}(\gamma\lambda)^{j-i}\Big(r_j + \gamma V(s_{j+1}) - V(s_j)\Big)\Big|s_i = s].$$

As learning is done over a finite trajectory, it is natural to introduce the following truncated operator, which considers samples until time n: $\forall s$, $T_n^\lambda V(s) = V(s) + E_\pi[\sum_{j=i}^{n}(\gamma\lambda)^{j-i}\Big(r_j + \gamma V(s_{j+1}) - V(s_j)\Big)\Big|s_i = s]$. To use it practically, we still need to remove the dependency to the model (*i.e.* the expectation) and to take into account the fact that we want to consider off-policy learning. Assume from now on that we have a trajectory $(s_1, a_1, r_1, s_2, \ldots, s_n, a_n, r_n, s_{n+1})$ sampled according to the behaviour policy π_0. As behavorial and target policies are different, estimates of T_n^λ need to be corrected through importance sampling [17]. For all s, a, let us introduce the following weight: $\rho(s, a) = \frac{\pi(a|s)}{\pi_0(a|s)}$. In our trajectory context, write $\rho_i^j = \prod_{k=i}^{j} \rho_k$ with $\rho_j = \rho(s_j, a_j)$. Now, consider the off-policy, sampled and truncated $\hat{T}_{i,n}^\lambda : \mathbb{R}^S \to \mathbb{R}$ operator as: $\hat{T}_{i,n}^\lambda V =$ $V(s_i) + \sum_{j=i}^{n}(\gamma\lambda)^{j-i}\Big(\rho_i^j \hat{T}_j V - \rho_i^{j-1} V(s_j)\Big)$. It can be seen that $\hat{T}_{i,n}^\lambda V$ is an unbiased estimate of $T_n^\lambda V(s_i)$ (see [16,21] for details).

Replacing $\hat{T}_j$ by $\hat{T}_{j,i}^\lambda$ in the optimization problem of Eq. (1), is a generic way to extend any parametric value function approximators to the *off-policy* setting and the use of *eligibility traces*: $\theta_i = \text{argmin}_{\omega \in \mathbb{R}^p} \sum_{j=1}^{i}(\hat{T}_{j,i}^\lambda \hat{V}_\xi - \hat{V}_\omega(s_j))^2$. In the rest of this section, by instantiating ξ to θ_i, θ_{i-1}, θ_{j-1} or ω, we derive the already existing algorithms off-policy LSTD(λ)/LSPE(λ) [21], and we extend two existing algorithms to eligibility traces and to off-policy learning, that we will naturally call FPKF(λ) and BRM(λ). With $\lambda = 0$, this exactly corresponds to the algorithms we described in the previous section. When $\lambda = 1$, it can be seen that $\hat{T}^\lambda_{i,n} V = \hat{T}^1_{i,n} V = \sum_{j=i}^n \gamma^{j-i} \rho^j_i r_j + \gamma^{n-i+1} \rho^n_i V(s_{n+1})$; thus, if $\gamma^{n-i+1} \rho^n_i$ tends to 0 when n tends to infinity[1] so that the influence of ξ in the definition of $\hat{T}^\lambda_{i,n} V_\xi$ vanishes, all algorithms should asymptotically behave the same.

Recall that a linear parameterization is chosen here, $\hat{V}_\xi(s_i) = \xi^T \phi(s_i)$. We adopt the following notations: $\phi_i = \phi(s_i)$, $\Delta\phi_i = \phi_i - \gamma\rho_i\phi_{i+1}$ and $\tilde{\rho}^{k-1}_j = (\gamma\lambda)^{k-j} \rho^{k-1}_j$. The generic cost function to be solved is therefore:

$$\theta_i = \operatorname*{argmin}_{\omega \in \mathbb{R}^p} \sum_{j=1}^i (\phi^T_j \xi + \sum_{k=j}^i \tilde{\rho}^{k-1}_j (\rho_k r_k - \Delta\phi^T_k \xi) - \phi^T_j \omega)^2. \tag{2}$$

3.1   Off-policy LSTD(λ)

The off-policy LSTD(λ) algorithm corresponds to instantiating Problem (2) with $\xi = \theta_i$. This can be solved by zeroing the gradient respectively to ω: $0 = \sum_{j=1}^i (\sum_{k=1}^j \phi_k \tilde{\rho}^{j-1}_k)(\rho_j r_j - \Delta\phi^T_j \theta_i)$. Introducing the (corrected) eligibility vector z_j:

$$z_j = \sum_{k=1}^j \phi_k \tilde{\rho}^{j-1}_k = \sum_{k=1}^j \phi_k (\gamma\lambda)^{j-k} \prod_{m=k}^{j-1} \rho_m = \gamma\lambda\rho_{j-1} z_{j-1} + \phi_j, \tag{3}$$

one obtains the following batch estimate:

$$\theta_i = (\sum_{j=1}^i z_j \Delta\phi^T_j)^{-1} \sum_{j=1}^i z_j \rho_j r_j = (A_i)^{-1} b_i. \tag{4}$$

A recursive implementation of this algorithm (where $M_i = (A_i)^{-1}$ is updated on-the-fly) has been proposed and analyzed recently by [21] and is described in Alg. 1. The author proves that if the *behavior* policy π_0 induces an irreducible Markov chain and chooses with positive probability any action that may be chosen by the *target* policy π, and if the compound (linear) operator $\Pi_0 T^\lambda$ has a unique fixed point[2], then off-policy LSTD(λ) converges to it almost surely. Formally, it converges to the solution θ^* of the so-called *projected fixed-point* equation:

$$V_{\theta^*} = \Pi_0 T^\lambda V_{\theta^*}. \tag{5}$$

Using the expression of the projection Π_0 and the form of the Bellman operator T^λ it can be seen that θ^* satisfies (see [21] for details) $\theta^* = A^{-1}b$ where

$$A = \Phi^T D_0 (I - \gamma P)(I - \lambda\gamma P)^{-1}\Phi \quad \text{and} \quad b = \Phi^T D_0 (I - \lambda\gamma P)^{-1} R. \tag{6}$$

The core of the analysis of [21] consists in showing that $\frac{1}{i} A_i$ and $\frac{1}{i} b_i$ defined in Eq. (4) respectively converge to A and b almost surely. Through Eq. (4), this implies the convergence of θ_i to θ^*.

---

[1] This is not always the case, see [21] and the discussion in Sec. 3.4.
[2] It is not always the case, see [20] or Sec. 4 for a counter-example.

Algorithm 1. LSTD(λ)

Initialization;
 Initialize vector θ_0 and matrix M_0 ;
 Set $z_0 = 0$;
 for $i = 1, 2, \ldots$ **do**
    Observe ϕ_i, r_i, ϕ_{i+1} ;
    Update traces ;
    $z_i = \gamma \lambda \rho_{i-1} z_{i-1} + \phi_i$;
    Update parameters ;
    $K_i = \dfrac{M_{i-1} z_i}{1 + \Delta \phi_i^T M_{i-1} z_i}$;
    $\theta_i = \theta_{i-1} + K_i(\rho_i r_i - \Delta \phi_i^T \theta_{i-1})$;
    $M_i = M_{i-1} - K_i(M_{i-1}^T \Delta \phi_i)^T$;

Algorithm 2. LSPE(λ)

Initialization;
 Initialize vector θ_0 and matrix N_0 ;
 Set $z_0 = 0$, $A_0 = 0$ and $b_0 = 0$;
 for $i = 1, 2, \ldots$ **do**
    Observe ϕ_i, r_i, ϕ_{i+1};
    Update traces ;
    $z_i = \gamma \lambda \rho_{i-1} z_{i-1} + \phi_i$;
    Update parameters ;
    $N_i = N_{i-1} - \dfrac{N_{i-1} \phi_i \phi_i^T N_{i-1}}{1 + \phi_i^T N_{i-1} \phi_i}$;
    $A_i = A_{i-1} + z_i \Delta \phi_i^T$;
    $b_i = b_{i-1} + \rho_i z_i r_i$;
    $\theta_i = \theta_{i-1} + N_i(b_i - A_i \theta_{i-1})$;

3.2   Off-policy LSPE(λ)

The off-policy LSPE(λ) algorithm corresponds to the instantiation $\xi = \theta_{i-1}$ in Problem (2). This can be solved by zeroing the gradient respectively to ω: $\theta_i = \theta_{i-1} + (\sum_{j=1}^i \phi_j \phi_j^T)^{-1} \sum_{j=1}^i z_j(\rho_j r_j - \Delta \phi_j^T \theta_{i-1})$, where we used the eligibility vector z_j defined Eq. (3). Write

$$N_i = (\sum_{j=1}^i \phi_j \phi_j^T)^{-1} = N_{i-1} - \frac{N_{i-1}\phi_i \phi_i^T N_{i-1}}{1 + \phi_i^T N_{i-1}\phi_i} \qquad (7)$$

where the second equality follows from the Sherman-Morrison formula. Using A_i and b_i as defined in the LSTD description in Eq. (4), one gets: $\theta_i = \theta_{i-1} + N_i(b_i - A_i\theta_{i-1})$. The overall computation is provided in Alg. 2. This algorithm, (briefly) mentionned by [21], generalizes the LSPE(λ) algorithm of [15] to off-policy learning. With respect to LSTD(λ), which computes $\theta_i = (A_i)^{-1}b_i$ (cf. Eq. (4)) at each iteration, LSPE(λ) is fundamentally recursive. Along with the almost sure convergence of $\frac{1}{i}A_i$ and $\frac{1}{i}b_i$ to A and b (defined in Eq. (6)), it can be shown that iN_i converges to $N = (\Phi^T D_0 \Phi)^{-1}$ (see for instance [15]) so that, asymptotically, LSPE(λ) behaves as: $\theta_i = \theta_{i-1} + N(b - A\theta_{i-1}) = Nb + (I - NA)\theta_{i-1}$, which is equivalent to (e.g. see [15]):

$$V_{\theta_i} = \Phi\theta_i = \Phi N b + \Phi(I - NA)\theta_{i-1} = \Pi_0 T^\lambda V_{\theta_{i-1}}. \qquad (8)$$

The behavior of this sequence depends on the spectral radius of $\Pi_0 T^\lambda$. Thus, the analyses of [21] and [15] (for the convergence of N_i) imply the following convergence result: under the assumptions required for the convergence of off-policy LSTD(λ), and the additional assumption that the operator $\Pi_0 T^\lambda$ has spectral radius smaller than 1 (so that it is contracting), LSPE(λ) also converges almost surely to the fixed point of the compound $\Pi_0 T^\lambda$ operator.

There are two sufficient conditions that can (independently) ensure such a desired contraction property. The first one is when one considers on-policy learning (see e.g. [15], where the authors studied the on-policy case and use this property in the proof). When the behavior policy π_0 is different from the target policy π, a sufficient condition for contraction is that λ be close enough to 1; indeed, when λ tends to 1, the spectral radius of T^λ tends to zero and can potentially balance an expansion of the projection Π_0. In the off-policy case, when γ is sufficiently big, a small value of λ can make $\Pi_0 T^\lambda$ expansive (see [20] for an example in the case $\lambda = 0$) and off-policy LSPE(λ) will then diverge. Eventually, Equations (5) and (8) show that when $\lambda = 1$, both LSTD(λ) and LSPE(λ) asymptotically coincide (as $T^1 V$ does not depend on V).

3.3   Off-policy FPKF(λ)

The off-policy FPKF(λ) algorithm corresponds to the instantiation $\xi = \theta_{j-1}$ in Problem (2). This can be solved by zeroing the gradient respectively to ω: $\theta_i = N_i(\sum_{j=1}^i \phi_j \phi_j^T \theta_{j-1} + \sum_{j=1}^i \sum_{k=1}^j \phi_k \tilde{\rho}_k^{j-1} (\rho_j r_j - \Delta\phi_j^T \theta_{k-1}))$ where N_i is the matrix introduced for LSPE(λ) in Eq. (7). With respect to the previously described algorithms, the difficulty here is that on the right side there is a dependence with all the previous terms θ_{k-1} for $1 \le k \le i$. Using the symmetry of the dot product $\Delta\phi_j^T \theta_{k-1} = \theta_{k-1}^T \Delta\phi_j$, it is possible to write a recursive algorithm by introducing the trace matrix Z_j that integrates the subsequent values of θ_k as follows: $Z_j = \sum_{k=1}^j \tilde{\rho}_k^{j-1} \phi_k \theta_{k-1}^T = Z_{j-1} + \gamma\lambda\rho_{j-1}\phi_j\theta_{j-1}^T$. With this notation we obtain: $\theta_i = N_i(\sum_{j=1}^i \phi_j \phi_j^T \theta_{j-1} + \sum_{j=1}^i (z_j \rho_j r_j - Z_j \Delta\phi_j))$. Using Eq. (7) and a few algebraic manipulations, we end up with: $\theta_i = \theta_{i-1} + N_i(z_i \rho_i r_i - Z_i \Delta\phi_i)$. This provides Alg. 3.

It generalizes the FPKF algorithm of [7] that was originally only introduced without traces and in the on-policy case. As LSPE(λ), this algorithm is fundamentally recursive. However, its overall behavior is quite different. As we discussed for LSPE(λ), iN_i asymptotically tends to $N = (\Phi^T D_0 \Phi)^{-1}$ and FPKF(λ) iterates eventually resemble: $\theta_i = \theta_{i-1} + \frac{1}{i} N(z_i \rho_i r_i - Z_i \Delta\phi_i)$. The term in brackets is a random component (that depends on the last transition) and $\frac{1}{i}$ acts as a learning coefficient that asymptotically tends to 0. In other words, FPKF(λ) has a *stochastic approximation* flavour. In particular, one can see FPKF(0) as a stochastic approximation of LSPE(0)[3]. When $\lambda > 0$, the situation is less clear (all the more that, as previously mentionned, we expect LSTD/LSPE/FPKF to asymptotically behave the same when λ tends to 1).

Due to its much more involved form (notably the matrix trace Z_j integrating the values of all the values θ_k from the start), we have not been able to obtain a formal analysis of FPKF(λ), even in the on-policy case. To our knowledge, there is no *full proof of convergence* for stochastic approximation algorithms with eligibility traces in the off-policy case[4], and a related result for FPKF(λ) thus seems difficult. Nevertheless, it is reasonable to conjecture that off-policy FPKF(λ) has the same asymptotic behavior as LSPE(λ).

---

[3] To see this, one can compare the asymptotic behavior of both algorithms. FPKF(0) does $\theta_i = \theta_{i-1} + \frac{1}{i} N(\rho_i \phi_i r_i - \phi_i \Delta\phi_i^T \theta_{i-1})$. One then notices that $\rho_i \phi_i r_i$ and $\phi_i \Delta\phi_i^T$ are samples of A and b to which A_i and b_i converge through LSPE(0).

[4] An analysis of TD(λ), with a simplifying assumption that forces the algorithm to stay bounded, is given in [21]. An analysis of a related algorithm, GQ(λ), is provided in [13], with an assumption on the second moment of the traces, which does not hold in general (see Propostion 2 in [21]). A full analysis thus remains to be done.

Algorithm 3. FPKF(λ)	**Algorithm 4.** BRM(λ)

Initialization;
 Initialize vector θ_0 and matrix N_0 ;
 Set $z_0 = 0$ and $Z_0 = 0$;
 for $i = 1, 2, \dots$ **do**
    Observe ϕ_i, r_i, ϕ_{i+1};
    Update traces ;
    $z_i = \gamma \lambda \rho_{i-1} z_{i-1} + \phi_i$;
    $Z_i = \gamma \lambda \rho_{i-1} Z_{i-1} + \phi_i \theta_{i-1}^T$;
    Update parameters ;
    $N_i = N_{i-1} - \dfrac{N_{i-1} \phi_i \phi_i^T N_{i-1}}{1 + \phi_i^T N_{i-1} \phi_i}$
    ;
    $\theta_i = \theta_{i-1} + N_i(z_i \rho_i r_i - Z_i \Delta \phi_i)$
    ;
	Initialization;
	 Initialize vector θ_0 and matrix C_0 ;
	 Set $y_0 = 0$, $\Delta_0 = 0$ and $z_0 = 0$;
	 for $i = 1, 2, \dots$ **do**
	    Observe ϕ_i, r_i, ϕ_{i+1};
	    Pre-update traces ;
	    $y_i = (\gamma \lambda \rho_{i-1})^2 y_{i-1} + 1$;
	    Compute ;
	    $U_i = \left(\sqrt{y_i} \Delta \phi_i + \dfrac{\gamma \lambda \rho_{i-1}}{\sqrt{y_i}} \Delta_{i-1} \quad \dfrac{\gamma \lambda \rho_{i-1}}{\sqrt{y_i}} \Delta_{i-1} \right)$;
	    $V_i = \left(\sqrt{y_i} \Delta \phi_i + \dfrac{\gamma \lambda \rho_{i-1}}{\sqrt{y_i}} \Delta_{i-1} \quad - \dfrac{\gamma \lambda \rho_{i-1}}{\sqrt{y_i}} \Delta_{i-1} \right)^T$;
	    $W_i = \left(\sqrt{y_i} \rho r_i + \dfrac{\gamma \lambda \rho_{i-1}}{\sqrt{y_i}} z_{i-1} \quad - \dfrac{\gamma \lambda \rho_{i-1}}{\sqrt{y_i}} z_{i-1} \right)^T$;
	    Update parameters ;
	    $\theta_i = \theta_{i-1} + C_{i-1} U_i (I_2 + V_i C_{i-1} U_i)^{-1} (W_i - V_i \theta_{i-1})$
	    ;
	    $C_i = C_{i-1} - C_{i-1} U_i (I_2 + V_i C_{i-1} U_i)^{-1} V_i C_{i-1}$;
	    Post-update traces ;
	    $\Delta_i = (\gamma \lambda \rho_{i-1}) \Delta_{i-1} + \Delta \phi_i y_i$;
	    $z_i = (\gamma \lambda \rho_{i-1}) z_{i-1} + r_i \rho_i y_i$;

3.4   Off-policy BRM(λ)

The off-policy BRM(λ) algorithm corresponds to the instantiation $\xi = \omega$ in Problem (2): $\theta_i = \operatorname{argmin}_{\omega \in \mathbb{R}^p} \sum_{j=1}^i (z_{j \to i} - \psi_{j \to i}^T \omega)^2$ where $\psi_{j \to i} = \sum_{k=j}^i \tilde{\rho}_j^{k-1} \Delta \phi_k$ and $z_{j \to i} = \sum_{k=j}^i \tilde{\rho}_j^{k-1} \rho_k r_k$. Thus $\theta_i = (\tilde{A}_i)^{-1} \tilde{b}_i$ where $\tilde{A}_i = \sum_{j=1}^i \psi_{j \to i} \psi_{j \to i}^T$ and $\tilde{b}_i = \sum_{j=1}^i \psi_{j \to i} z_{j \to i}$. The recursive implementation of θ_i is somewhat tedious, but space restriction precludes its description. The resulting algorithm, which is based on 3 traces — 2 reals (y_i and z_i) and 1 vector (Δ_i) — and involves the (straightforward) inversion of a 2×2 matrix, is described in Alg. 4.

GPTD and KTD, which are close to BRM, have also been extended with some trace mechanism; however, GPTD(λ) [8], KTD(λ) [9] and the just described BRM(λ) are different algorithms. Briefly, GPTD(λ) mimics the on-policy LSTD(λ) algorithms and KTD(λ) uses a different Bellman operator[5]. As BRM(λ) builds a linear systems of which it updates the solution recursively, it resembles LSTD(λ). However, the system it builds is different. The following theorem characterizes the behavior of BRM(λ) and its potential limit.

Theorem 1. *Assume that the stochastic matrix P_0 of the* behavior *policy is irreducible and has stationary distribution μ_0. Further assume that*

$$\text{there exists a coefficient } \beta < 1 \text{ such that } \forall (s, a), \quad \lambda \gamma \rho(s, a) \leq \beta, \qquad (9)$$

then $\frac{1}{i} \tilde{A}_i$ and $\frac{1}{i} \tilde{b}_i$ respectively converge almost surely to
$\tilde{A} = \Phi^T [D - \gamma D P - \gamma P^T D + \gamma^2 D' + S(I - \gamma P) + (I - \gamma P^T) S^T] \Phi$ *and* $\tilde{b} = \Phi^T [(I - \gamma P^T) Q^T D + S] R^\pi$ *where we wrote: $D = \operatorname{diag}((I - (\lambda \gamma)^2 \tilde{P}^T)^{-1} \mu_0)$, $D' = \operatorname{diag}(\tilde{P}^T (I - (\lambda \gamma)^2 \tilde{P}^T)^{-1} \mu_0)$, $Q = (I - \lambda \gamma P)^{-1}$, $S = \lambda \gamma (D P - \gamma D') Q$, and where $\tilde{P}$ is the matrix of coordinates $\tilde{p}_{ss'} = \sum_a \pi(s, a) \rho(s, a) T(s, a, s')$.*

---

[5] Actually, the corresponding loss is $(\hat{T}_{j,i}^0 \hat{V}(\omega) - \hat{V}_\omega(s_j) + \gamma \lambda (\hat{T}_{j+1,i}^1 \hat{V}(\omega) - \hat{V}_\omega(s_{j+1})))^2$. With $\lambda = 0$ it gives $\hat{T}_{j,i}^0$ and with $\lambda = 1$ it provides $\hat{T}_{j,i}^1$.

As a consequence the BRM(λ) algorithm converges with probability 1 to $\tilde{A}^{-1}\tilde{b}$. The assumption given by Eq. (9) trivially holds in the on-policy case (in which $\rho(s,a) = 1$ for all (s,a)) and in the off-policy case when $\lambda\gamma$ is small with respect to the mismatch between policies $\rho(s,a)$. The matrix $\tilde{P}$, which is in general not a stochastic matrix, can have a spectral radius bigger than 1; Eq. (9) ensures that $(\lambda\gamma)^2\tilde{P}$ has spectral radius smaller than β so that D and D' are well defined. Finally, note that there is probably no hope to remove the assumption of Eq. (9) since by making $\lambda\gamma$ big enough, one may force the spectral radius of $(\lambda\gamma)^2\tilde{P}$ to be as close to 1 as one may want, which would make $\tilde{A}$ and $\tilde{b}$ diverge.

The proof of this Theorem follows the general lines of that of Proposition 4 in [3]. Due to space constraints, we only provide its sketch: Eq. (9) implies that the traces can be truncated at some depth l, of which the influence on the potential limit of the algorithm vanishes when l tends to ∞. For all l, the l-truncated version of the algorithm can easily be analyzed through the ergodic theorem for Markov chains. Making l tend to ∞ allows to tie the convergence of the original arguments to that of the truncated version. Eventually, the formula for the limit of the truncated algorithm is (tediously) computed and one derives the limit.

The fundamental idea behind the Bellman Residual approach is to address the computation of the fixed point of T^λ differently from the previous methods. Instead of computing the projected fixed point as in Eq. (5), one considers the overdetermined system: $\Phi\theta \simeq T^\lambda\Phi\theta \quad\Leftrightarrow\quad \Phi\theta \simeq (I - \lambda\gamma P)^{-1}(R + (1 - \lambda)\gamma P\Phi\theta) \quad\Leftrightarrow\quad \Phi\theta \simeq QR + (1 - \lambda)\gamma PQ\Phi\theta \quad\Leftrightarrow\quad \Psi\theta \simeq QR$ with $\Psi = \Phi - (1-\lambda)\gamma PQ\Phi$ and solves it in a least-squares sense, that is by computing $\theta^* = \bar{A}^{-1}\bar{b}$ with $\bar{A} = \Psi^T\Psi$ and $\bar{b} = \Psi^T QR$. One of the motivation for this approach is that, contrary to the matrix A of LSTD/LSPE/FPKF, $\bar{A}$ is invertible for all values of λ, and one can always guarantee a finite error bound with respect to the best projection [18]. If the goal of BRM(λ) is to compute $\bar{A}$ and $\bar{b}$ from samples, what it actually computes ($\tilde{A}$ and $\tilde{b}$) will in general be biased because it is based on a single trajectory[6]. Such a bias adds an uncontrolled variance term to $\bar{A}$ and $\bar{b}$ (e.g. see [1]) of which an interesting consequence is that $\tilde{A}$ remains non singular[7]. More precisely, there are two sources of bias in the estimation: one results from the non Monte-carlo evaluation (the fact that $\lambda < 1$) and the other from the use of the correlated importance sampling factors (as soon as one considers off-policy learning). The interested reader may check that in the on-policy case, and when λ tends to 1, $\tilde{A}$ and $\tilde{b}$ coincide with $\bar{A}$ and $\bar{b}$. However, in the strictly off-policy case, taking $\lambda = 1$ does not prevent the bias due to the correlated importance sampling factors. If we have argued that LSTD/LSPE/FPKF asymptotically coincide when $\lambda = 1$, we see here that BRM may generally differ in an off-policy situation.

---

[6] It is possible to remove the bias when $\lambda = 0$ by using double samples. However, in the case where $\lambda > 0$, the possibility to remove the bias seems much more difficult.

[7] $\bar{A}$ is by construction positive definite, and $\tilde{A}$ equals $\bar{A}$ plus a positive term (the variance term), and is thus also positive definite.

4   Illustration of the Algorithms

In this section, we briefly illustrate the behavior of all the algorithms we have described so far. In a first set of experiments, we consider random Markov chains involving 3 states and 2 actions and projections onto random spaces[8] of dimension 2. The discount factor is $\gamma = 0.99$. For each experiment, we have run all algorithms (plus TD(λ) with stepsize $\alpha_t = \frac{1}{t+1}$) 50 times with initial matrix (M_0, N_0, C_0) equal to[9] $100I$, with $\theta_0 = 0$ and during $100,000$ iterations. For each of these 50 runs, the different algorithms share the same samples, that are generated by a random uniform policy π_0 (*i.e.* that chooses each action with probability 0.5). We consider two situations: *on-policy*, where the policy to evaluate is $\pi = \pi_0$, and *off-policy*, where the policy to evaluate is random (*i.e.* , it picks the actions with probabilities p and $1 - p$, where p is chosen uniformly at random). In the curves we are about to describe, we display on the abscissa the iteration number and on the ordinate the median value of the distance (quadratic, weighted by the stationary distribution of P) between the computed value $\Phi\theta$ and the real value $V = (I - \gamma P)^{-1} R$ (*i.e.* the lower the better).

For each of the two situations (*on-* and *off-policy*), we present data in two ways. To appreciate the influence of λ, we display the curves on one graph per algorithm with different values of λ (Fig. 1 and 2). To compare the algorithms for solving the Bellman equation $V = T^\lambda V$, we show on one graph per value of λ the error for the different algorithms (Fig. 3 and 4). **In the on-policy setting,** LSTD and LSPE have similar performance and convergence speed for all values of λ. They tend to converge much faster than FPKF, which is slightly faster than TD. BRM is usually in between LSTD/LSPE and FPKF/TD, though for small values of λ, the bias seems significative. When λ increases, the performance of FPKF and BRM improves. At the limit when $\lambda = 1$, all algorithms (except TD) coincide (confirming the intuition for $\lambda = 1$, the influence of the choice ξ vanishes in Eq. (2)). **In the off-policy setting,** LSTD and LSPE still share the same behavior. The drawbacks of the other algorithms are amplified with respect to the on-line situation. As λ increases, the performance of FPKF catches that of LSTD/LSPE. However, the performance of BRM seems to worsen while λ is increased from 0 to 0.99 and eventually approaches that of the other algorithms when $\lambda = 1$ (though it remains different, *cf.* the discussion in the previous section). **Globally,** the use of eligibility traces allows to significantly improve the performance of FPKF(λ) over FPKF [7] in both on- and off-policy cases, and that of BRM(λ) over BRM/GPTD/KTD of [8,10] in the on-policy case. The performance of BRM(λ) in the off-policy case is a bit disappointing, probably because of its inherent bias, which deserves further investigation. However, LSTD(λ)/LSPE(λ) appear to be in general the best algorithms.

---

[8] For each action, rewards are uniform random vectors on $(0, 1)^3$, transition matrices are random uniform matrices on $(0, 1)^{3 \times 3}$ normalized so that the probabilities sum to 1. Random projections are induced by random uniform matrices Φ of size 3×2.

[9] This matrix acts as an L_2 regularization and is used to avoid numerical instabilities at the beginning of the algorithms. The bigger the value, the smaller the influence.

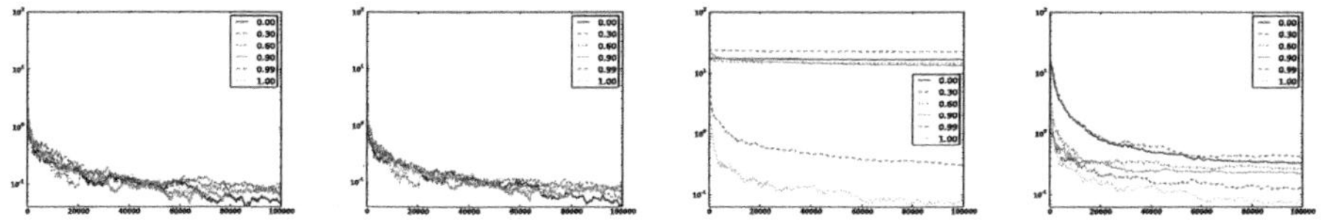

Fig. 1. Influence of λ, *on-policy* (LSTD, LSPE, FPKF and BRM)

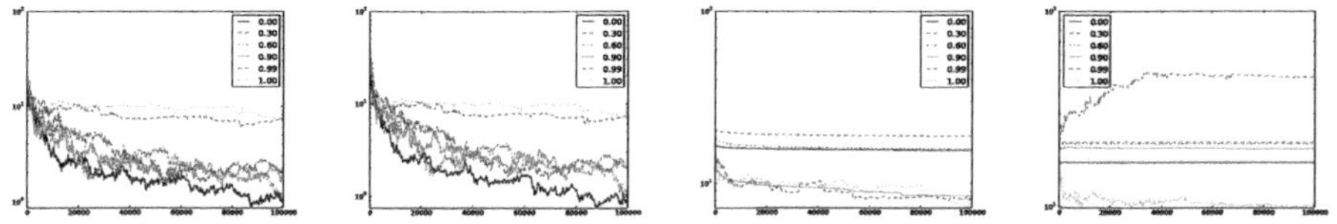

Fig. 2. Influence of λ, *off-policy* (LSTD, LSPE, FPKF and BRM)

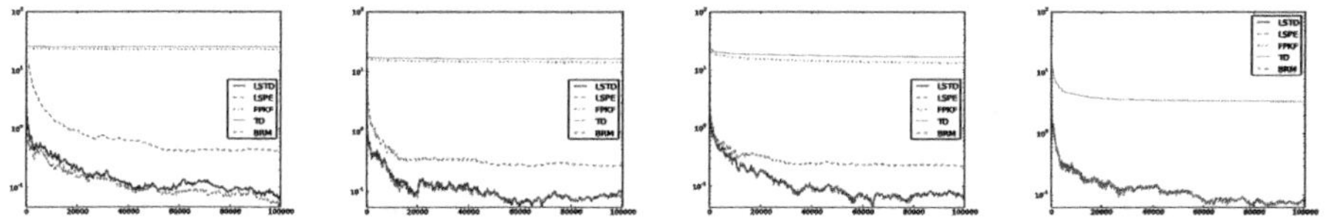

Fig. 3. Comparison of the algorithms, *on-policy* ($\lambda \in \{0.3, 0.6, 0.9, 1\}$)

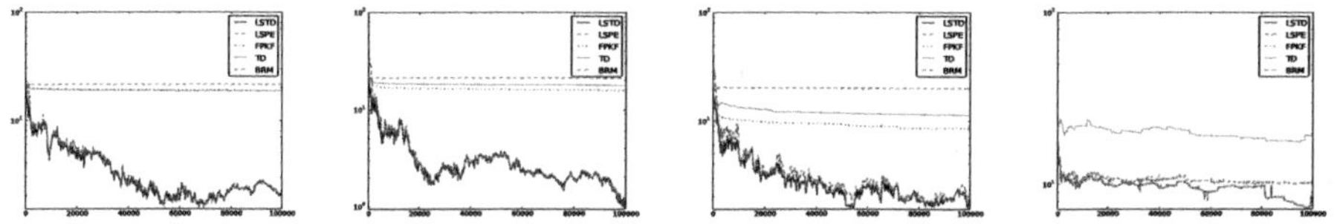

Fig. 4. Comparison of the algorithms, *off-policy* ($\lambda \in \{0.3, 0.6, 0.9, 1\}$)

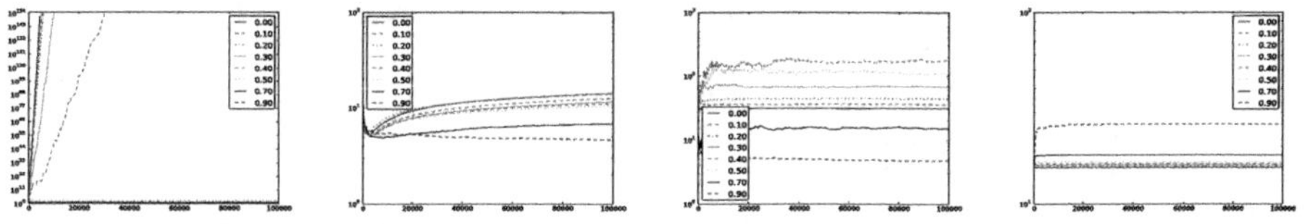

Fig. 5. Pathological situation where LSPE and FPKF diverge, while LSTD converges (LSPE, FPKF, LSTD and BRM)

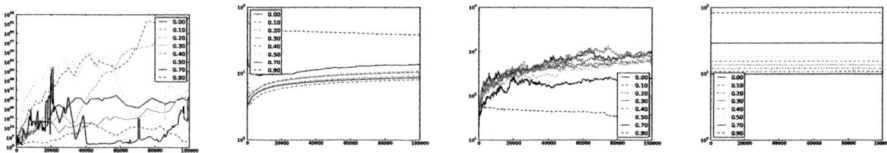

Fig. 6. Pathological situation where LSPE, FPKF and LSTD all diverge (LSPE, FPKF, LSTD and BRM)

Eventually, we consider 2 experiments involving an MDP and a projection due to [20], in order to illustrate possible numerical issues when solving the projected fixed-point Eq. (5). In the first experiment one sets (λ, γ) such that $\Pi_0 T^\lambda$ is expansive; as expected one sees (Fig. 5) that LSPE and FPKF both diverge. In the latter experiment, one sets (λ, γ) so that the spectral radius of $\Pi_0 T^\lambda$ is 1 (so that A is singular), and in this case LSTD also diverges (Fig. 6). In both situations, BRM is the only one not to diverge[10].

5   Conclusion

We considered LS algorithms for value estimation in an MDP context. Starting from the on-policy case with no trace, we recalled that several algorithms (LSTD, LSPE, FPKF and BRM/GPTD/KTD) optimize similar cost functions. Substituting the original Bellman operator by an operator that deals with traces and off-policy samples leads to the state-of-the-art off-policy trace-based versions of LSTD and LSPE, and suggests natural extensions of FPKF and BRM. We described recursive implementations of these algorithms, discussed their convergence properties, and illustrated their behavior empirically. Overall, our study suggests that even if the use of eligibility traces generally improves the efficiency of the algorithms, LSTD(λ) and LSPE(λ) remain in general better than FPKF(λ) (that is much slower) and BRM(λ) (that may suffer from high bias). Furthermore, since LSPE(λ) requires more conditions for stability, LSTD(λ) probably remains the best choice in practice.

References

1. Antos, A., Szepesvári, C., Munos, R.: Learning Near-Optimal Policies with Bellman-Residual Minimization Based Fitted Policy Iteration and a Single Sample Path. In: Lugosi, G., Simon, H.U. (eds.) COLT 2006. LNCS (LNAI), vol. 4005, pp. 574–588. Springer, Heidelberg (2006)
2. Baird, L.C.: Residual Algorithms: Reinforcement Learning with Function Approximation. In: ICML (1995)
3. Bertsekas, D.P., Yu, H.: Projected Equation Methods for Approximate Solution of Large Linear Systems. J. Comp. and Applied Mathematics 227(1), 27–50 (2009)
4. Bertsekas, D.P., Tsitsiklis, J.N.: Neuro-Dynamic Programming. Athena Scientific (1996)
5. Boyan, J.A.: Technical Update: Least-Squares Temporal Difference Learning. Machine Learning 49(2-3), 233–246 (1999)
6. Bradtke, S.J., Barto, A.G.: Linear Least-Squares algorithms for temporal difference learning. Machine Learning 22(1-3), 33–57 (1996)
7. Choi, D., Van Roy, B.: A Generalized Kalman Filter for Fixed Point Approximation and Efficient Temporal-Difference Learning. DEDS 16, 207–239 (2006)

---

[10] Note that this adverserial setting is meant to illustrate the fact that for the problem considered, some values (λ, γ) may be problematic for LSTD/LSPE/FPKF. In practice, λ can be chosen big enough so that these algorithms will be stable.

8. Engel, Y.: Algorithms and Representations for Reinforcement Learning. Ph.D. thesis, Hebrew University (2005)
9. Geist, M., Pietquin, O.: Eligibility Traces through Colored Noises. In: ICUMT (2010)
10. Geist, M., Pietquin, O.: Kalman Temporal Differences. JAIR 39, 483–532 (2010)
11. Geist, M., Pietquin, O.: Parametric Value Function Approximation: a Unified View. In: ADPRL (2011)
12. Kearns, M., Singh, S.: Bias-Variance Error Bounds for Temporal Difference Updates. In: COLT (2000)
13. Maei, H.R., Sutton, R.S.: GQ(λ): A general gradient algorithm for temporal-difference prediction learning with eligibility traces. In: Conference on Artificial General Intelligence (2010)
14. Munos, R.: Error Bounds for Approximate Policy Iteration. In: ICML (2003)
15. Nedić, A., Bertsekas, D.P.: Least Squares Policy Evaluation Algorithms with Linear Function Approximation. DEDS 13, 79–110 (2003)
16. Precup, D., Sutton, R.S., Singh, S.P.: Eligibility Traces for Off-Policy Policy Evaluation. In: ICML (2000)
17. Ripley, B.D.: Stochastic Simulation. Wiley & Sons (1987)
18. Scherrer, B.: Should one compute the Temporal Difference fix point or minimize the Bellman Residual? The unified oblique projection view. In: ICML (2010)
19. Sutton, R.S., Barto, A.G.: Reinforcement Learning: An Introduction (Adaptive Computation and Machine Learning), 3rd edn. MIT Press (1998)
20. Tsitsiklis, J., Van Roy, B.: An analysis of temporal-difference learning with function approximation. IEEE Transactions on Automatic Control 42(5), 674–690 (1997)
21. Yu, H.: Convergence of Least-Squares Temporal Difference Methods under General Conditions. In: ICML (2010)

Value Function Approximation through Sparse Bayesian Modeling

Nikolaos Tziortziotis and Konstantinos Blekas

Department of Computer Science, University of Ioannina,
P.O. Box 1186, 45110 Ioannina, Greece
{ntziorzi,kblekas}@cs.uoi.gr

Abstract. In this study we present a sparse Bayesian framework for value function approximation. The proposed method is based on the on-line construction of a dictionary of states which are collected during the exploration of the environment by the agent. A linear regression model is established for the observed partial discounted return of such dictionary states, where we employ the Relevance Vector Machine (RVM) and exploit its enhanced modeling capability due to the embedded sparsity properties. In order to speed-up the optimization procedure and allow dealing with large-scale problems, an incremental strategy is adopted. A number of experiments have been conducted on both simulated and real environments, where we took promising results in comparison with another Bayesian approach that uses Gaussian processes.

Keywords: Value function approximation, Sparse Bayesian modeling, Relevance Vector Machine, Incremental learning.

1 Introduction

Reinforcement learning (RL) [13] aims at controlling an autonomous agent in an environment which is usually unknown. The agent is only aware of a reward signal that is applied to it when acting with the environment. In this manner, the actions are evaluated and the learning process is designed on choosing the action with the optimum expected return. The goal of RL is to discover an optimal policy, where in most cases this is equivalent to estimating the value function of states. A plethora of methods has been proposed in the last decades using a variety of value-function estimation techniques [5]. Algorithms such as the Q-learning [18] and Sarsa [9,12] try to estimate the long-term expected value of each possible action given a particular state by choosing actions with the maximum value. However, these methods have some drawbacks that prevent them from using in large or continuous state spaces of real-world applications. Value function approximation approaches offer a nice solution to this problem. Least-squares temporal-difference (LSTD) learning [2] is a widely used algorithm for value function learning of a fixed policy. Also, the least-squares policy-iteration (LSPI) method [6] extends the LSTD by using it in the policy evaluation step of policy estimation.

Recently, kernelized reinforcement learning methods have been paid a lot of attention by employing all the benefits of kernel techniques [14]. In this manner, standard RL methods have been extended by mapping to kernel spaces, see for example [20,19]. One

S. Sanner and M. Hutter (Eds.): EWRL 2011, LNCS 7188, pp. 128–139, 2012.

particularly elegant Bayesian RL formulation is the Gaussian Process Temporal Difference (GPTD) [3], that constitutes an efficient adaptation of the Gaussian processes to the problem of online value-function estimation. The GPTD employs a probabilistic generative model for the state value function, and the solution to the inference problem is given by the posterior distribution conditioned on the observed sequence of rewards. An on-line kernel sparsification algorithm has also been proposed in [3], by incrementally constructing an appropriate dictionary of representative states. Finally, the Kalman Temporal Differences (KTD) framework has been introduced only recently [4], where the value function approximation is stated as a filtering problem and nonstationarities are allowed through the specification of some evolution model for parameters.

In this study an alternative Bayesian scheme for value function approximation is presented. The key aspects of our method are the creation of a state dictionary and the partial discounted return which corresponds to the accumulated reward between two states that get placed in the dictionary. The advantages of this approach are threefold. First, it achieves a reduced computational complexity, since our analysis deals only with the states which are stored in the dictionary. At a second level, it manages to avoid making approximations when dealing with large-scale problems, as in the case of GPTD method for calculating the kernel covariance matrix. Finally, it offers enhanced modeling capabilities due to the embedded sparsity model properties. More specifically, the proposed method addresses the problem of value function approximation by appropriately creating a linear regression model. Training this model is achieved through a sparse Bayesian methodology [15,11] that offers many advantages in regression. Enforcing sparsity is a fundamental machine learning regularization principle that causes to obtain more flexible inference methods. In sparse Bayesian regression we employ models having initially many degrees of freedom, where we apply a heavy tail prior over coefficients. After training, only few coefficients will be maintained, since they will be automatically considered as significant. This is equivalent to retaining only a part of the dictionary which will be responsible for estimating the value function and designing the optimum policy. Furthermore, we have used a computationally efficient incremental strategy that presented in [16], in order to accelerate the optimization procedure. The proposed method was tested on a suite of benchmarks including known simulated environments, as well as real environments using a PeopleBot mobile robot. Comparison has been made using the sparse on-line version of the GPTD algorithm.

In section 2, we briefly describe the Markov Decision Processes (MDPs) and the GPTD method as a Bayesian framework for value function approximation. The proposed sparse regression model is then presented in section 3, along with an incremental learning procedure. To assess the performance of our methodology we present in section 4 numerical experiments with artificial and real test environments. Finally, in section 5 we give conclusions and suggestions for future research.

2   Markov Decision Processes and GPTD

In the most standard formulation of the problem, the environment where an agent acts, is modeled as a Markov Decision Process (MDP) [13]. A MDP is denoted as a tuple $\{\mathcal{S}, \mathcal{A}, R, P, \gamma\}$, where $\mathcal{S}$ and $\mathcal{A}$ are the state and action spaces, respectively; R is a reward function that specifies the immediate reward for each transition; P is the state transition distribution; and $\gamma \in [0, 1]$ is a discount factor that determines the importance of current and future rewards. A stationary policy π defines a probability distribution over the action space condition on the states and can be seen as a mapping $\pi : \mathcal{S} \times \mathcal{A} \rightarrow [0, 1]$. The discounted return $D(s)$ for a state s under a policy π, having a policy dependent state transition probability distribution $p^{\pi}(\cdot|s_t)$, is given by

$$D(s) = \sum_{t=0}^{\infty} \gamma^t R(s_t)|s_0 = s . \tag{1}$$

This can be written more concisely as

$$D(s) = R(s) + \gamma D(s') , \text{ where } s' \sim p^{\pi}(\cdot|s). \tag{2}$$

The objective of RL problems is to estimate an optimal policy π^* which maximize the expected discounted return, $V^{\pi}(s) = E_{\pi}[D(s)]$. This can be translated into a value function approximation problem, according to the following recursive formulation:

$$V^{\pi}(s) = E_{\pi}\left[R(s_t) + \gamma V^{\pi}(s_{t+1})|s_t = s\right] . \tag{3}$$

Equation 3 is the Bellman equation for V^{π} which expresses a relationship between the values of current and next state. Alternatively, the state-action value function usually used to facilitate policy improvement. This is the expected discounted return starting from state s, taking the action a and then following the policy π:

$$Q^{\pi}(s, a) = E_{\pi}\left[\sum_{t=0}^{\infty} \gamma^t R(s_t)|s_0 = s, a_0 = a\right] . \tag{4}$$

Having found the optimal action-state value function Q^*, the optimal policy is given by $\pi^*(s) = \arg\max_a Q^*(s, a)$.

Gaussian Processes [8] have recently been used as a Bayesian framework for modeling RL tasks. Gaussian Process Temporal Difference (GPTD) [3] is based on describing the value function as a Gaussian process. In particular, a decomposition of the discounted return $D(s)$ is first considered into its mean value and a zero mean residual:

$$D(s) = V(s) + (D(s) - V(s)) = V(s) + \Delta V(s) . \tag{5}$$

By combining Eqs. 5, 2 we obtain the following rule:

$$R(s) = V(s) - \gamma V(s') + N(s, s') , \tag{6}$$

where $N(s, s') = \Delta V(s) - \gamma \Delta V(s')$ is the difference between residuals. Given a sample trajectory of states $\{s_1, \ldots, s_t\}$, the model results in a set of $t - 1$ linear equations

$$R_t = H_t V_t + N_t, \tag{7}$$

where R_t, V_t, N_t are vectors of rewards, value functions and residuals, respectively. Additionally, the H_t is a matrix of size $(t-1) \times t$ and is given by

$$H_t = \begin{bmatrix} 1 & -\gamma & 0 & \cdots & 0 \\ 0 & 1 & -\gamma & \cdots & 0 \\ \vdots & & & & \vdots \\ 0 & 0 & \cdots & 1 & -\gamma \end{bmatrix}. \tag{8}$$

By considering the above equation as a Gaussian Process, a zero-mean Gaussian prior is assumed over the value functions V_t, i.e., $V_t \sim \mathcal{N}(\mathbf{0}, K_t)$, where K_t is a kernel covariance matrix over states. Also, the residuals N_t is assumed to be zero-mean Gaussian, $N_t \sim \mathcal{N}(\mathbf{0}, \Sigma_t)$, where the covariance matrix is calculated as $\Sigma_t = \sigma_t^2 H_t H_t^\top$. In this way, at each time a state s is visited, the value function of the state is given by the posterior distribution, which is also Gaussian, $(V(s)|R_t) \sim \mathcal{N}(\hat{V}(s), p_t(s))$, where

$$\hat{V}(s) = k_t(s)^\top \alpha_t, \qquad \alpha_t = H_t^\top (H_t K_t H_t^\top + \Sigma_t)^{-1} R_t,$$

and $\quad p_t(s) = k(s,s) - k_t(s)^\top C_t k_t(s), \qquad C_t = H_t^\top (H_t K_t H_t^\top + \Sigma_t)^{-1} H_t.$

A limitation to the application of the GPTD is the computational complexity that increases linearly with time t. To solve this problem, an on-line kernel sparsification algorithm has been proposed in [3] which is based on the construction of a dictionary of representative states, $\mathcal{D}_{t-1} = \{\tilde{s}_1, \ldots, \tilde{s}_{d_{t-1}}\}$. An approximate linear dependence (ALD) analysis is performed in order to examine whether or not a visited state s_t must be entered into the dictionary. This is achieved according to a least squares problem, where we test if the image of the candidate state, $\phi(s_t)$, can be adequately approximated by the elements of the current dictionary [3], i.e.

$$\delta_t = \min_{\mathbf{a}} \left\| \sum_j a_j \phi(\tilde{s}_j) - \phi(s_t) \right\|^2 \leq \nu, \tag{9}$$

where ν is a positive threshold that controls the level of sparsity. The sparse on-line version of GPTD makes further approximations for calculating the kernel matrix K_t, where it uses only the dictionary members for this purpose; for more details see [3].

3 The Proposed Method

An advanced methodology to the task of control learning is presented in this section that employs the Relevance Vector Machine (RVM) [15] generative model for value function approximation. We start our analysis by taking into account the stationarity property of the MDP, which allows us to rewrite the discounted return of Eq. 1 at time step t as

$$D(s_t) = \mathcal{R}(s_t) + \gamma^{k_t} D(s_{t+k_t}), \tag{10}$$

where $\mathcal{R}(s_t) = \sum_{j=0}^{k_t-1} \gamma^j R(s_{t+j})$ is the *partial discounted return* of a state-reward subsequence. The term k_t denotes the time difference between two states that have been observed. According to the Eq. 5, the discounted return can be decomposed into its mean and a zero-mean residual. Considering this assumption and substituting Eq. 5 into Eq. 10 leads us to the following rule:

$$\mathcal{R}(s_t) = V(s_t) - \gamma^{k_t} V(s_{t+k_t}) + N(s_t, s_{t+k_t}), \tag{11}$$

where $N(s_t, s_{t+k_t}) = \Delta V(s_t) - \gamma^{k_t} \Delta V(s_{t+k_t})$ is the residuals difference.

Thus, assuming a dictionary of n states $\mathcal{D}_t = \{\tilde{s}_1, \ldots, \tilde{s}_n\}$, we obtain a set of $n-1$ equations:

$$\mathcal{R}(\tilde{s}_i) = V(\tilde{s}_i) - \gamma^{k_i} V(\tilde{s}_{i+1}) + N(\tilde{s}_i, \tilde{s}_{i+1}), \quad \text{for } i = 1, \ldots, n-1, \tag{12}$$

which can be written more concisely as

$$\mathcal{R}_n = H_n V_n + N_n, \tag{13}$$

where $\mathcal{R}_n = (\mathcal{R}(\tilde{s}_1), \ldots, \mathcal{R}(\tilde{s}_{n-1}))^T$, $V_n = (V(\tilde{s}_1), \ldots, V(\tilde{s}_n))^T$ and $N_n = (N(\tilde{s}_1, \tilde{s}_2), \ldots, N(\tilde{s}_{n-1}, \tilde{s}_n))^T$. The matrix H_n is of size $(n-1) \times n$ and has the following form

$$H_n = \begin{bmatrix} 1 & -\gamma^{k_1} & 0 & \cdots & 0 \\ 0 & 1 & -\gamma^{k_2} & \cdots & 0 \\ \vdots & & & & \vdots \\ 0 & 0 & \cdots & 1 & -\gamma^{k_{n-1}} \end{bmatrix}. \tag{14}$$

Moreover, we assume that the (hidden) vector of the value functions is described with the functional form of a linear model

$$V_n = \Phi_n \mathbf{w}_n, \tag{15}$$

where $\mathbf{w}_n$ is the vector of the n unknown model regression coefficients. $\Phi_n = [\phi_1^T \ldots \phi_n^T]$ is a kernel 'design' matrix that contains n basis functions ϕ_i, where the values of their components have been calculated using a kernel function $\phi_i(\tilde{s}_j) \equiv k(\tilde{s}_i, \tilde{s}_j)$. It must be noted that during our experimental study we have considered Gaussian type of kernels governed by a scalar parameter (kernel width). Thus, Eq. 13 can be written as

$$\mathcal{R}_n = H_n \Phi_n \mathbf{w}_n + N_n, \tag{16}$$

that can be further simplified as:

$$\mathbf{y}_n = \Phi_n \mathbf{w}_n + \mathbf{e}_n. \tag{17}$$

The above equation describes a linear regression model that fits the modified observations, $\mathbf{y}_n = (H_n^T H_n)^{-1} H_n^T \mathcal{R}_n$. The term $\mathbf{e}_n$ plays the role of the stochastic model noise and is assumed to be a zero-mean Gaussian with precision β_n, i.e. $\mathbf{e}_n \sim \mathcal{N}(0, \beta_n^{-1} I)$. Under this prism, the conditional probability density of the sequence $\mathbf{y}_n$ is also Gaussian, i.e.

$$p(\mathbf{y}_n | \mathbf{w}_n, \beta_n) = \mathcal{N}(\mathbf{y}_n | \Phi_n \mathbf{w}_n, \beta_n^{-1} I). \tag{18}$$

An important issue is how to define the optimal order of the above regression model. Sparse Bayesian methodology offers an advanced solution to this problem by penalizing large order models. This is the idea behind the Relevance Vector Machines (RVM) [15]. More specifically, a heavy-tailed prior distribution, $p(\mathbf{w}_n)$, is imposed over the regression coefficients $\mathbf{w}_n$ to zero out most of the weights w_{ni} after training. This is achieved in an hierarchical way: First, a zero-mean Gaussian distribution is considered

$$p(\mathbf{w}_n|\boldsymbol{\alpha}_n) = \mathcal{N}(\mathbf{w}_n|0, A_n^{-1}) = \prod_{i=1}^{n} \mathcal{N}(w_{ni}|0, \alpha_{ni}^{-1}), \tag{19}$$

where A_n is a diagonal matrix containing the n elements of the precision vector $\boldsymbol{\alpha}_n = (\alpha_{n1}, \ldots, \alpha_{nn})^\top$. At a second level, a Gamma hyperprior is imposed over each hyperparameter, α_{ni},

$$p(\boldsymbol{\alpha}_n) = \prod_{i=1}^{n} Gamma(\alpha_{ni}|a, b). \tag{20}$$

It must be noted that both Gamma parameters a, b, are *a priori* set to zero in order to make these priors uninformative.

This two-stage hierarchical prior is actually a Student's-t distribution that provides sparseness to the model [15], since it enforces most of the parameters α_{ni} to become large and as a result the corresponding weights w_{ni} are set to zero. In this way, the complexity of the regression models is controlled automatically, while at the same time over-fitting is avoided. Furthermore, we can obtain the marginal likelihood distribution of sequence $\mathbf{y}_n$ by integrating out the weights $\mathbf{w}_n$. This gives a zero mean Gaussian:

$$p(\mathbf{y}_n|\boldsymbol{\alpha}_n, \beta_n) = \int p(\mathbf{y}_n|\mathbf{w}_n, \beta_n)p(\mathbf{w}_n|\boldsymbol{\alpha}_n)dw = \mathcal{N}(0, C_n), \tag{21}$$

where the covariance matrix has the form, $C_n = \Phi_n A_n^{-1}\Phi_n^\top + \beta_n^{-1}I$.

From the Bayes rule, the posterior distribution of the weights can be also obtained as [15]:

$$p(\mathbf{w}_n|\mathbf{y}_n, \boldsymbol{\alpha}_n, \beta_n) = \mathcal{N}(\mathbf{w}_n|\boldsymbol{\mu}_n, \Sigma_n), \tag{22}$$

where

$$\boldsymbol{\mu}_n = \beta_n \Sigma_n \Phi_n^\top \mathbf{y}_n, \quad \Sigma_n = (\beta_n \Phi_n^\top \Phi_n + A_n)^{-1}. \tag{23}$$

The maximization of the log-likelihood function of Eq. 21, leads to the following update rules for the model parameters [15]:

$$\alpha_{ni} = \frac{\gamma_i}{\mu_{ni}^2}, \tag{24}$$

$$\beta_n^{-1} = \frac{\|\mathbf{y}_n - \Phi_n \boldsymbol{\mu}_n\|^2}{n - \sum_{i=1}^{n} \gamma_i}, \tag{25}$$

where $\gamma_i = 1 - \alpha_{ni}[\Sigma_n]_{ii}$ and $[\Sigma_n]_{ii}$ is the i-th diagonal element of the matrix Σ_n. Thus, Equations 23, 24 and 25 are applied iteratively until convergence. The mean values of weights, $\boldsymbol{\mu}_n$, are finally used for the value function approximation of a state s, i.e. $\tilde{V}(s) = \boldsymbol{\phi}(s)^T \boldsymbol{\mu}_n$, where $\boldsymbol{\phi}(s) = (k(s, \tilde{s}_1), \ldots, k(s, \tilde{s}_n))^T$.

3.1   Incremental Optimization

The application of RVM in large scaling problem is problematic, since it requires the computation of matrix Σ_n in Eq. 23. In our case this is happening when the size of the dictionary (n) becomes large. To deal with this problem we can follow an incremental learning algorithm that has been proposed in [16]. The method initially assumes that all states in the dictionary (all basis functions) have been pruned due to the sparsity constraint. This is equivalent to assuming that $\alpha_{ni} = \infty, \forall i = \{1, \ldots, n\}$. Then, at each iteration a basis function is examined whether to be either added to the model, or removed from the model, or re-estimated. When adding a basis function, the value of the hyperparameter α_{ni} is estimated according to the maximum likelihood criterion.

In particular, it is easily to show that the term of the marginal likelihood of Eq. 21 which referred to the single parameter α_{ni} is [16]

$$\ell(\alpha_{ni}) = \frac{1}{2}\left(\log \alpha_{ni} - \log(\alpha_{ni} + s_{ni}) + \frac{q_{ni}^2}{\alpha_{ni} + s_{ni}}\right), \tag{26}$$

where

$$s_{ni} = \frac{\alpha_{ni}S_{ni}}{\alpha_{ni} - S_{ni}}, \qquad q_{ni} = \frac{\alpha_{ni}Q_{ni}}{\alpha_{ni} - S_{ni}}, \tag{27}$$

and $S_{ni} = \phi_i^\top C_n^{-1}\phi_i$, $Q_{ni} = \phi_i^\top C_n^{-1}\mathbf{y}_n$, $\phi_i = (\phi_i(\tilde{s}_1), \ldots, \phi_i(\tilde{s}_n))^\top$. Note that in this manner the matrix inversion is avoided by using the Woodbury identity [16]. It has been shown in [16] that the log-likelihood has a single maximum at:

$$\alpha_{ni} = \frac{s_{ni}^2}{q_{ni}^2 - s_{ni}}, \quad \text{if } q_{ni}^2 > s_{ni}, \tag{28}$$

$$\alpha_{ni} = \infty, \qquad \text{if } q_{ni}^2 \leq s_{ni}. \tag{29}$$

Thus, a basis function ϕ_i is added (and so the corresponding state of the dictionary becomes active) in the case of $q_{ni}^2 > s_{ni}$. In the opposite case, this basis function is removed.

3.2   Working in Episodic Tasks and Unknown Environments

So far we have focused on solving continuing tasks, where the agent is placed initially to a random state and then is let to wander-off indefinitely. Since most of the RL tasks are episodic, a modification to our approach is needed to meet these requirements. During episodic tasks the agent will reach a terminal state within a finite number of steps. After that, a new episode (epoch) begins by placing the agent to a random initial position. An absorbing state may be thought as a state that only zero-rewards are received and that each action plays no role. In this case, the partial discounted return $\mathcal{R}(\tilde{s}_n)$ of the last inserted state ($\tilde{s}_n$) in an episode, is actually the discounted return $D(\tilde{s}_n)$, itself. Also, the discount factor γ for the subsequent dictionary state ($\tilde{s}_{n+1}$) will be set to zero. Thus, the matrix H_{n+1} will take the following form

$$H_{n+1} = \begin{bmatrix} 1 & -\gamma^{k_1} & 0 & \cdots & 0 & 0 \\ 0 & 1 & -\gamma^{k_2} & \cdots & 0 & 0 \\ \vdots & \vdots & & & \vdots & \vdots \\ 0 & 0 & \cdots & 1 & -\gamma^{k_{n-1}} & 0 \\ 0 & 0 & \cdots & 0 & 1 & 0 \end{bmatrix}. \tag{30}$$

This new form of the matrix H is the only modification to our approach in order to deal with episodic tasks.

Finally, in our study we have considered transitions between state-action pairs instead of single states, since the model environment is completely unknown. This is achieved by determining the kernel function as a product of state kernel k_s and action kernel k_a, i.e. $k(s, a, s', a') = k_s(s, s') k_a(a, a')$ (legitimate kernel [10,1]). Therefore, we have considered the approximation of the optimal state-value function Q.

4   Experimental Results

The performance of our model (called as RVMTD) has been studied to several simulated and real environments. In all cases, two evaluation criteria have been used: the mean number of steps, as well as the mean return with respect to episodes. Comparison has been made with the on-line GPTD algorithm [3]. It must be noted that both methods use the same criterion for adding a new state to the dictionary (Eq. 9) with the same threshold parameter ν. Also, the proper value for the scalar parameter of the Gaussian kernel in each problem was found experimentally. However, in some cases (mostly on simulated environments) the sensitivity of the performance of both approaches to this parameter was significant. Finally, in all cases, the decay parameter γ was set to 0.99.

4.1   Experiments on Simulated Environments

The first series of experiments was made using two well-known benchmarks [1]. The first one is the mountain car [7], where the objective of this task is to drive an underpowered car up a steep mountain road from a valley to tophill, as illustrated in the Fig. 1(a). Due to the force of gravity, the car cannot accelerate up to the tophill and thus it must go to the opposite slope to acquire enough momentum, so as to reach the goal on the right slope. The environmental states consist of two continuous variables: the position ($p_t \in [-1.5, +0.5]$) and the current velocity ($v_t \in [-0.07, 0.07]$) of the car. Initially, the car is standing motionless ($v_0 = 0$) at the position $p_0 = -0.5$. At each time step, it receives a negative reward $r = -1$. Three are the possible actions: +1 (full throttle forward), -1 (full throttle reverse) and 0 (zero throttle). An episode is terminated either when the car reaches the goal at the right tophill, or the total number of steps exceeds a maximum allowed value (1000). The state kernel k_s was set as $k_s = k(s, s') = \exp\left(-\sum_{i=1}^{2}(s_i - s'_i)^2/(2\sigma_i^2)\right)$, where $\sigma_1^2 = 5 \times 10^{-2}$ and $\sigma_2^2 = 5 \times 10^{-4}$. On the other hand we have used a simple action kernel of type: 1 when the actions are the same, 0.5 when differs by one and 0 otherwise. Finally, the parameter ν that

---

[1] Both simulators have been downloaded from http://www.dacya.ucm.es/jam/download.htm

specifies the sparsity of our model was set to $\nu = 0.001$, resulting in a dictionary that contains about 150 states.

Another test environment is the famous cart pole shown in Fig. 2(a), where the objective is to keep the pole balanced and the cart within its limits by applying a fixed magnitude force either to the left or to the right. The states consists of four continuous variables: the horizontal position (x) and the velocity $(\dot{x})$ of the cart, and the angle (θ) and the angular velocity $(\dot{\theta})$ of the pole. There are 21 possible discrete actions from -10 to 10, while the reward received by the environment takes the form $r = 10 - 10|10\theta|^2 - 5|x| - 10\dot{\theta}$. The cart is initially positioned in the middle of the track having zero velocity, and the pole is parallel to the vertical line having velocity $\dot{\theta} = 0.01$. An episode terminates when either the cart moves off the track, or the pole falls, or the pole is successfully balanced for 1000 time steps. Similarly, we have considered a Gaussian state kernel with different scalar parameter (σ_i^2) per variable: $\sigma_1^2 = 2, \sigma_2^2 = 0.5, \sigma_3^2 = 0.008$ and $\sigma_4^2 = 0.1$. The action kernel was also Gaussian with variance, $\sigma^2 = 1$. Finally, the parameter ν was set to 0.1, resulting in a dictionary of size 100.

The depicted results on these problems are illustrated in Figs. 1 and 2, respectively. As it is obvious our method achieves to find the same or improved policy in comparison

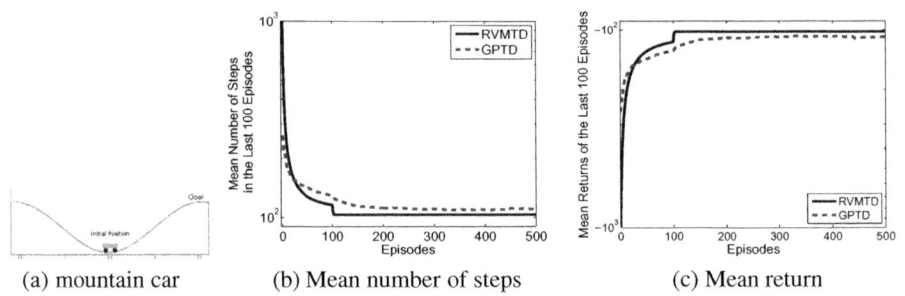

(a) mountain car   (b) Mean number of steps   (c) Mean return

Fig. 1. Experimental results with the mountain car simulator

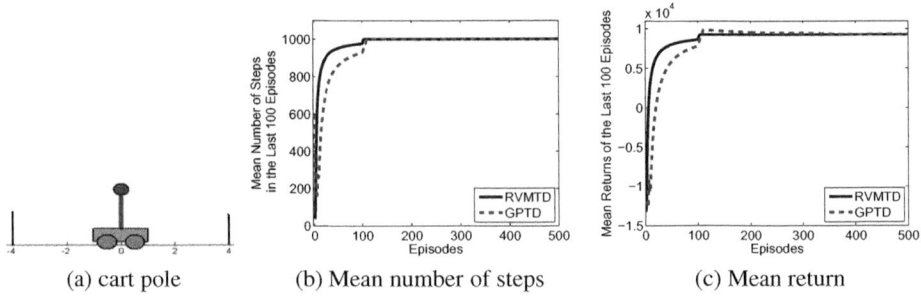

(a) cart pole   (b) Mean number of steps   (c) Mean return

Fig. 2. Experimental results with the cart pole simulator

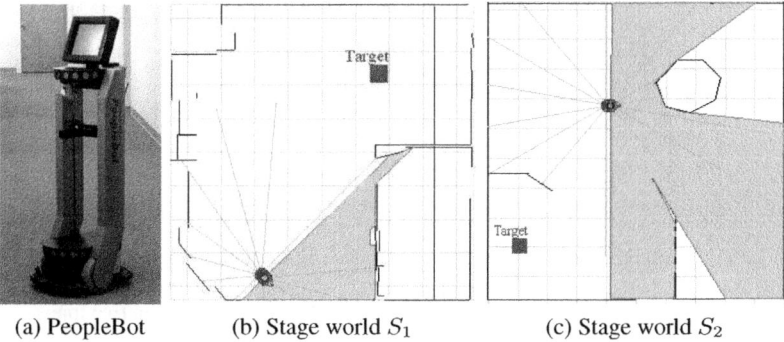

(a) PeopleBot   (b) Stage world S_1 (c) Stage world S_2

Fig. 3. The mobile robot and the 2D-grid maps used in our experiments. These are two snapshots from the simulator with visualization of the robot's laser and sonar range scanner.

with the online GPTD. However, our approach has the tendency to convergence to the optimum solution much faster than GPTD. Especially in the case of the cart-pole, only a few episodes were capable of reaching the optimum policy with a smaller in size dictionary. It is interesting to note that, although both approaches converged to almost the same policy, the GPTD method requires a larger dictionary (almost double size), as well as higher execution time.

4.2   Experiments on a Mobile Robot

The performance of the proposed method have also been studied to a PeopleBot mobile robot, shown in Fig. 3, which is based on the robust P3-DX base. This is a wheeled mobile robot occupied with advanced tools for communication, through the ARIA (Advanced Robot Interface for Applications) library and various sensors, such as sonar, laser and a pan-tilt camera. In this work, only the sonar and laser sensors were used for obstacle avoidance. There is also available the MobileSim simulation environment built on the Stage platform which manages to simulate the real environment with satisfactory precision[2].

Two different grid maps (stage worlds) have been selected during our experiments, as shown in Fig.3. Note that the first one was obtained by mapping our laboratory using the PeopleBot robot and the MobileEyes software. In this study, the objective is to find a steady landmark (shown with a rectangular box in both maps of Fig.3) starting from any position in the world with the minimum number of steps. The robot receives a reward of -1 per time step, except when it finds an obstacle where the reward is -100. In our study we have discretized the action space into the 8 major compass winds, while the length of each step was $0.5m$. Also, the maximum allowed number of steps per episode was set to 100. Finally, we have used a Gaussian type of kernel for the environmental state with a scalar parameter value $\sigma^2 = 1$. The action kernel takes 4 possible values: 1 (when actions are the same), 0.6 (when differs $45°$), 0.3 (when differs $90°$), 0 (otherwise). Note that, while we have used the same state kernel for both methods, in the case of the

---

[2] more details can be found at http://robots.mobilerobots.com/wiki/MobileSim

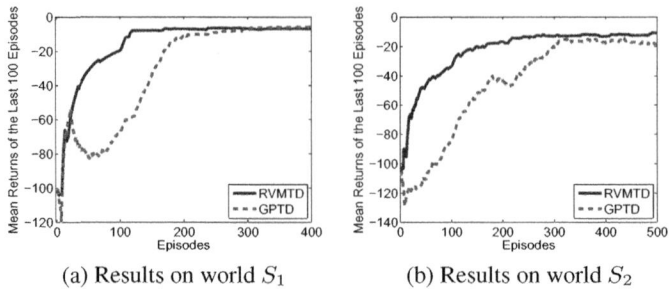

(a) Results on world S_1 (b) Results on world S_2

Fig. 4. Plots of the mean return as estimated by both methods in two maps

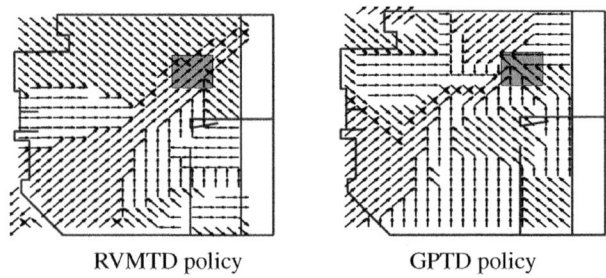

RVMTD policy   GPTD policy

Fig. 5. Learned policies by both comparative methods in the case of test world S_1

online GPTD we have adopted the action kernel function described in [3], since it gave better performance.

The experimental results of the two worlds are shown in Fig. 4 which gives the plots of the mean returns received by the agent in the last 100 episodes. Obviously, the proposed method manages to discover a more optimal policy in a higher rate in comparison with the GPTD. This is more apparent in Fig. 5, where we show the estimated trajectories following the learned policy of each method during studying the stage word S_1 (Fig. 3b). In the proposed method, the robot is close to the optimal path between any start point and destination (target), and as a result it reaches the destination reliably and faster. We took the same performance behavior with the other stage word S_2 of Fig. 3.

5   Conclusions

In this paper we have proposed an advanced methodology for model-free value function approximation using the RVM regression framework as the generative model. The key aspect of the proposed technique lies on the introduction of the partial discounted returns that are observed during the creation of a state dictionary. This sequential data of rewards is modeled using a sparse Bayesian framework as employed by the RVM method that incorporates powerful modeling properties. We have also applied an incremental learning strategy that accelerates the optimization procedure and makes the method to be practical for large scale problems. As experiments have shown, our method is able to achieve better performance and to learn significantly more optimal policies. A future research direction to our study is to further improve the regression method and the kernel design matrix specification, by incorporating a mechanism for adapting the scale parameter of the Gaussian kernel function [17]. Another interesting topic for future study is to work on different schemes for the on-line dictionary construction that allowing the dictionary to be dynamically adjusted during learning.

References

1. Bishop, C.M.: Pattern Recognition and Machine Learning. Springer, Heidelberg (2006)
2. Bradtke, S.J., Barto, A.G.: Linear least-squares algorithms for temporal difference learning. Machine Learning 22, 33–57 (1996)
3. Engel, Y., Mannor, S., Meir, R.: Reinforcement learning with gaussian process. In: International Conference on Machine Learning, pp. 201–208 (2005)
4. Geist, M., Pietquin, O.: Kalman Temporal Differences. Journal of Artificial Intelligence Research 39, 483–532 (2010)
5. Kaelbling, L.P., Littman, M.L., Moore, A.W.: Reinforcement learning: A survey. Journal of Artificial Inteligence Research 4, 237–285 (1996)
6. Lagoudakis, M.G., Parr, R.: Least-squares policy iteration. Journal of Machine Learning Research 4, 1107–1149 (2003)
7. Moore, A.: Variable resolution dynamic programming: Efficiently learning action maps in multivariate real-valued state-spaces. In: Machine Learning: Proceedings of the Eighth International Conference. Morgan Kaufmann (June 1991)
8. Rasmussen, C., Williams, C.: Gaussian Processes for Machine Learning. MIT Press (2006)
9. Rummery, G.A., Niranjan, M.: On-line q-learning using connectionist systems. Tech. rep., Cambridge University Engineering Department (1994)
10. Scholkopf, B., Smola, A.: Learning with Kernels. MIT Press (2002)
11. Seeger, M.: Bayesian Inference and Optimal Design for the Sparse Linear Model. Journal of Machine Learning Research 9, 759–813 (2008)
12. Singh, S., Sutton, R.S., Kaelbling, P.: Reinforcement learning with replacing eligibility traces. Machine Learning, 123–158 (1996)
13. Sutton, R.S., Barto, A.G.: Reinforcement Learning: An Introduction. MIT Press, Cambridge (1998)
14. Taylor, G., Parr, R.: Kernelized value function approximation for reinforcement learning. In: International Conference on Machine Learning, pp. 1017–1024 (2009)
15. Tipping, M.E.: Sparse bayesian learning and the relevance vector machine. Journal of Machine Learning Research 1, 211–244 (2001)
16. Tipping, M.E., Faul, A.C.: Fast marginal likelihood maximization for sparse bayesian models. In: Proceedings of the Ninth International Workshop on Artificial Intelligence and Statistics (2003)
17. Tzikas, D., Likas, A., Galatsanos, N.: Sparse Bayesian modeling with adaptive kernel learning. IEEE Trans. on Neural Networks 20(6), 926–937 (2009)
18. Watkins, C., Dayan, P.: Q-learning. Machine Learning 8(3), 279–292 (1992)
19. Xu, X., Hu, D., Lu, X.: Kernel-based least squares policy iteration for reinforcement learning. IEEE Transactions on Neural Networks 18(4), 973–992 (2007)
20. Xu, X., Xie, T., Hu, D., Lu, X.: Kernel least-squares temporal difference learning. International Journal of Information Technology 11(9), 54–63 (2005)

Automatic Construction of Temporally Extended Actions for MDPs Using Bisimulation Metrics

Pablo Samuel Castro and Doina Precup

School of Computer Science
McGill University
{pcastr,dprecup}@cs.mcgill.ca

Abstract. Temporally extended actions are usually effective in speeding up reinforcement learning. In this paper we present a mechanism for automatically constructing such actions, expressed as options [24], in a finite Markov Decision Process (MDP). To do this, we compute a bisimulation metric [7] between the states in a small MDP and the states in a large MDP, which we want to solve. The *shape* of this metric is then used to completely define a set of options for the large MDP. We demonstrate empirically that our approach is able to improve the speed of reinforcement learning, and is generally not sensitive to parameter tuning.

1 Introduction

Temporally extended actions are a well-studied approach for speeding up stochastic planning and reinforcement learning [2; 5; 16; 24]. Much attention has been devoted to the problem of learning such courses of action (e.g. [10; 12; 13; 27; 30]). This is still a challenge, as it goes back to the fundamental problem of intelligence: how can one learn useful *representations* of an environment from data?

In this paper we propose a new approach to the problem of learning extended actions, which is based on similar intuitions to learning by analogy, an approach used in cognitive architectures such as ACT-R [1] and SOAR [11]. The idea is to find similarities between a new problem and older, solved problems, and adapt the previous solution to the new situation. Here, we use small example systems, compute the similarity of states from such systems to states in larger problems, then use this similarity measurement in order to produce new temporally extended actions for the large system. It is important to note that, unlike in other research such as transfer learning (see [26] for a survey), we do not attempt to provide an entire solution to the new system. Instead, "pieces" of knowledge are constructed, and they will be used further in learning and planning for the new problem. This task is less daunting, as we will illustrate in the paper; as a result, the solved problem and the new problem could actually be quite different, and the approach still succeeds in providing useful representations.

We develop this idea in the context of the *options* framework for modeling temporally extended actions in reinforcement learning [17; 24]. We use bisimulation metrics [7] to relate states in the two different problems, from the point

S. Sanner and M. Hutter (Eds.): EWRL 2011, LNCS 7188, pp. 140–152, 2012.

of view of behavior similarity. Previous work [3] explored the use of bisimulation metrics for transferring the optimal policy from a small system to a large system. Here, we instead identify subsets of states in which an option should be defined, based on the "shape" of the bisimulation metric. We use the metric to define all aspects of the option: the initiation set, the termination probabilities, and its internal policy. These options can then be used for reinforcement learning. In the experiments, we use these options instead of primitive actions, although they could also be used in addition to the primitives. We use a sampling-based approximant of the bisimulation metric [6] to compute the distances, removing the requirement that a model of the system be known beforehand.

This paper is organized as follows. In section 2 we present the necessary background and introduce the notation. In section 3 we describe the procedure for constructing options. We empirically evaluate the quality of the options obtained in section 4. We present concluding remarks and avenues of future work in section 5.

2   Background and Notation

2.1   MDPs and Q-Learning

A Markov Decision Process (MDP) is defined as a 4-tuple $\langle S, A, P, R \rangle$ where S is a finite set of states, A is a finite set of actions available from each state, $P : S \times A \rightarrow Dist(S)^1$ specifies the probabilistic transition function and $R : S \times A \rightarrow \mathbb{R}$ is the reward function. A policy $\pi : S \rightarrow Dist(A)$ indicates the action choice at each state. The value of a state $s \in S$ under a policy π is defined as $V^\pi(s) = \mathbb{E}^\pi \left[\sum_{\tau=1}^\infty \gamma^{\tau-1} r_\tau | s_0 = s \right]$, where r_τ is a random variable denoting the reward received at time step τ and $\gamma \in (0, 1)$ is the discount factor. The optimal state-action value function $Q^* : S \times A \rightarrow \mathbb{R}$ gives the maximal expected return for each state-action pair, given that the optimal policy is followed afterwards. It obeys the following set of Bellman optimality equations: $Q^*(s, a) = R(s, a) + \gamma \sum_{s' \in S} P(s, a)(s') \max_{a' \in A} Q^*(s', a')$.

Q-learning is a popular reinforcement learning algorithm which maintains an estimate of Q^* based on samples. After performing action a from state s, landing in state s' and receiving reward r, the estimate for $Q(s, a)$ is updated as follows: $Q(s, a) = Q(s, a) + \alpha \left[r + \gamma \max_{a' \in A} Q(s', a') - Q(s, a) \right]$, where $\alpha \in (0, 1)$ is a learning rate parameter. Watkins & Dayan (1992) proved that the above method converges to Q^* as long as all actions are repeatedly taken from all states (and some conditions on the learning rate α). To guarantee this, a simple strategy is choosing an action randomly with probability ϵ and choosing the action that appears best (according to the current estimates) the rest of the time; this is called ϵ-greedy exploration. For a more detailed treatment of these and other reinforcement learning algorithms, please see [23].

---

[1] $Dist(X)$ is defined as the set of distributions over X.

2.2   Options

The options framework [24; 17] allows the definition of temporally extended actions in an MDP. Formally, an option o is a triple $\langle \mathcal{I}_o, \pi_o, \beta_o \rangle$, where $\mathcal{I}_o \subseteq S$ is the set of states where the option is available, $\pi_o : S \to Dist(A)$ is the option's policy and $\beta_o : S \to [0, 1]$ is the probability of the option terminating at each state. When an option o is started in state $s \in \mathcal{I}_o$, the policy π_o is followed until the option is terminated, as dictated by β_o. The *model* of an option consists of the *discounted* transition probabilities $Pr(s'|s, o), \forall s, s' \in S$ and the expected reward received while executing the option: $\mathfrak{R}(s, o)$.

More formally, cf. [24; 17], let $Pr(s', k|s, o)$ be the probability that option o terminates after k time steps in state s', after being started in s. Then, $Pr(s'|s, o)$ is defined as $Pr(s'|s, o) = \sum_{k=1}^{\infty} Pr(s', k|s, o)\gamma^k$. Note that because the transition model of an option incorporates the discount factor, it actually consists of *sub-probabilities* (*i.e.* $\sum_{s'} Pr(s'|s, o) < 1$).

Similarly, the reward model of an option is defined as:
$\mathfrak{R}(s, o) = \mathbb{E}^o \left[\sum_{\tau=1}^{k} \gamma^{\tau-1} r_\tau | s_0 = s \right]$, where k is the time at which the option terminates.

With these model definitions, planning and reinforcement learning algorithms can be defined for an MDP with given options in a manner very similar to algorithms involving just primitive actions. Options have also been shown to speed up learning and planning in application domains, especially in robotics (e.g. [22]).

A significant amount of research has been devoted to methods for learning options (and temporal abstractions in general). Much of the existing work is based on the notion of *subgoals*, i.e. important states that may be beneficial to reach. Subgoals can be identified based on a learned model of the environment, using graph theoretic techniques [12; 27], or based on trajectories through the system, without learning a model [13; 21]. Some existing work targets specifically the problem of learning subgoals from trajectories when the state space is factored and state abstractions need to be acquired at the same time as options [9; 14; 30; 15]. Other work is focused on learning options from demonstrations [10]. The notion of subgoals is conceptually attractive, as it creates abstractions that are easy to interpret, but it is also somewhat brittle: options defined by subgoals may not be useful, if the subgoals are not well identified. Comanici & Precup (2010) define an option construction algorithm where the general probabilities of termination are learned instead. This is also the approach we take here, but the learning algorithm is not gradient-based, as in their case.

The class of option construction methods most related to our approach is based on MDP homomorphisms [18; 29; 19; 20]. In this case, states from one MDP are mapped into states from a different MDP, based on the transition dynamics and reward functions. The existing work attempts to learn such mappings from data; Taylor et al. (2009) relate the notion of approximate MDP homomorphisms to a notion of MDP similarity called lax bisimulation, which allows for the construction of approximate homomorphisms with provable guarantees. Our approach builds on their metrics (as detailed below). Note also that, in contrast with existing work, we will construct *all* parts of the option: initiation set, policy and termination conditions.

2.3 Bisimulation Metrics

Bisimulation captures behavioral equivalence between states in an MDP [8]. Roughly speaking, it relates two states when they have equal immediate rewards and equivalent dynamics.

Definition 1. *An equivalence relation E is a bisimulation relation if for any $s, t \in S$, sEt implies that for all $a \in A$ (i): $R(s, a) = R(t, a)$; and (ii): for all $C \in S/_E$, $\sum_{s' \in C} P(s, a)(s') = \sum_{s' \in C} P(t, a)(s')$, where $S/_E$ is the set of all equivalence classes in S w.r.t. E. Two states $s, t \in S$ are called bisimilar, denoted $s \sim t$, if there exists a bisimulation relation E such that sEt.*

Ferns et al. (2004) defined bisimulation metrics, a quantitative analogue of bisimulation relations. A metric d is said to be a bisimulation metric if for any $s, t \in S$, $d(s, t) = 0 \Leftrightarrow s \sim t$.

The bisimulation metric is based on the Kantorovich probability metric $T_K(d)(P, Q)$, where d is a pseudometric[2] on S and P, Q are two state distributions. The Kantorovich probability metric is defined as a linear program, which intuitively computes the cost of "converting" P into Q under metric d. The dual formulation of the problem is an instance of the minimum cost flow (MCF) problem, for which there are many efficient algorithms. Ferns et al. (2004) proved that the functional $F(d)(s, t) = \max_{a \in A} (|R(s, a) - R(t, a)| + \gamma T_K(d)(P(s, a), P(t, a))$ has a greatest fixed point, $d_\sim$, and $d_\sim$ is a bisimulation metric. It can be computed to a desired degree of accuracy δ by iteratively applying F for $\left\lceil \frac{\ln \delta}{\ln \gamma} \right\rceil$ steps. However, each step involves the computation of the Kantorovich metric for all state-state-action triples, which requires a model and has worst case running time of $O(|S|^3 \log |S|)$. We will refer to this bisimulation metric as the *exact metric*. Ferns et al. (2006) introduced a more efficient way of computing the metric by replacing the MCF problem with a weighted matching problem. The state probability distributions P and Q are estimated using statistical samples. More precisely, let $X_1, X_2, \ldots, X_N$ and $Y_1, Y_2, \ldots, Y_N$ be N points independently sampled from P and Q, respectively. The *empirical distribution* P_N is defined as: $P_N(s) = \frac{1}{N} \sum_{i=1}^{N} \mathbb{1}_{X_i = s}, \forall s$, where $\mathbb{1}$ is the indicator function. Then, for any metric d, $T_K(d)(P_N, Q_N) = \min_\sigma \frac{1}{N} \sum_{i=1}^{N} d(X_i, Y_{\sigma(i)})$, where the minimum is taken over all permutations σ on N elements. This is an instance of the weighted assignment problem, which can be solved in $O(N^3)$ time. In this approach, one can control the computation cost by varying the number of samples taken. Ferns et al. (2006) proved that $\{T_K(d)(P_N, Q_N)\}$ converges to $T_K(d)(P, Q)$ almost surely, and the metric $d_{\sim, N}(s, t)$ computed using $T_K(d)(P_N, Q_N)$ converges to $d_\sim$. We refer to this approximant as the *sampled metric*.

---

[2] A pseudometric is similar to a metric but $d(x, y) = 0 \not\Rightarrow x = y$.

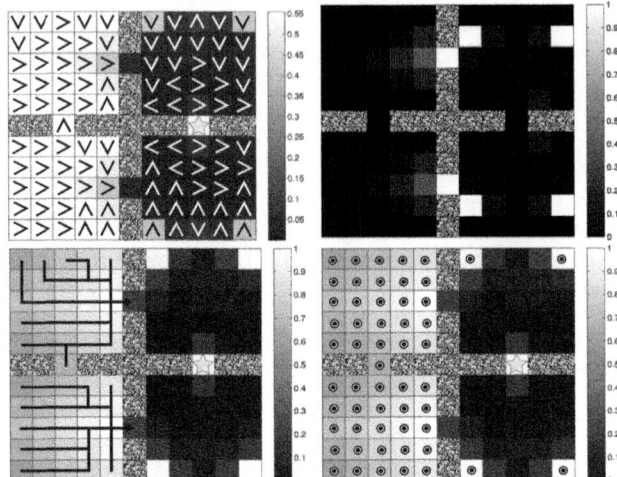

Fig. 1. Top left: Distances to leftmost source state and π_o; Top right: Computed values of β_o; Bottom left: Sum of discounted probabilities with chains used to find $\diamond$; Bottom right: Sum of discounted probabilities, dotted states are in $\mathcal{I}_o$

A shortcoming of bisimulation relations and metrics as defined above is that the action sets of two states under comparison must be the same. Taylor et al. (2009) overcome this problem by introducing lax-bisimulation relations.

Definition 2. *An equivalence relation E is a lax-bisimulation relation if for any $s, t \in S$, sEt implies that $\forall a \in A, \exists b \in A$ such that (i): $R(s, a) = R(t, b)$; (ii): for all $C \in S/E$, $\sum_{s' \in C} P(s, a)(s') = \sum_{s' \in C} P(t, b)(s')$; and vice versa. Two states s, t are called lax-bisimilar, denoted $s \sim_L t$, if there exists a lax-bisimulation relation E such that sEt.*

Taylor et al. (2009) also extend lax-bisimulation relations to lax-bisimulation metrics by first defining the metric $J(d)$ between state-action pairs, given any metric d, as follows:

$$J(d)((s, a), (t, b)) = |R(s, a) - R(t, b)| + \gamma T_K(d)(P(s, a), P(t, a)) \qquad (1)$$

They show that the functional

$$F_L(d)(s, t) = \max \left(\begin{array}{l} \max_{a \in A} \min_{b \in A} J(d)((s, a), (t, b)), \\ \max_{b \in A} \min_{a \in A} J(d)((s, a), (t, b)) \end{array} \right) \qquad (2)$$

has a greatest fixed point d_L, and d_L is a lax-bisimulation metric. As for the original bisimulation metric, we can approximate T_K using statistical sampling.

The lax bisimulation metric (or its approximants) can be used to establish homomorphisms between MDPs. Castro & Precup (2010) used lax-bisimulation metrics to transfer an optimal policy from a small source MDP to a large target

MDP. Their approach was to pair every state in the target MDP with its "nearest" source state. The optimal action is used in the source state, and its "best matching" action counterpart is used in the target.

3   Option Construction

We are now ready to present our approach to automatically construct options using a small source MDP for which an optimal policy has been found. The intuition behind the approach is that we will try to define options in a new, large task based on identifying "patches" of states that are similar to some particular state from the small problem. Each state from the small problem, with its corresponding optimal action, will be translated into a full option in the larger task. Only states that are similar enough to the original one should be included as part of the initiation set for the option. Large changes in similarity indicate large changes in either rewards or dynamics, and we use this as a cue that the option should be discontinued.

To assist in the description of the algorithms, we will use a running example in which a 7-state chain with a unique goal state in the rightmost position acts as a source system M_1, and a domain with four 5x5 rooms and 4 hallways is the target system M_2 (similar to [24]); one of the hallways is the goal state, and the only positive reward is obtained when entering this goal; transitions succeed with 90% probability. We will focus the example on the option arising from the leftmost source state for simplicity. Note that the algorithm will construct one option corresponding to every state in the small source task, but in general one could decide to only construct options from a subset of states in the source. For each source state s we denote the resulting option as o_s. We first construct the policy π_{o_s}, then the termination function β_{o_s}, and finally the initialization set $\mathcal{I}_{o_s}$.

3.1   Constructing π_{o_s}

The policy for o_s is obtained initially by the method used in [3] for knowledge transfer. More precisely, given MDPs M_1 and M_2, the following metric is computed, where π^* is the optimal policy for MDP M_1:

$$ J(d)(s,(t,b)) = |R_1(s,\pi^*(s)) - R_2(t,b)| \quad + \gamma T_K(d)(P_1(s,\pi^*(s)), P_2(t,a)) \quad (3) $$

Note that this is similar to equation (1), but the source state s is restricted to an optimal action given by $\pi^*(s)$. This was suggested in [3] to speed up the computation of the metric. Now, for any $t \in S_2$: $\pi_{o_s}(t) = \arg\min_{b \in A_2} J(d_L)(s,(t,b))$, where d_L is the lax-bisimulation metric. The top left image in Figure 1 displays the d_L distance between the states in the target system and the leftmost state in the source system, as well as the resulting policy. Note that the policy is defined, at the moment, for all states; it is optimal in the rooms on the left but sub-optimal in the rooms on the right. This is expected as we are comparing against the leftmost state in the source system. The policy generated by the source states that are closer to their goal is better for the rooms on the right than for those on the left (not shown for lack of space).

3.2   Constructing β_{o_s}

From the left panel of Fig. 1, it is clear that the bisimulation distance has areas of smooth variation, among states that are at the same relative position to the goal (i.e. the source of reward) and areas of sharp variation around the hallways. Intuitively, if one considers the temporal neighborhood of states, areas in which the bisimulation distance varies drastically indicate a big behavioral change, which suggests a *boundary*. This is a similar idea to the change point detection of [10], but here there is no human demonstration involved; the changes arise automatically from the similarities. These *areas of interest* will be used to define β_{o_s}.

To formalize this idea, for all state-action pairs (t, a) from M_2, we consider the most probable next state: $\bigcirc(t, a) = \arg\max_{t'} P_2(t, a)(t')$. For any state t' and option o_s, let $\zeta(t', o_s)$ be the set of states that have t' as their most probable next state:

$$\zeta(t', o_s) = \{t \in S_2 | \bigcirc (t, \pi_{o_s}(t)) = t'\}$$

The change in distance at each target state t' can be computed as follows:

$$\Delta(t', o_s) = \left| d_L(s, t') - \frac{1}{|\zeta(t', o_s)|} \sum_{t \in \zeta(t', o_s)} d_L(s, t) \right|$$

We now convert Δ into a termination probability:

$$\beta_{o_s}(t') = \frac{\Delta(t', o_s)}{\max_{t'' \in S_2} \Delta(t'', o_s)}$$

This definition ensures that options have a greater probability of terminating at states with the greatest change in the distance function. The top right panel of Fig. 1 displays the termination probabilities at each state.

3.3   Constructing the Initiation set $\mathcal{I}_{o_s}$

The question of learning good initiation sets is perhaps the least studied in the option construction literature. Intuitively, one desirable property of options is that they terminate within a reasonable amount of time, in order to allow a reconsideration of the course of action. Note that the model of an option provides this information readily, as its probabilities are discounted (*i.e.* $\sum_{t' \in S_2} Pr(t'|t, o) < 1$). The bottom panel of Fig. 1 displays the sum of discounted probabilities under the transferred option. Note that low values result from two different types of behavior: either states in which the policy of the option causes the agent to enter a cycle (i.e. by running into a wall), or states where the policy is good, but the option takes longer to terminate. We would like to distinguish between these two situations, and exclude from the initiation set mainly the states where the policy can lead to high-probability cycles. To achieve this, we define for all $t \in S_2$ and option o the most likely *ending state*, $\Diamond(t, o)$, in a recursive manner:

$$\Diamond(t, o) = \begin{cases} t & \text{If } \bigcirc(t, o) = t \\ \Diamond(\bigcirc(t, o), o) & \text{If } \sum_{t' \in S_2} Pr(t'|t, o) \leq \sum_{t' \in S_2} Pr(t'| \bigcirc (t, o), o) \\ t & \text{Otherwise} \end{cases}$$

Thus, in order to determine $\Diamond(t, o)$ we follow the chain of most likely states (using the $\bigcirc$ operator) as long as the sum of discounted probabilities continues to increase (the second condition above). In the bottom left panel of figure 1 we display the chain of states followed to find the most likely states. States without a "path" are those states for which the same state is the most likely state (either by condition 1 or 3 above). We determine whether t should be in $\mathcal{I}_{o_s}$ based on the sum of discounted probabilities of the most likely ending state. Let τ be a threshold parameter. Formally, we define the initialization set as follows:

$$\mathcal{I}_{o_s} = \left\{ t \in S_2 \,\middle|\, \sum_{t' \in S_2} Pr(t'|\Diamond(t, o_s), o_s) \geq \tau \right\}$$

In the bottom right panel of figure 1 the dotted cells are the states in $\mathcal{I}$.

It is possible that some states will not be included in the initialization set for any option. If such a state is encountered, we add it to the $\mathcal{I}_{\hat{o}}$ for the option $\hat{o}$ that has the highest probability of termination when started from t. Note that we could also simply allow such states to use only primitive actions. We did not encounter any such states in our experiments.

4   Empirical Evaluation

In this section we evaluate how effective the constructed options are in speeding up learning. Given an MDP, we use a small source MDP to construct options in the larger MDP, using a threshold parameter of $\tau = 0.7^3$ for all environments. Then we perform standard SMDP Q-learning using *only* the options, and compare its results to Q-learning with just primitive actions. The Q-value functions were initialized to 0 in all domains. We use the sampled version of the metric, as we do not assume that the environment model is known; all needed quantities for the construction algorithm are computed based on these samples. The sampling is performed once at the beginning of the algorithm, resulting in $N * |S| * |A|$ total samples. To allow a fair comparison with Q-learning, we grant Q-learning a "burn-in" phase, where it is allowed to take $N * |S| * |A|$ steps and update its Q-value function estimate. For all our experiments we set $N = 10$. As a baseline,

---

[3] The size of the initiation set is inversely proportional to τ. There was not much difference in performance when $\tau \in [0.6, 0.8]$.

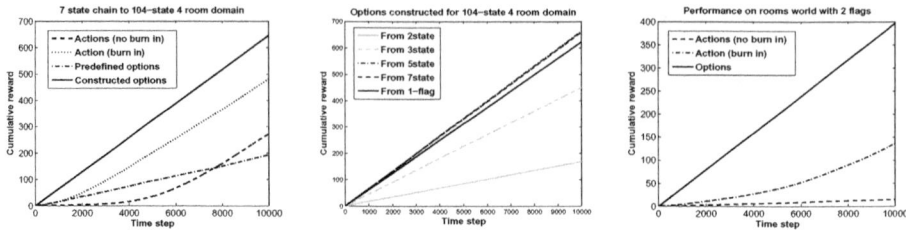

Fig. 2. Rooms world. Top left: No flags with 7-state chain as source. Top right: Various source domains with no flag target domain. Bottom: 2 flags added to target domain.

we also compared with the performance when using the exact metric for option construction. The exact metric was not able to handle the larger domains, and in the domains where the exact metric was computed, the difference with the performance of the sampled metric was negligible. Thus, for clarity we have excluded the plots with the exact metric.

We allow the learning algorithms to run for 10,000 time steps and plot the cumulative reward received, averaged over 30 runs. Additionally, for the computation of the sampled metric we take 30 different sets of samples and average over the performance resulting from the respective metrics. The default settings are ϵ-greedy exploration policy with $\epsilon = 0.1$, and a learning rate of $\alpha = 0.1$ for all algorithms; in some experiments we vary these parameters to study their effect.

4.1 Rooms World

We use the 7-state chain and the rooms world domain described above as source and target, respectively. The starting state is fixed throughout the experiments. In the left panel of figure 2 we display the performance of the different algorithms. We also ran learning using the four predefined options defined in [3], which are the options one would intuitively define for this domain. Despite this, the options constructed by our method have a better performance. This is most likely because our method automatically disables "bad" options for certain states. In the middle panel of figure 2, we compare the performance of using chains of varying sizes as the source domains, as well as a tiny 3-state domain with one flag (resulting in 6 overall states). As we can see, the performance is not affected much by changing source domains. This behaviour was observed in all of the experiments presented below.

To increase the complexity of the problem, we added two flags, which increases the state space to 416 states. Upon entering the goal state, the agent receives a reward of +1 for each flag picked up. The source system used for this problem is the one-flag system mentioned above. In the right panel of figure 2 we display the performance on the flagged domain. We note that the flagged domain was too large for the exact metric.

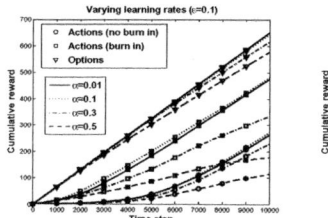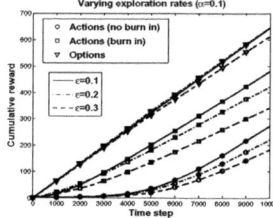

Fig. 3. Effect of varying parameters on rooms world

In Fig. 3 we evaluate the effect of varying parameters on the rooms world domain (without flags). Varying the learning and exploration rates has little effect on our method, which suggests that the constructed options are nearly optimal. This behavior was observed in all the problems discussed below.

So far we have been using the same reward "structure" for the source and target domain (*i.e.* a reward of 1 is received upon entering the goal). A natural question is whether our method is sensitive to differences in reward. In figure 4 it can be seen that this is not the case. We modified the target task so a reward of 5 is received upon entering the goal, and varied the goal reward in the source domain. In the right panel, we added a penalty of 0.1 in the target domain for each step that did not end in a goal state. Learning using the constructed options yields much better results than learning with primitive actions. It is interesting to note that the performance is improved when we add a penalty to the target domain. This is due to the fact that bisimulation distinguishes first based on differences in reward, so adding a non-zero reward at each state provides the bisimulation metric with more information.

4.2   Maze Domain

To test our method on a domain with a different topology and that does not have clear "subgoals", we used a maze domain with 36 states. The results are displayed in figure 5. This is a difficult domain and standard Q-learning performs quite poorly.

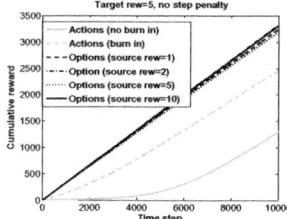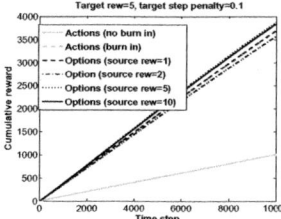

Fig. 4. Varying reward structures. Left: Reward at target fixed at 5. Right: Target domain has a step penalty of 0.1

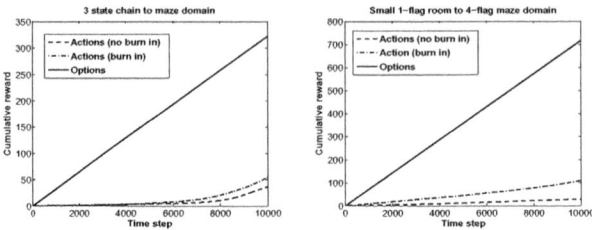

Fig. 5. Left: Performance without flags. Right: Performance with 4 flags (576 states).

Using the constructed options has a far better performance, even when 4 flags are added to the maze (right image in figure 5). For the maze without flags we only used a 3-state chain; despite this small source domain, the resulting options yield a far superior performance. For the maze with flags, we used the small one flag system mentioned above.

5   Conclusion and Future Work

We presented a novel approach for option construction using bisimulation metrics. The most interesting aspect of our results is the minimal amount of prior knowledge required, which comes in the form of very simple, small prior tasks. We note that the target domains we use bear very little resemblance to the source systems; they do not have similar topology, for example. Yet the bisimuation metric is powerful enough to define good options by comparison to these cases. We have conducted preliminary experiments in larger domains, and the results are generally consistent with those presented here. When using the 4-room domain with very large rooms and no intermediate rewards, however, there is a decrease in performance if the source chain is too small. This is probably due to the fact that bisimulation uses either immediate rewards or differences in dynamics to distinguish states. In very large rooms there is a lot of "open space" with no reward, so most of the states are clustered together, resulting in sub-optimal options. By using chains that are approximately the length of the longest "open space" in the target domain, the performance is as above. Furthermore, as mentioned above, adding a penalty at each step leads to improved performance, as it gives the metric more information.

One objection against using bisimulation is its computational cost. We avoid it here by using statistical sampling, which has a two-fold advantage: an exact model of the system is not required and we can control the computational cost by choosing number of samples.

We believe that our method can also be used to perform a hierarchical decomposition of an MDP by clustering "similar" states. Specifically, two states $t_1, t_2 \in S_2$ are clustered together if for all options o, $t_1 \in \mathcal{I}_o \Leftrightarrow t_2 \in \mathcal{I}_o$ and $\Diamond(t_1, o) == \Diamond(t_2, o)$. Initial results on the rooms world domain are promising, and return a clustering generally consistent with the room topology. A different way of approaching this problem would be to estimate the reward model of the option explicitly and incorporate it in the option construction. This line of work requires further investigation.

Acknowledgements. The authors would like to thank NSERC and ONR for providing financial support.

References

[1] Anderson, J.R.: Act: A simple theory of complex cognition. American Psychologist 51, 355–365 (1996)
[2] Barto, A.G., Mahadevan, S.: Recent advances in hierarchical reinforcement learning. Discrete Event Dynamic Systems 13(4), 341–379 (2003)
[3] Castro, P.S., Precup, D.: Using bisimulation for policy transfer in MDPs. In: Proceedings of the 24th AAAI Conference on Artificial Intelligence (AAAI 2010), pp. 1065–1070 (2010)
[4] Comanici, G., Precup, D.: Optimal policy switching algorithms in reinforcement learning. In: Proceedings of AAMAS (2010)
[5] Dietterich, T.G.: Hierarchical reinforcement learning with the maxq value function decomposition. Journal of Artificial Intelligence Research 13, 227–303 (2000)
[6] Ferns, N., Castro, P.S., Precup, D., Panangaden, P.: Methods for computing state similarity in Markov Decision Processes. In: Proceedings of the 22nd Annual Conference on Uncertainty in Artificial Intelligence (UAI 2006), pp. 174–181 (2006)
[7] Ferns, N., Panangaden, P., Precup, D.: Metrics for finite Markov decision processes. In: Proceedings of the 20th Annual Conference on Uncertainty in Artificial Intelligence (UAI 2004), pp. 162–169 (2004)
[8] Givan, R., Dean, T., Greig, M.: Equivalence Notions and Model Minimization in Markov Decision Processes. Artificial Intelligence 147(1-2), 163–223 (2003)
[9] Jonsson, A., Barto, A.G.: Causal graph based decomposition of factored MDPs. Journal of Machine Learning Research 7, 2259–2301 (2006)
[10] Konidaris, G., Kuindersma, S., Barto, A.G., Grupen, R.A.: Constructing skill trees for reinforcement learning agents from demonstration trajectories. In: Advances in Neural Information Processing Systems 23, pp. 1162–1170 (2010)
[11] Laird, J., Bloch, M.K., the Soar Group: Soar home page (2011)
[12] Mannor, S., Menache, I., Hoze, A., Klein, U.: Dynamic abstraction in reinforcement learning via clustering. In: Proceedings of the 21st International Conference on Machine Learning, ICML 2004 (2004)
[13] McGovern, A., Barto, A.G.: Automatic discovery of subgoals in reinforcement learning using diverse density. In: Proceedings of the 18th International Conference on Machine Learning, ICML 2001 (2001)
[14] Mehta, N., Ray, S., Tapadalli, P., Dietterich, T.: Automatic discovery and transfer of maxq hierarchies. In: Proceedings of the 25th International Conference on Machine Learning, ICML 2008 (2008)
[15] Mugan, J., Kuipers, B.: Autonomously learning an action hierarchy using a learned qualitative state representation. In: Proceedings of the 21st International Joint Conference on Artificial Intelligence (2009)
[16] Parr, R., Russell, S.: Reinforcement learning with hierarchies of machines. In: Advances in Neural Information Processing Systems, NIPS 1998 (1998)

[17] Precup, D.: Temporal Abstraction in Reinforcement Learning. PhD thesis, University of Massachusetts, Amherst (2000)

[18] Ravindran, B., Barto, A.G.: Relativized options: Choosing the right transformation. In: Proceedings of the 20th International Conference on Machine Learning, ICML 2003 (2003)

[19] Soni, V., Singh, S.: Using Homomorphism to Transfer Options across Reinforcement Learning Domains. In: Proceedings of AAAI Conference on Artificial Intelligence, AAAI 2006 (2006)

[20] Sorg, J., Singh, S.: Transfer via Soft Homomorphisms. In: Proceedings of the 8th International Conference on Autonomous Agents and Multiagent Systems, AAMAS 2009 (2009)

[21] Stolle, M., Precup, D.: Learning Options in Reinforcement Learning. In: Koenig, S., Holte, R.C. (eds.) SARA 2002. LNCS (LNAI), vol. 2371, p. 212. Springer, Heidelberg (2002)

[22] Stone, P., Sutton, R.S., Kuhlmann, G.: Reinforcement learning for robocup-soccer keepaway. Adaptive Behavior 13(3), 165–188 (2005)

[23] Sutton, R.S., Barto, A.G.: Reinforcement Learning: An Introduction. MIT Press, Cambridge (1998)

[24] Sutton, R.S., Precup, D., Singh, S.: Between MDPs and semi-MDPs: A framework for temporal abstraction in reinforcement learning. Artificial Intelligence 112, 181–211 (1999)

[25] Taylor, J., Precup, D., Panangaden, P.: Bounding performance loss in approximate MDP homomorphisms. In: Proceedings of the Conference on Advances in Neural Information Processing Systems, NIPS 2009 (2009)

[26] Taylor, M.E., Stone, P.: Transfer learning for reinforcement learning domains: A survey. Journal of Machine Learning Research 10, 1633–1685 (2009)

[27] Šimšek, Ö., Wolfe, A.P., Barto, A.G.: Identifying useful subgoals in reinforcement learning by local graph partitioning. In: Proceedings of the 22nd International Conference on Machine Learning, ICML 2005 (2005)

[28] Watkins, C.J.C.H., Dayan, P.: Q-learning. Machine Learning 8, 279–292 (1992)

[29] Wolfe, A.P., Barto, A.G.: Defining object types and options using MDP homomorphisms. In: Proceedings of the ICML 2006 Workshop on Structural Knowledge Transfer for Machine Learning (2006)

[30] Zang, P., Zhou, P., Minnen, D., Isbell, C.: Discovering options from example trajectories. In: Proceedings of the 26th International Conference on Machine Learning, ICML 2009 (2009)

Unified Inter and Intra Options Learning Using Policy Gradient Methods

Kfir Y. Levy and Nahum Shimkin

Faculty of Electrical Engineering, Technion, Haifa 32000, Israel
{kfiryl@tx,shimkin@ee}.technion.ac.il

Abstract. Temporally extended actions (or macro-actions) have proven useful for speeding up planning and learning, adding robustness, and building prior knowledge into AI systems. The options framework, as introduced in Sutton, Precup and Singh (1999), provides a natural way to incorporate macro-actions into reinforcement learning. In the subgoals approach, learning is divided into two phases, first learning each option with a prescribed subgoal, and then learning to compose the learned options together. In this paper we offer a unified framework for concurrent inter- and intra-options learning. To that end, we propose a modular parameterization of intra-option policies together with option termination conditions and the option selection policy (inter options), and show that these three decision components may be viewed as a unified policy over an augmented state-action space, to which standard policy gradient algorithms may be applied. We identify the basis functions that apply to each of these decision components, and show that they possess a useful orthogonality property that allows to compute the natural gradient independently for each component. We further outline the extension of the suggested framework to several levels of options hierarchy, and conclude with a brief illustrative example.

1 Introduction

In complex planning problems it is often useful to utilize macro-actions, where every macro-action is a restricted plan or a policy, and compose these macro-actions in order to form an overall plan or a policy [1,2,3]. In the terminology of Sutton et al. [4], a macro action is called an *option* and the overall policy, composing these options, is called a *policy over options*. That article suggested how to compose existing options into a policy over options in a manner analogous to standard Reinforcement Learning (RL) algorithms, employing the Semi-Markov Decision process (SMDP) framework. An important problem is how to devise a good option. A popular solution is to define a subgoal task, which is a task that terminates if it arrives at a subgoal state, and an artificial reward is given for arriving there. A plan for the subgoal task can be learned using conventional RL methods, and the optimal policy, which is learned for the subgoal task is then used as an option which halts in a subgoal state. Thus, the learning procedure originally suggested in [4] divides the learning procedure into an intra-option

S. Sanner and M. Hutter (Eds.): EWRL 2011, LNCS 7188, pp. 153–164, 2012.

learning phase, where every option is learned using subgoals, and to inter-option learning phase, where the composition of the options is learned, using standard RL algorithms applied to the SMDP framework. A possible shortcoming of this approach is that options are rigid and do not change during the inter-option learning phase. Sutton et al. [4] suggest a partial solution which they call interruption, which corresponds essentially to a single step of greedy policy improvement. Another interesting approach is to find subgoals online, during the learning phase, relying for example on reward or visit statistics of states [5,6].

The options framework provides temporal abstraction of the planning and learning problem. A different approach towards abstraction of RL problems is to look for the best policy within a parameterized family $\{\pi_\theta\}_{\theta \in \Theta}$. The restriction to a parameterized family allows to introduce prior knowledge, limits the search space, and it allows use of gradient methods, [7,8,9] , in order to find a local maximum. Policy gradient algorithms have been successfully applied to various reinforcement learning problems in complex domains [8,10,11].

Previous work on parametric options includes [1,2,3]. In [1] gradient learning of the stopping condition is derived through embedding the options into the state space. In [2] the task is first manually decomposed into subtasks which are learned separately and only then the higher decision hierarchy is being optimized. In [3] the SMDP framework is utilized together with a certain regression method in order to learn parametric intra-policies. All of these works do not consider the problem of simultaneous optimization of the intra-options together with the overall policy, and only the first addresses the problem of learning the stopping conditions of options.

In this paper we are interested in schemes that can jointly learn the options policies (intra-option learning) and the policy-over-options (inter-option learning). We first formulate a framework in which the three components of the overall policy: intra-option policies, option termination conditions and the option selection policy, are viewed as a unified stationary policy over an augmented state-action space, to which standard RL algorithms may be applied. We next propose a modular parameterization of these three components, and identify the basis functions that apply to each of these policy components, we then show that these basis functions enjoy a useful orthogonality property that allows to compute the natural gradient independently for each component, leading to great computational savings.

The paper is constructed as follows: In Section 2 we describe the basic MDP model, and provide a brief overview of natural gradient methods and the options framework. In Section 3 we define the augmented state-action spaces and the augmented hierarchial policy (AHP). In Section 4 we propose a parameterization of the AHP for which we prove the orthogonality property of the parameterized base functions, and in Section 5 we briefly consider extending our framework to hierarchical multi-level options, Section 6 presents an illustrative example and in Section 7 we conclude the paper. Due to space limitations we leave some details of the example in Section 6 to the extended online version[1].

---

[1] http://webee.technion.ac.il/people/shimkin/PREPRINTS/
LevyShimkinEWRL2011.pdf

2   Model and Background

A discounted-reward Markov Decision Process (MDP) M may be defined by a 5 tuple $< S, A, r, P, \gamma >$. At time t, the system is in state $s_t \in S$, the agent chooses an action $a_t \in A$, a reward $r(s_t, a_t)$, is given, and the system transfers to a new state s_{t+1}, randomized according to a transition probability function, $P(\cdot|s_t, a_t)$. A policy $\pi = \{\pi_t\}_{t=0}^{\infty}$ is a series of mappings from the collection of possible histories to a probability function over the action space. A stationary policy depends only on the current state, and is therefore defined by the map $\pi : S \rightarrow \Delta(A)$, where $\Delta(A)$ is the probability simplex over the set A. For a given stationary policy π, the value function V^{π} and the action-value function Q^{π} are defined as:

$$V^{\pi}(s) = E^{\pi}[\sum_{t=0}^{\infty} \gamma^t r(s_t, a_t)|s_0 = s] \tag{1}$$

$$Q^{\pi}(s, a) = r(s, a) + E^{\pi}[\sum_{t=1}^{\infty} \gamma^t r(s_t, a_t)|s_0 = s, a_0 = a] \tag{2}$$

We assume that the initial state s_0 is chosen according to a fixed initial distribution $\eta(s)$. The general goal is to maximize the discounted return, defined as:

$$J(\pi) = \sum_{s \in S} \eta(s) V^{\pi}(s) = \sum_{s \in S} d^{\pi}(s) \sum_{a \in A} \pi(a|s) r(s, a) \tag{3}$$

where $d^{\pi}(s) = E_{s_0 \sim \eta}[\sum_{t=0}^{\infty} \gamma^t p^{\pi}(s_t = s|s_0)]$.

2.1   Natural Policy Gradient

In the policy gradient framework [7], the policy itself is parameterized as $\pi(a|s) = \pi(a|s, \theta)$, and the vector θ is modified in the direction of gradient of expected return $J(\pi_\theta) \triangleq J(\theta)$. Plain methods of gradient learning estimate the value of $\nabla_\theta J(\theta)$ during the run, and modify the value of θ greedily in this direction. Natural gradient methods do not follow the steepest direction in parameter space, but rather the steepest direction with respect to the Fisher metric $G(\theta)$ [12,9,8], which leads to improved convergence properties. The relation between the standard gradient $\nabla_\theta J(\theta)$ and the natural gradient $\widetilde{\nabla}_\theta J(\theta)$ is given by:

$$\widetilde{\nabla}_\theta J(\theta) = G^{-1}(\theta) \nabla_\theta J(\theta) \tag{4}$$

It is shown in [13] that the Fisher metric captures the geometric structure which is induced by the parametric family of policies. As shown in [7], for a stationary policy π, the gradient of the discounted return is given by:

$$\nabla_\theta J(\theta) = \sum_{s \in S} d^{\pi}(s) \sum_{a \in A} \pi(a|s) Q^{\pi}(s, a) \psi(s, a) \tag{5}$$

where $\psi(s, a) = \nabla_\theta \log \pi(a|s)$. Assuming $\theta \in R^N$, for every element $\theta^{(m)}$ in the parameter vector θ define:

$$\psi_m(s, a) = \frac{\partial \log \pi(a|s)}{\partial \theta^{(m)}} \tag{6}$$

We shall refer to $\{\psi_m\}_{m=1}^N$ as the θ *base functions*.
For functions $f, g : S \times A \to R$, consider the following inner product:

$$\langle f, g \rangle = \sum_{s \in S} d^\pi(s) \sum_{a \in A} \pi(a|s) f(s, a) g(s, a) \tag{7}$$

Using the notation of this inner product, equation (5) can be written as:

$$\frac{\partial J(\pi)}{\partial \theta^{(m)}} = \langle Q^\pi, \psi_m \rangle \tag{8}$$

Hence, if we denote $\Psi_\theta = \text{span}\{\psi_{\theta_1}, \ldots, \psi_{\theta_N}\}$, and $Q_{\Psi_\theta}^\pi(s, a)$ as the orthogonal projection of $Q^\pi(s, a)$ onto Ψ_θ, the function Q^π in equation (8) can be replaced with $Q_{\Psi_\theta}^\pi$. The orthogonal projection $Q_{\Psi_\theta}^\pi(s, a)$, by definition, could be written as

$$Q_{\Psi_\theta}^\pi(s, a) = \sum_{i=1}^N w_i \psi_i(s, a) = w^T \nabla_\theta \log \pi(a|s) \tag{9}$$

From the orthogonality principle we know that $\langle Q^\pi - Q_{\Psi_\theta}^\pi, \psi_i \rangle = 0$ for every i. From the last two equations one can obtain a linear equation for w

$$G_\theta w = c_\theta \tag{10}$$

where

$$\begin{aligned} G_\theta(i, j) &= \langle \psi_i, \psi_j \rangle, & G_\theta \in R^{N \times N} \\ c_\theta(i) &= \langle Q^\pi, \psi_i \rangle, & c_\theta \in R^N \end{aligned} \tag{11}$$

It is shown in [12,9,8] that w is the natural gradient of $J(\theta)$. So, if we manage to estimate G_θ and c_θ, we can calculate w from equation (10) and gradient update θ. In order to estimate the entries of the matrix G_θ, we can simply use the temporal average of $\psi_i(s_t, a_t)\psi_j(s_t, a_t)$. As for the entries of c_θ, it is demonstrated in [7] that if we define the advantage function $A^\pi(s, a) \triangleq Q^\pi(s, a) - V^\pi(s)$, we can replace the Q^π function in equation (11) with the advantage function. The latter may be estimated with the time difference (TD) error:

$$\delta_t = r_t + V^\pi(s_{t+1}) - V^\pi(s_t) \tag{12}$$

which is an unbiased estimate of the advantage function. It is suggested in [8] to estimate the entries $c_\theta(i)$ with the temporal average of $\psi_i(s_t, a_t)\delta_t$. We still require an estimate of the value function V^π, which can be done with TD methods as in [9] or with the least square methods [14,15]. Once we have an estimate of G_θ and c_θ, we can calculate the natural gradient: $w = G_\theta^{-1} c_\theta$. We refer the reader to [9] for a complete description of the natural gradient learning algorithm.

2.2   The Options Framework

An option o is characterized by a 3-tuple $< \mathcal{I}, \pi, \beta >$, where $\mathcal{I} \subseteq S$ is the set from which the option can be initiated, $\beta : S \to [0,1]$ is a termination probability, and π is the intra-option policy, which in general, may depend on the entire history since the option was initiated (but not before). Here we restrict attention to stationary intra-option policies. Note that a primitive action can be considered as a single step option. A policy-over-options is defined to be a mapping $\mu : S \to \Delta(\mathcal{O})$, where $\mathcal{O}$ is the set of all options, determines which option should be initiated in the current state. Given the the policy-over-options μ and options set $\mathcal{O}$ we refer to their combination as the overall policy (OP). We refer to intra-option policies, stopping conditions and policy-over-options as the three *decision components* of the OP. In [4] the SMDP framework is utilized in order to optimize the policy-over-options, μ.

3   The Augmented Options Model

In this section we formulate an augmented MDP model that will enable us to utilize existing RL algorithms in order to learn simultaneously the three decision components of the overall policy (OP). In Subsection 3.1 we describe the decisions process made by the OP and in the following Subsection we define augmented state-action spaces and show that the OP in the original state-action spaces is equivalent to a *stationary* policy in the augmented spaces.

We consider a given OP $\{\mu, \mathcal{O}\}$, where μ is the policy-over-options and $\mathcal{O} = \{\pi_i, \beta_i\}_{i=1}^n$ is the options set. We index the individual option by i or j, where $i, j \in \mathcal{O}$, and denote by π_i and β_i the intra policy and stopping condition of option i. Note the use of the same notation i for the option itself and its index.

3.1   Overall Policy (OP) Description

The following is an outline of the decision process made by an OP:
1. At time t the process arrives at a state s_t with the option i_t, chosen at the previous step. We divide the choices of the policy to three decision phases:

- **Stopping decision phase** (sp): Choose whether or not to stop the current option, with the choice made according to the stopping probability $\beta_{i_t}(s_t)$. We relate to the decision to stop or not as an action, chosen from the binary action set $A_{\text{stop}} \triangleq \{stop, cont\}$.
- **Option decision phase** (op): Choose a new option j_t. If in the previous phase stopping was not chosen, the former option persists ($j_t = i_t$), otherwise, the policy-over-options chooses a new option j_t according to $j_t \sim \mu(\cdot|s_t)$.
- **Action decision phase** (ap): Choose a new action a_t, according to the option j_t chosen at the previous phase, i.e. $a_t \sim \pi_{j_t}(\cdot|s_t)$.

2. At time $t + 1$ the process arrives at a new state s_{t+1} under the policy i_{t+1} chosen in the previous step. Again, three phases of decision will take place as in the previous step. Note that the following holds by definition: $i_{t+1} = j_t$, i.e., at the current step we arrive with the option we chose in the previous step.

We will relate to the option i_t with which we arrive at a state s_t as the *arrive-option* at time t, and the option j_t which we choose in time t as the *act-option*.

3.2   The Augmented Model

Given the original state and action spaces, S, A, and the overall policy $\{\mu, \mathcal{O}\}$, we define the augmented state and actions spaces, $\tilde{S}, \tilde{A}$:

- **Augmented state** $\tilde{s} = (i, s)$, where $i \in \mathcal{O}$ is the the option with which we *arrive* at the state s.
- **Augmented action** $\tilde{a} = (\varphi, j, a)$, where $\varphi \in A_{\text{stop}}$ is the decision whether to stop the current *arrive*-option i. The action $j \in \mathcal{O}$ is the choice of the *act*-option, and $a \in A$ is the primitive action chosen by the *act*-option j.

State and option transitions in the original state space translate to state transitions in the augmented space. Given the original transition probabilities $P(s'|s, a)$ we can calculate the transition probability function in the augmented one:

$$P(\tilde{s}'|\tilde{s}, \tilde{a}) = P((i', s')|(i, s), (\varphi, j, a)) = 1_{\{i'=j\}} P(s'|s, a) \qquad (13)$$

Hence, the transition probability is the original $P(s'|s, a)$ if the next *arrive-option* i' is equal to the current *act*-option j, otherwise it is 0. Note that $\varphi = cont$ implies $i = j$.

Given a overall policy $\{\mu, \mathcal{O}\}$ in the MDP M, we define the following policy Π in the augmented space:

$$\Pi(\tilde{a}|\tilde{s}) = \Pi((\varphi, j, a)|(i, s)) = P_{sp}(\varphi|i, s) P_{op}(j|\varphi, i, s) P(a|j, s) \qquad (14)$$

where

$$P_{sp}(\varphi|i, s) = 1_{\{\varphi=stop\}} \beta_i(s) + 1_{\{\varphi=cont\}} (1 - \beta_i(s)) \qquad (15)$$
$$P_{op}(j|\varphi, i, s) = 1_{\{\varphi=cont\}} 1_{\{j=i\}} + 1_{\{\varphi=stop\}} \mu(j|s) \qquad (16)$$
$$P(a|j, s) = \pi_j(a|s) \qquad (17)$$

We relate to Π as the augmented hierarchical policy (AHP). Notice that $\Pi(\tilde{a}|\tilde{s})$ depends only on $(\tilde{a}, \tilde{s})$ and is therefore stationary. The term $P_{sp}(\varphi|i, s)$ is the probability that the current *arrive*-option i will stop at the current state s. The next term $P_{op}(j|\varphi, i, s)$ is the probability that an *act*-option j will be chosen given the current state and stopping action, and the last term $P(a|j, s)$ is the probability that a new action a is chosen in state s, which is done according to the policy of the *act*-option, j. In equations (15)-(17), we relate the equivalent policy Π to the policy-over-options μ and options set $\mathcal{O} = \{\pi_i, \beta_i\}_{i=1}^n$. Notice that we can design μ to be also dependent at the *arrive*-option, i.e., $\mu = \mu(j|i, s)$.

It is easy to show that the policy Π over the augmented spaces is equivalent to the overall policy $\{\mu, \mathcal{O}\}$, over the original MDP, in the sense that both induce the same probability measure over augmented state-actions trajectories (which include stopping decisions and option choices), $\{i_t, s_t, \varphi_t, j_t, a_t\}_{t=0}^{T}$ particularly:

$$E^{\{\mu,\mathcal{O}\}}[\sum_{t=0}^{\infty} \gamma^t r_t | s_0, i_0] = E^{\Pi}[\sum_{t=0}^{\infty} \gamma^t r_t | (i_0, s_0)] \tag{18}$$

Therefore, in order to simultaneously learn the three decision components of the overall policy (stopping, intra policies, policy-over-options), we can directly apply standard RL methods to the augmented hierarchical policy Π. In particular, we can apply natural gradient methods which we mentioned in Subsection 2.1, in the next Section we investigate natural gradient learning of the AHP, Π.

4   Natural Gradient of the AHP

In this Section we offer a modular parameterization of the AHP Π, for which we prove an orthogonality property (Proposition 1), which enables us to substantially reduce the computational burden of calculating the natural gradient for the parameterized AHP (Corollary 1).

The special structure of the AHP Π in equations (14)-(17) suggests the following parameterization:

$$\beta_i(s) = \beta_i(s, \lambda_i) \tag{19}$$
$$\mu(j|s) = \mu(j|s, \chi)$$
$$\pi_j(a|s) = \pi_j(a|s, \theta_j)$$

with the parameter vector

$$\Theta = (\theta_1, \theta_2 \ldots, \theta_N, \lambda_1, \lambda_2 \ldots, \lambda_N, \chi) \tag{20}$$

The parameter vector Θ is composed of three types of sub-vector parameters: θ_i controls the stationary policy of option i, λ_i controls the stopping of the ith policy, and χ controls the choices of the policy over options μ.

Generally, in order to calculate the natural gradient (equation (10)), we should invert an $N \times N$ matrix, where $N = \dim\{\Theta\}$.

In what follows we denote by $N_{\theta_m}, N_{\lambda_k}, N_\chi$ the dimensions of the vectors $\theta_m, \lambda_k, \chi$, and by $\theta_j^{(h)}$ the hth element of the sub-vector θ_j.

Proposition 1. *Let $\{\mu, \mathcal{O}\}$ be an overall policy with stationary options set $\mathcal{O} = \{\pi_i, \beta_i\}_{i=1}^n$, and let Π be its equivalent augmented hierarchical policy as in equations (14)-(17), parameterized as in equations (19)-(20). Define the following linear subspaces of the Θ-base functions from equation (6):*

$$\Psi_{\theta_m} = \text{span}\{\psi_{\theta_m^{(h)}}\}_{h=1}^{N_{\theta_m}} \qquad \forall m \in \mathcal{O}$$

$$\Psi_{\lambda_k} = \text{span}\{\psi_{\lambda_k^{(h)}}\}_{h=1}^{N_{\lambda_k}} \qquad \forall k \in \mathcal{O}$$

$$\Psi_\chi = \text{span}\{\psi_{\chi^{(h)}}\}_{h=1}^{N_\chi} \tag{21}$$

Then these subspaces are orthogonal under the inner product defined in (7).

Proof: We may calculate the base function for the elements of the parameter vector Θ as follows:

$$\psi_{\lambda_k^{(h)}} = \frac{\partial \log \Pi((\varphi, j, a)|(i, s), \Theta)}{\partial \lambda_k^{(h)}} = 1_{\{i=k\}} \frac{1}{P_{sp}(\varphi|i, s, \lambda_i)} \frac{\partial P_{sp}(\varphi|i, s, \lambda_i)}{\partial \lambda_i^{(h)}} \quad (22)$$

$$\psi_{\chi^{(h)}} = \frac{\partial \log \Pi((\varphi, j, a)|(i, s), \Theta)}{\partial \chi^{(h)}} = 1_{\{\varphi = stop\}} \frac{1}{P_{op}(j|\varphi, i, s)} \frac{\partial \mu(j|s, \chi)}{\partial \chi^{(h)}} \quad (23)$$

$$\psi_{\theta_m^{(h)}} = \frac{\partial \log \Pi((\varphi, j, a)|(i, s), \Theta)}{\partial \theta_m^{(h)}} = 1_{\{j=m\}} \frac{\partial \log \pi_j(a|s, \theta_j)}{\partial \theta_j^{(h)}} \quad (24)$$

It is easy to see that the indicator functions in (22)(24) imply that the following orthogonality relations hold:

$$\langle \psi_{\lambda_m^{(f)}}, \psi_{\lambda_k^{(h)}} \rangle = 0 \qquad \forall m \neq k \quad (25)$$

$$\langle \psi_{\theta_m^{(f)}}, \psi_{\theta_k^{(h)}} \rangle = 0 \qquad \forall m \neq k \quad (26)$$

Less trivial are the following relations:

$$\langle \psi_{\theta_m^{(f)}}, \psi_{\lambda_k^{(h)}} \rangle = 0, \quad \langle \psi_{\theta_m^{(f)}}, \psi_{\chi^{(h)}} \rangle = 0 \qquad \forall k, m \quad (27)$$

$$\langle \psi_{\chi^{(f)}}, \psi_{\lambda_k^{(h)}} \rangle = 0 \qquad \forall k \quad (28)$$

Let us prove (27) first :

$$\langle \psi_{\theta_m^{(f)}}, \psi_{\lambda_k^{(h)}} \rangle = \sum_{\tilde{s} \in \tilde{S}} d^{\Pi}(\tilde{s}) \sum_{\tilde{a} \in \tilde{A}} \Pi(\tilde{a}|\tilde{s}) \psi_{\theta_m^{(f)}} \psi_{\lambda_k^{(h)}} =$$

$$\sum_{i,s} d^{\Pi}(i, s) \sum_{\varphi, j, a} \Pi(\tilde{a}|\tilde{s}) 1_{\{j=m\}} \frac{\partial \log \pi_j(a|s, \theta_j)}{\partial \theta_j^{(f)}} 1_{\{i=k\}} \frac{\partial \log P_{sp}(\varphi|i, s, \lambda_i)}{\partial \lambda_i^{(h)}} =$$

$$\sum_s d^{\Pi}(k, s) \sum_{\varphi, a} P_{op}(m|\varphi, k, s, \chi) \frac{\partial P_{sp}(\varphi|k, s, \lambda_k)}{\partial \lambda_k^{(h)}} \frac{\partial \pi_m(a|s, \theta_m)}{\partial \theta_m^{(f)}} =$$

$$\sum_s d^{\Pi}(k, s) \sum_{\varphi} P_{op}(m|\varphi, k, s, \chi) \frac{\partial P_{sp}(\varphi|k, s, \lambda_k)}{\partial \lambda_k^{(h)}} \frac{\partial}{\partial \theta_m^{(f)}} \sum_a \pi_m(a|s, \theta_m) = 0$$

where in the last step we used the identity $\sum_a \pi_m(a|s) = 1$. The proof of $\langle \psi_{\theta_m^{(f)}}, \psi_{\chi^{(h)}} \rangle = 0$ is similar. We next prove equation (28):

$$\langle \psi_{\chi^{(f)}}, \psi_{\lambda_k^{(h)}} \rangle = \sum_{\tilde{s} \in \tilde{S}} d^{\Pi}(\tilde{s}) \sum_{\tilde{a} \in \tilde{A}} \Pi(\tilde{a}|\tilde{s}) \psi_{\chi^{(f)}} \psi_{\lambda_k^{(h)}} =$$

$$\sum_{i,s} d^{\Pi}(i,s) \sum_{\varphi,j,a} \Pi(\tilde{a}|\tilde{s}) \frac{1_{\{\varphi=stop\}}}{P_{op}(j|\varphi,i,s)} \frac{\partial \mu(j|s,\chi)}{\partial \chi^{(h)}} 1_{\{i=k\}} \frac{\partial \log P_{sp}(\varphi|i,s,\lambda_i)}{\partial \lambda_i^{(h)}} =$$

$$\sum_{s \in S} d^{\Pi}(k,s) \sum_{j,a} \pi_j(a|s,\theta_j) \frac{\partial \mu(j|s,\chi)}{\partial \chi^{(h)}} \frac{\partial P_{sp}(stop|k,s,\lambda_k)}{\partial \lambda_k^{(h)}} =$$

$$\sum_{s} d^{\Pi}(k,s) \frac{\partial P_{sp}(stop|k,s,\lambda_k)}{\partial \lambda_k^{(h)}} \frac{\partial}{\partial \chi^{(h)}} \{ \sum_j \mu(j|s,\chi) \sum_a \pi_j(a|s,\theta_j) \} = 0$$

where in the last step we used the identities $\sum_a \pi_j(a|s) = 1$, $\sum_j \mu(j|s) = 1$. $\square$
Given the last orthogonality result, it is clear that the matrix G_Θ in (11) becomes block diagonal with each block corresponding to a component of Θ.

Corollary 1. *Given a parameterization of the AHP Π as in equations (19)-(20), we can estimate each sub-vector of the natural gradient w independently as follows: we can write an independent equation with the form of (10) for the relevant sub vector* $w_{\theta_m}, w_{\lambda_k}, w_\chi$. *The* w_{θ_m} *equations:*

$$G_{\theta_m} w_{\theta_m} = c_{\theta_m} \qquad (29)$$

where:

$$G_{\theta_m}(i,j) = \langle \psi_{\theta_m^{(i)}}, \psi_{\theta_m^{(j)}} \rangle \qquad G_{\theta_m} \in R^{N_{\theta_m} \times N_{\theta_m}} \qquad (30)$$

$$c_{\theta_m}(i) = \langle Q^\pi(s,a), \psi_{\theta_m^{(i)}} \rangle \qquad c_{\theta_m} \in R^{N_{\theta_m}} \qquad (31)$$

Similar equations apply to w_{λ_k}, w_χ.

Thus, in order to calculate the natural gradient vector, it is sufficient to invert $2n+1$ Fisher matrices, where n denotes the number of options, each matrix has the dimension of the corresponding sub-vector.

5 Multilevel Decision Hierarchies

In the previous sections we have considered a hierarchical structure with one level of options below the top (policy-over-options) level, so that every low level option chooses primitive actions. In this section we briefly consider a multi-level hierarchical structure, in which higher-level options can treat options from a lower level as extended actions. We describe options with only two levels of hierarchy, as the extension to structures with more levels of hierarchy is similar.
In figure 1 we illustrate the hierarchical structure of an overall policy with two levels of hierarchy, where root option is denoted by μ_0, the first level option-nodes are denoted by μ_1, μ_2 and beneath every node i in the 1st level lie the leaf options $o_{i,n}$ whose policies are stationary. Similarly to section 3, we can divide the choices made by this hierarchical policy into 5 decision phases as follows. Suppose at time t the process arrives a state s_t with a leaf option o_{i_t,n_t}:

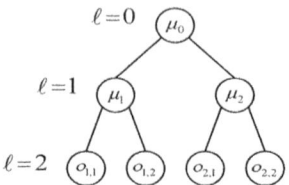

Fig. 1. Hierarchical options structure

1. Decide whether to stop the current leaf option.
2. The hierarchy above, μ_{i_t}, makes a decision whether to stop the execution of the current level 1 decision node.
3. The root policy μ_0 chooses a new level 1 decision node, μ_{j_t}.
4. The level 1 hierarchy, μ_{j_t} chooses a new leaf option o_{j_t, m_t}
5. The new leaf option chooses a primitive action $a_t \sim \pi_{j_t, m_t}(\cdot | s_t)$

As in section 3, we can define corresponding augmented state and action spaces and an equivalent augmented hierarchical policy above these augmented spaces, it can be shown that the orthogonality property between the basis function of the various policy components (Section 4) will be maintained. We omit the details here due to space constraints.

6   Experimental Results – Inverted Pendulum

The inverted pendulum task is a known RL benchmark in which we have to swing up an inverted pendulum from the down position to the up-position and keep it stable. We have two state variables, the angle and it derivative,namely $s = (\theta, \dot{\theta})$. We can apply a limited torque $|a| \leq a_{max}$ at the rotary joint. Motivated by [16] where multiple controllers are designed based on traditional control theory and then combined using an RL scheme, we chose to use the following three *parameterized options* in order to learn an effective control policy:

1. Swinging Option: $a_1 \sim \mathcal{N}(k_1 \text{sign}(\dot{\theta}), \sigma_1^2)$.
2. Decelerating Option: $a_2 \sim \mathcal{N}(-k_2 \text{sign}(\dot{\theta}), \sigma_2^2)$.
3. Stabilizing Option $a_3 \sim \mathcal{N}(k_3\theta + k_4\dot{\theta}, \sigma_3^2)$.

Here $\mathcal{N}(\mu, \sigma^2)$ is gaussian random variable with a fixed variance σ^2. We further parameterized the options stopping conditions. For instance, the stopping condition of the swinging option is: $\beta_1(s) = \Phi(\frac{\theta - (\pi + \lambda_1)}{\sigma_{sp}}) + \Phi(\frac{-(\theta - (\pi - \lambda_1))}{\sigma_{sp}})$, where Φ is the cumulative normal distribution[2]. We optimize over the action parameters and stopping parameters:$\{k_i, \lambda_i\}_{i=1}^4$. Results are illustrated in Fig.2. From Fig. 2a, we can see that our algorithm converges within 90 episodes, where in most of the cases it converges within less 40. Interestingly our algorithm eventually neglects the decelerating option, as shown in Figure 2b.

---

[2] For the stopping conditions of the other options as well as further simulation details see the extended online version.

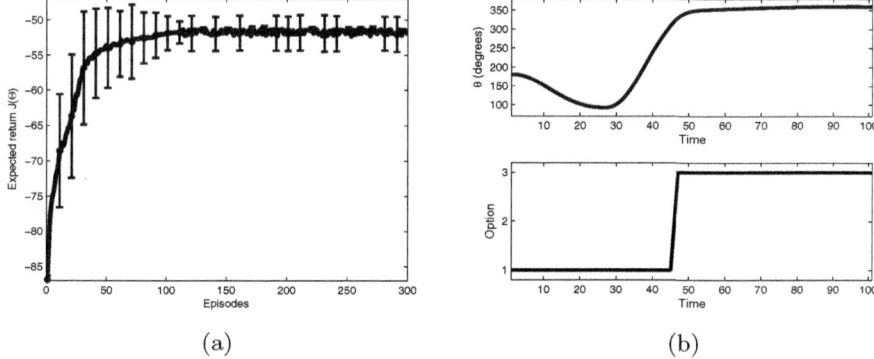

(a)   (b)

Fig. 2. Learning curves for the pendulum task. In (a) we can see the expected return averaged over 20 runs. In (b),we can see the options selection and θ as a function of time for the converged parameters .

7   Concluding Remarks

The unified framework that we propose enables to take a structure of overall policy and simultaneously optimize the intra-option policies, the stopping conditions and the policy-over-options. However, the simultaneous optimization of all parameters is only optional, and our framework allows one to freeze some of the parameters while learning the others. Comparing simultaneous verses alternating optimization schedules is an interesting direction for further research.

The possibility to tune the stopping conditions enables us to obtain "smoother" transitions in between the options, as opposed to the subgoals approach, where the termination conditions are rigid.

As shown, the orthogonality property between the basis function of the different policy components leads to significant simplification in the computation of the natural gradient. In future research, one could look for similar computational leverages in other learning algorithms such as second-order methods for parameter tuning, and least squares methods for the value function evaluation [14,15]. More generally, additional theoretical and empirical work on parametric options learning is evidently called for.

Acknowledgements. This work was supported in part by the IST Programme of the European Community, under the PASCAL2 Network of Excellence, IST-2007-216886. This publication only reflects the authors' views.

References

1. Comanici, G., Precup, D.: Optimal policy switching algorithms for reinforcement learning. In: Proceedings of the 9th International Conference on Autonomous Agents and Multiagent Systems, pp. 709–714 (2010)

2. Ghavamzadeh, M., Mahadevan, S.: Hierarchical policy gradient algorithms. In: Twentieth ICML, pp. 226–233 (2003)
3. Neumann, G., Maass, W., Peters, J.: Learning complex motions by sequencing simpler motion templates. In: ICML (2009)
4. Sutton, R.S., Precup, D., Singh, S.: Between MDPs and semi-MDPs: A framework for temporal abstraction in reinforcement learning. Artificial intelligence 112, 181–211 (1999)
5. Simsek, O., Barto, A.: Using relative novelty to identify useful temporal abstractions in reinforcement learning. In: ICML, vol. 21, p. 751. Citeseer (2004)
6. Menache, I., Mannor, S., Shimkin, N.: Q-Cut - Dynamic Discovery of Sub-goals in Reinforcement Learning. In: Elomaa, T., Mannila, H., Toivonen, H. (eds.) ECML 2002. LNCS (LNAI), vol. 2430, pp. 295–306. Springer, Heidelberg (2002)
7. Sutton, R.S., McAllester, D., Singh, S., Mansour, Y.: Policy gradient methods for reinforcement learning with function approximation. In: Advances in Neural Information Processing Systems, vol. 12 (2000)
8. Peters, J., Schaal, S.: Natural actor-critic. Neurocomputing 71(7-9), 1180–1190 (2008)
9. Bhatnagar, S., Sutton, R.S., Ghavamzadeh, M., Lee, M.: Natural actor-critic algorithms. Automatica 45, 2471–2482 (2009)
10. Richter, S., Aberdeen, D., Yu, J.: Natural actor-critic for road traffic optimisation. In: Advances in Neural Information Processing Systems, vol. 19, p. 1169 (2007)
11. Buffet, O., Dutech, A., Charpillet, F.: Shaping multi-agent systems with gradient reinforcement learning. In: Autonomous Agents and Multi-Agent Systems (2007)
12. Kakade, S.: A natural policy gradient. In: Advances in Neural Information Processing Systems 14, vol. 2, pp. 1531–1538 (2002)
13. Bagnell, J., Schneider, J.: Covariant policy search. In: International Joint Conference on Artificial Intelligence, vol. 18, pp. 1019–1024. Citeseer (2003)
14. Boyan, J.A.: Technical update: Least-squares temporal difference learning. Machine Learning 49, 233–246 (2002)
15. Nedić, A., Bertsekas, D.: Least squares policy evaluation algorithms with linear function approximation. Discrete Event Dynamic Systems 13 (2003)
16. Yoshimoto, J., Nishimura, M., Tokita, Y., Ishii, S.: Acrobot control by learning the switching of multiple controllers. Artificial Life and Robotics 9 (2005)

Options with Exceptions

Munu Sairamesh and Balaraman Ravindran

Indian Institute of Technology Madras, India

Abstract. An option is a policy fragment that represents a solution to a frequent subproblem encountered in a domain. Options may be treated as temporally extended actions thus allowing us to reuse that solution in solving larger problems. Often, it is hard to find subproblems that are exactly the same. These differences, however small, need to be accounted for in the reused policy. In this paper, the notion of options with exceptions is introduced to address such scenarios. This is inspired by the Ripple Down Rules approach used in data mining and knowledge representation communities. The goal is to develop an option representation so that small changes in the subproblem solutions can be accommodated without losing the original solution. We empirically validate the proposed framework on a simulated game domain.

Keywords: Options framework, Transfer Learning, Maintenance of skills.

1 Introduction

One of the main advantages of Hierarchical Reinforcement Learning(HRL) is that HRL frameworks allow us to learn *skills* required to solve a task and transfer that knowledge to solving other related problems. This allows us to considerably cut down on learning time and achieve better success rates in solving new problems. But the main difficulty in such transfer is that a problem is seldom encountered in exactly the same form. There are some changes in the task, such as a different obstacle configuration, or minor changes in the dynamics of the world. We need a framework where we can accommodate such changes without compromising the quality of the solution of the original sub-task that the skill was learned on. In fact we would like to cascadingly accommodate more such changes, without degrading any of the solutions learned along the way.

The problem is exacerbated when we look at spatial abstractions that are specific to the skill being learnt. There has been a lot of interest in recent times on deriving skill specific representations (MAXQ[3], VISA [4] and Relativized option [13]). Dieterich, while proposing the MaxQ framework for value function decomposition has discussed safe state abstraction conditions that would minimally affect the quality of the sub-task policies that are being learned. Jonsson and Barto have looked at the problem of jointly determining spatial and temporal abstractions in factored MDPs. Ravindran and Barto have proposed relativized options that use the notion of partial homomorphisms to determine lossless option specific state representations that satisfy some form of safe state abstraction conditions.

S. Sanner and M. Hutter (Eds.): EWRL 2011, LNCS 7188, pp. 165–176, 2012.

Such approaches yield more compact representations of the skills and allow for greater transfer since they ignore all aspects of the state representation not required for that sub-task. This also renders them more fragile since they can fail if there are small changes to the system dynamics that require that different spatial abstractions be used to represent the modified solutions. For example, in the simulated game used in [13], the introduction of an additional obstacle might require changing the policy only around the states with the obstacle. But this would require us to represent the location of the obstacle as well as the original information that we were considering.

In order to accommodate such scenarios we have proposed the notion of *options with exceptions* (OWE). The knowledge management community have long dealt with the problem of incremental maintenance of rule bases using Ripple Down Rule(RDR) [5]. Suppose that some new training instances contradict a compact rule that was derived from the original training data, the traditional approach would be to either accept the decreased performance or try to re-learn the rule base from scratch. The ripple down rules representation allows us to minimally modify the existing rule by adding exceptions that are derived from only the contradicting instances. Taking inspiration from this concept, we propose a framework that allows us to modify an option's policy only using instances where the existing policy seems to fail.

There are several key challenges to address here. How to represent the policy so that modifications can be made without affecting the base policy when we add the exceptions? In a stochastic domain, how do we even detect that the old policy has failed or does not perform up to expectations? In a deterministic environment this is somewhat easy to address, since we can form models of expected behavior of the world. How to compactly represent such models in stochastic environments? How to detect which of the new features of the state space should we add to the option representation in order to trigger the exception?

Our framework has several components that extend and augment the traditional option representation. The first is a more structured representation of the option policy that allows for easy addition of exceptions. The second is a network of *landmarks* or way points that acts as a compact model of the environment and quickly allows us to narrow down to the region where the option failed. The third is a path model for the original policy that allows us to pinpoint the features that we need to include in the exception representation.

We test the ability of the framework to identify exceptions and to learn new policies to address the exceptions. We also show that when the option with exception is applied to the task it was originally learnt on, there is no degradation of performance.

The next section describes the necessary notations and background work. Section 3 describes the landmark network and the transition time model, while Section 4 describes the identification of the actual exceptions and presents our experimental results. Section 5 concludes with some discussion and directions for future work.

2   Notation and Background

2.1   Notation

A trajectory is a sequence of states through which an agent moves. The length of the trajectory is denoted by l. Let s_t and a_t be the state and action at time t. Let η_s be the total number of times that a state s has occurred in all the trajectories. A transition instance is the tuple $\langle s_t, a_t \rangle$ and a transition chain is an ordered sequence of transition instances.

2.2   Option

An option [1] is represented as $\langle I, \pi, \beta \rangle$, where I is the initiation set of the option, i.e., the set of states in which the option can be invoked, π is the policy according to which option selects the actions and β is the termination function where $\beta(s)$ gives the probability that the option terminates in state s. An option can be thought of as representing a skill.

2.3   Policy Representation

To explicitly represent the policies of the option, we choose a specific representation. Suppose, the policies are represented using Q values, reusing them in another task would result in changes in the Q values. This could modify the original policy adversely. Hence a representation is required in which changes in these values do not alter the existing policy. For example, in order to accommodate exceptions, this can be achieved if the representation of the new policy is conditioned on the feature that caused the exception. Also, the exception may affect the policy in its neighboring states, hence requiring the use of a small amount of history in representing the new policy. This can be achieved using a suffix tree, which divides the history space.

Suffix trees are used to build a discrete internal state action representation e.g. U-Tree. The internal nodes of the suffix tree are from the set { s_{t-1}, a_{t-1}, s_{t-2}, $a_{t-2} \cdots s_{t-H}$, a_{t-H} }, where H is the history index. It denotes the number of time steps in the past from the current time. The Suffix tree is trained using a transition chain.

3   Transition Time Model

One of the chief components of *Option With Exceptions* (OWE) framework is a transition time model that records the time of transition between certain distinguished states or *landmarks*.

Landmarks are places that are visited often and are used to find the way back or through an area. Using the same notion we define a landmark for a region as a state that is visited on most of the paths (successful or unsuccessful) through that region. A sub-goal may be a landmark whereas a landmark need not be a sub-goal. To build an efficient transition time model, landmarks need to be uniformly distributed throughout the state space.

The landmark network is a weighted *Directed Acyclic Graph(DAG)*, where nodes are landmarks and an edge $i \rightarrow j$ indicates that node j is followed by node i in a majority of the sampled trajectories, with the edge weight denoting average transition time between i and j. The agent, constrained with the topology of the landmark network has to follow a sequence of landmarks(l_{sequence}) starting with the start landmark and ending with the terminal landmark. If the landmark in the l_{sequence} is not visited within the transition time stored in the time model then a failure is ascertained. Failure leads to learning a new path between the landmarks or learning a new average transition time between existing paths of the landmarks. This new transition time is added as an element to already existing set of average transition times.

3.1   Identification of Landmark

We propose two heuristics to identify landmarks :

☐ Spiked state: While traveling most of the people try to associate landmarks with some junctions, where they want to leave the road and move onto another road. Spiked state heuristic is based on this idea and the junction of many trajectories is called a spiked state. Here the spiked state has highest frequency of visits across trajectories as compared to other neighboring states. Once a policy is learnt, trajectories are sampled with random start states. For a given trajectory let s_{t-1}, s_t, s_{t+1} be state that occurred at successive time instances. If $\eta_{s_{t-1}} < \eta_{s_t}$ and $\eta_{s_t} > \eta_{s_{t+1}}$ then s_t is a spiked state. If two states s and s' spiked in the same trajectory, they are called co-occurring spiked states. From this set of spiked states we choose the set of states which spiked in maximum number of trajectories. Since we need landmarks to be uniformly distributed in the domain, we define a lower bound on the count i.e., number of times state spiked in all the trajectories. This bound can be estimated empirically as it depends upon the domain.

One of the problems with this heuristic is that it requires the domain to be highly connected so that there are various paths through the states which enables us to find the pattern required to identify the spiked states.

☐ Mean-to-variance ratio: This heuristic uses a simpler approach as compared to the spiked state heuristic. Trajectories are sampled as in spiked state heuristic, but here all trajectories start from the same state. A state is said to have been accessed at time t if it is encountered in the trajectory at step count t. The time step at which each state is accessed in the trajectory is extracted for all states in the trajectories and its mean and variance. The states are sorted in decreasing order of mean-to-variance ratio. Here variance is considered in order to avoid the effect of states closer to initial states and the terminal states in the trajectory as these states will always be accessed. The analysis is done for the states which appears to occur in

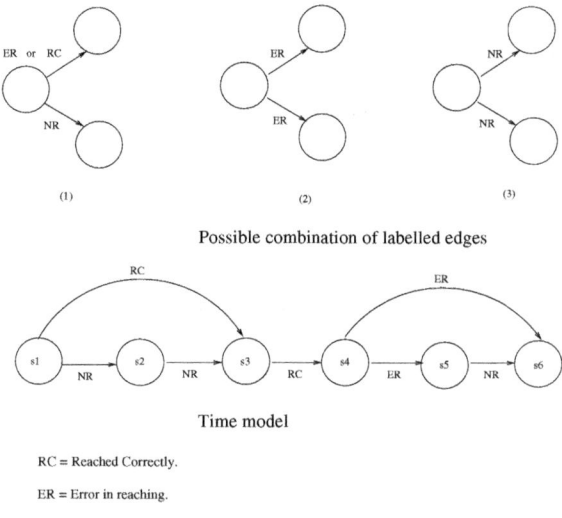

Possible combination of labelled edges

Time model

RC = Reached Correctly.

ER = Error in reaching.

NR = Path not taken.

Fig. 1. Identification of exception landmark

maximum number of trajectories. Distance between the landmarks is defined as the average number of time steps taken to traverse between the landmarks. Landmarks are selected sequentially from the sorted list of states so that a criteria of minimum distance between the landmarks is satisfied.

3.2   Construction and Updation of Transition Time Model

Transition time model stores the transition time between the source and destination landmarks. This is estimated from sampled trajectories for all landmark pairs of the landmark sequence(l_{sequence}), if they are part of the same trajectory. Here $l_{sequence}$ is a topological ordering of landmarks in the landmark network. Let the old average transition time between the landmarks p and q stored in the transition model be called as t^{pq}_{old}.

In order to find a failure, a new instance of the transition time model is created. It is populated with the average new transition time, extracted from the new sampled trajectories. These new trajectories are sampled in batches using the old policies. If there is no significant change in the transition probability of the states in the batch, sampling is stopped. Let t^{pq}_{new} be a new average transition time between the landmark p and q stored in the new instance of transition model. Percentage increase in the average transition time between any two landmark, p and q is called Change In Transition Time($CITT$).

$$CITT = \frac{t^{pq}_{\text{new}} - t^{pq}_{\text{old}}}{t^{pq}_{\text{old}}} \times 100$$

Failure is ascertained if $CITT$ exceeds the empirically estimated threshold percentage. Since each source and destination landmark may store the set of average transition times. This enforces computation of $CITT$ between t^{pq}_{new} and the set of t_{old}^{pq}s. Hence minimum $CITT$ for each edge is compared with the threshold percentage, leading to edges being labeled as shown in Figure 1. There are three types of scenarios which usually occur in the transition time model. In the first scenario shown in Figure 1, one of the destination landmarks is reached from the source landmark. If at least one of the $CITT$ value is within the threshold percentage, the edge is labeled as "RC" (Reached Correctly), else it is labeled as "ER"(Error In Reaching). The rest of the edges leading to other landmarks are labeled as "NR"(Path Not Taken). An edge is labeled as "NR" only if the source and destination landmark pair does not occur simultaneously within the specified threshold count of trajectories. There cannot be a node with two out edges labeled as "RC" and "ER".

The transition time model with the labeled edges is used to find the exception. If the task is successfully completed then there exists a unique path consisting of a sequence of edges labeled as "RC" between the start and terminal landmarks. This indicates that an exception has not occurred. Failure is ascertained when at least one of the outgoing edges from the current landmark is labeled as "ER" or all the outgoing edges are labeled as "NR" as shown in second scenario in Figure 1. The current landmark is then called an exception landmark. For example, in Figure 1, s_4 is the exception landmark in the sequence of landmarks s_1, s_2, s_3, s_4, s_5 and s_6

Topological ordering of landmarks($l_{sequence}$) helps to identify a set of landmarks called *potential destination landmarks*, that come topologically after the exception landmark. A new path is learnt between the exception landmark and some potential destination landmarks using Q-learning. The transition time model is updated to include the new transition time between the landmarks. While updating, the newly learnt path might add edges between landmarks which were not connected earlier or it might add new task times to existing set of task times between already connected landmarks. The landmark network is topologically reordered to reflect the new ordering of the landmarks.

4   Identification of Exception State

An exception happens when the policy fails. The state in the which failure first happens is called the exception state. Failure can happen because of changes in the dynamics of the domain. The reasons for this to happen is either introduction of a new object in the domain and a change in the values that features of the domain can take. In either case, there are some features of the exception state which reflect the change in the domain. The new policy is represented in the suffix tree by conditioning on these features.

In order to identify the exceptions, transition probabilities of all the state action pairs need to be stored. But rather than storing it for all the actions, a model called as *minimum model* can be used to store only the optimal policy action for all the states. Exception state is found by comparing the minimum model learnt before and after the failure is ascertained by the time model. Rather than comparing for all the states in the trajectory, exception landmarks can be used

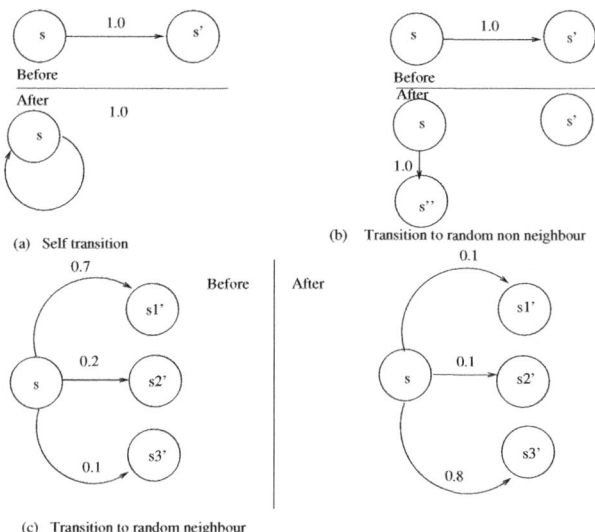

(a) Self transition

(b) Transition to random non neighbour

(c) Transition to random neighbour

Fig. 2. Different types of exceptions

as reference point from where comparison could start. Exception landmark is the landmark after which an exception occurs. Figure 2 shows basic types of exceptions, which change the dynamics of the domain. These are self transitions, transitions to random neighboring states and transitions to random non-neighboring states . Other forms of exceptions that are possible are combinations of these basic types of exceptions.

After the exception policy is learnt, the model is updated to reflect the changes in transition probabilities of exception states.

5   Experiment and Results

In Figure 3, the agent's goal is to collect the diamond(o) in every room by occupying the same square as the diamond. Each of the rooms is a 11 by 11 grid with certain obstacles in it. The actions available to the agent are N, S, E, W. An action would result in transition to the expected state with probability 0.9. It results in a transition in any one of the unintended directions with a probability 0.1.

The state is described by the following features: the room number the agent is in, with 0 denoting the corridor, the x and y co-ordinates within the room or corridor and boolean variables $have(i)$, $i = 1,..., 5$, indicating possession of diamond in room i. The goal is any state of the form $\langle 0, ..1, 1, 1, 1, 1 \rangle$, where o indicates the corridor and 1 indicates possession of diamonds in the respective rooms. The state abstraction for the option is $\langle x, y, have \rangle$. Exceptions are caused by the objects C, F (self transition) , D (transition to random neighbor) and E (transition to random non-neighbor).

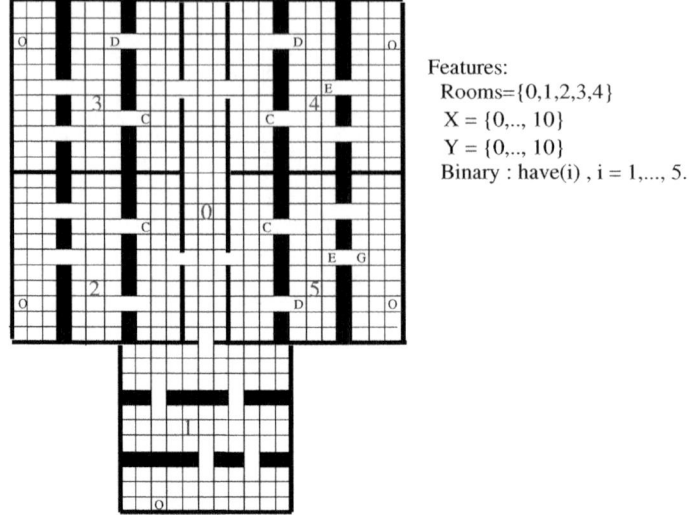

Features:
Rooms={0,1,2,3,4}
X = {0,.., 10}
Y = {0,.., 10}
Binary : have(i) , i = 1,..., 5.

Fig. 3. A simple room domain with different objects. The task is to collect the diamond, o in the environment.

Table 1. Selected Landmarks

Experiment		
1	2	3
(10, 5, 0), (0, 3, 1), (10, 5, 1)	(10, 5, 0),(5, 5, 0),(1, 2, 0) (4, 5, 1),(10, 5, 1)	(10 ,5, 0),(7 ,7 ,0),(3 ,5, 0) (1, 3, 0),(0, 3, 1),(1, 5, 1) (4, 5, 1),(5, 7, 1),(9, 7, 1) (10, 5, 1)

The agent used a learning rate of .05, discount rate of 1.0 and ϵ- greedy exploration, with an ϵ of 0.1. The result were averaged over 25 independent runs. The trials were terminated either on completion of the task or after 1000 steps. The landmarks were selected by the mean to variance ratio. States that occur in more than 80% of the trials only were considered as candidate landmarks. Three sets of experiments were performed with different desired minimum distance (in time) between the landmarks, resulting in a different number of landmarks for each experiment. The distance between the landmarks is 18 steps in experiment 1, 8 steps for experiment 2 and 4 steps for experiment 3. The selected landmarks are shown in table 1. To compare the transition time between the landmarks, threshold value is taken as 25% for experiment 1, 30% for experiment and 35% for experiment 3. A landmark is considered as not reachable("NR") if the number of trajectories is less than 75%.

The objects were introduced in the domain in the order C, D, E and F. These objects can be introduced randomly in the domain but for the purpose of learning nested OWE, they were introduced in the specific order. Exception caused by the

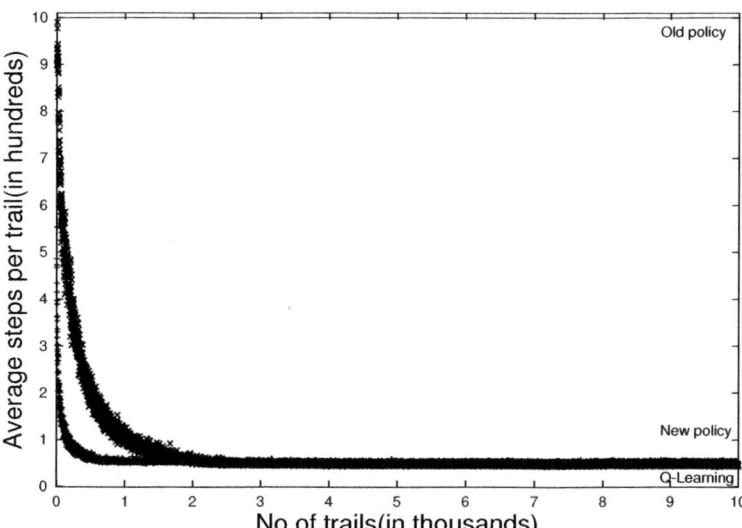

Fig. 4. Comparison of old policy, new policy and policy learnt using Q-learning where the exception is caused by the object C

object "D" is not identified in most of the experiments because the maximum change in the number of steps caused by the exception is not more than 1.5 on an average. If the threshold value is adjusted to account for this small change, it leads to detection of exceptions caused by the stochasticity of the environment. Hence exception caused due to D in the domain remains unidentified. Whereas, if the same object is placed in the grid (8,7), exception is identified because the most of trajectories do not go through (8,6).

In Figure 4,5 and 6, the new policies are plotted for various obstacles. The new policy is a concatenation of the path given by the suffix tree from the start state to exception state or terminal state, followed by the path given by Q-learning from exception state to potential terminal landmarks, followed by a path given by the suffix tree from potential terminal landmark to terminal state. Graphs have been plotted for Experiment 3. Due to lack of space, we have not plotted the graphs for other experiments although the results agree with our expectations.

When the OWE policies were used in the old domain, i.e., without exceptions, the average number of steps taken to complete the task does not change appreciably, as shown in Table 2. The average was taken over 1000 trails. In Table 2, the rows indicate transfer of policies to old domain after particular object was introduced in the domain and columns indicate different objects present in the domain at that time.

In the column "Object C,D,E", the average number of steps for experiment 3 is higher than the others because the new policy is constrained to go through extra landmarks which requires 3 extra steps in comparison to that required for experiment 1 and experiment 2. Therefore for experiment 3, all the values under the columns are greater than corresponding values of the other experiments.

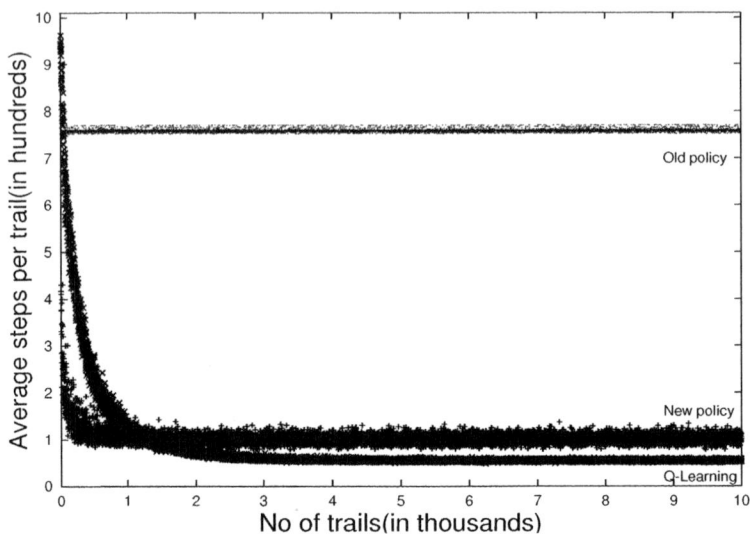

Fig. 5. Comparison of old policy, new policy and policy learnt using Q-learning where the exception is caused by the object E

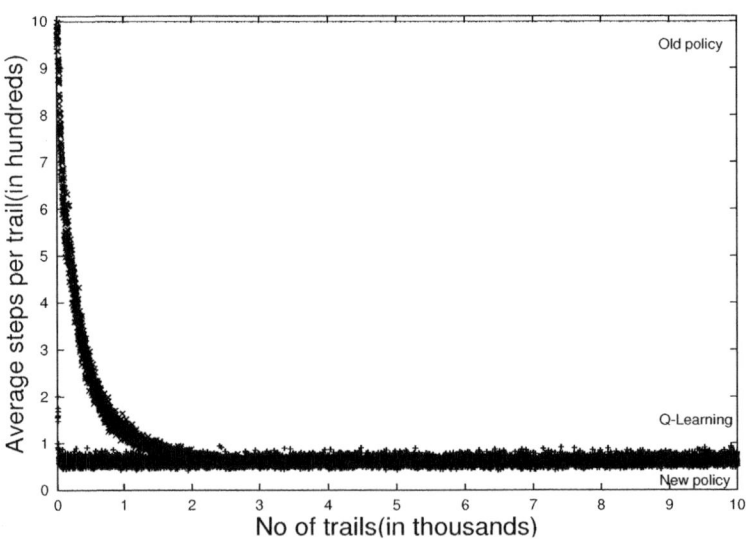

Fig. 6. Comparison of old policy, new policy and policy learnt using Q-learning where the exception is caused by the object F

Table 2. Transfer back of the policy to old domain(Avg No of steps)

		No Object	Object C	Object C,D	Object C,D,E
Experiment 1	No Object	44.43			
	After Object C	44.07	52		
	After Object D	44.32	52.29	55.12	
	After Object E	44.55	52.71	55.02	63.12
	After Object F	44.58	52.36	55.21	63.87
Experiment 2	No Object	44.45			
	After Object C	44.23	52.16		
	After Object D	44.14	52.30	54.42	
	After Object E	43.90	52.40	54.84	63.4
	After Object F	43.62	52.55	54.69	63.54
Experiment3	No Object	44.33			
	After Object C	44.39	53.19		
	After Object D	43.86	53.20	55.30	
	After Object E	44.76	53.78	55.17	66.56
	After Object F	44.27	53.19	55.60	66.24

6 Conclusion

In this work we draw attention to the problem of efficient incremental maintenance of library of skills. We propose a framework in which exceptions can be
added to already learnt options without disturbing the original policies. While
the framework succeeds in identifying small changes required to the options,
we still have some distance to go before building a complete skill management
system. We are currently working on the problem of automatically deriving the
minimal set of features required to represent exceptions. We also want to automatically identify when an exception would suffice and when a new option is
warranted. While we can think of several heuristics, based on the size of the
suffix tree for instance, we would like a more principled approach.

References

1. Barto, A.G., Mahadevan, S.: Recent Advances in Hierarchical Reinforcement
   Learning. Discrete Event Dynamic Systems 13(1-2) (2003)
2. Taylor, M.E., Stone, P.: Transfer Learning for Reinforcement Learning Domains:
   A Survey. Journal of Machine Learning Research 10, 1633–1685 (2009)
3. Dietterich, T.G.: Hierarchical Reinforcement Learning with the MAXQ Value Function Decomposition. Journal of Artificial Intelligence Research 13, 227–303 (2000)
4. McCallum, A.K.: Reinforcement Learning with Selective Perception and Hidden
   State, Ph.D. Thesis, Department of Computer Science, The College of Arts and
   Science, University of Rocheater, USA (1995)
5. Asadi, M., Huber, M.: Autonomous Subgoal Discovery and Hierarchical Abstraction Learned Policies. In: FLAIRS Conference, pp. 346–350 (2003)
6. Gaines, B.R., Compton, P.: Induction of Ripple-Down Rules Applied to Modeling
   Large Database. Knowledge Acquisition 2(3), 241–258 (1995)

7. McGovern, A.: Autonomous Discovery of Temporal Abstraction from Interaction with An Environment, Ph.D. Thesis, Department of Computer Science, University of Massachusetts, Amherst, USA (2002)
8. Precup, D.: Temporal Abstraction in Reinforcement Learning, Ph.D. Thesis, Department of Computer Science, University of Massachusetts, Amherst, USA (2000)
9. McGovern, A., Barto, A.G.: Automatic Discovery of Subgoals in Reinforcement Learning using Diverse Density. In: Proc. 18th International Conf. on Machine Learning, pp. 361–368. Morgan Kaufmann, San Francisco (2001)
10. Bradtke, S.J., Duff, M.O.: Reinforcement Learning Methods for Continuous-Time Markov Decision Problems. In: Tesauro, G., Touretzky, D., Leen, T. (eds.) Advances in Neural Information Processing Systems, vol. 7, pp. 393–400. The MIT Press (1995)
11. Sutton, R.S., Precup, D.: Intra-option learning about temporally abstract actions. In: Proceedings of the Fifteenth International Conference on Machine Learning, pp. 556–564. Morgan Kaufman (1998)
12. Kaelbling, L.P.: Hierarchical learning in stochastic domains: Preliminary results. In: Proceedings of the Tenth International Conference on Machine Learning, pp. 167–173 (1993)
13. Ravindran, B., Barto, A.G.: Relativized Options: Choosing the Right Transformation. In: Proceedings of the Twentieth International Conference on Machine Learning, pp. 608–615 (2003)

Robust Bayesian Reinforcement Learning through Tight Lower Bounds

Christos Dimitrakakis

EPFL, Lausanne, Switzerland
christos.dimitrakakis@epfl.ch

Abstract. In the Bayesian approach to sequential decision making, exact calculation of the (subjective) utility is intractable. This extends to most special cases of interest, such as reinforcement learning problems. While utility bounds are known to exist for this problem, so far none of them were particularly tight. In this paper, we show how to efficiently calculate a lower bound, which corresponds to the utility of a near-optimal *memoryless* policy for the decision problem, which is generally different from both the Bayes-optimal policy and the policy which is optimal for the expected MDP under the current belief. We then show how these can be applied to obtain robust exploration policies in a Bayesian reinforcement learning setting.

1 Setting

We consider decision making problems where an agent is acting in a (possibly unknown to it) environment. By choosing actions, the agent changes the state of the environment and in addition obtains scalar rewards. The agent acts so as to maximise the expectation of the utility function: $U_t \triangleq \sum_{k=t}^{T} \gamma^k r_k$, where $\gamma \in [0,1]$ is a discount factor and where the instantaneous rewards $r_t \in [0, r_{\max}]$ are drawn from a Markov decision process (MDP) μ, defined on a state space $\mathcal{S}$ and an action space $\mathcal{A}$, both equipped with a suitable metric and σ-algebra, with a set of transition probability measures $\left\{ \mathcal{T}_\mu^{s,a} \mid s \in \mathcal{S}, a \in \mathcal{A} \right\}$ on $\mathcal{S}$, and a set of reward probability measures $\left\{ \mathcal{R}_\mu^{s,a} \mid s \in \mathcal{S}, a \in \mathcal{A} \right\}$ on $\mathbb{R}$, such that:

$$r_t \mid s_t = s, a_t = a \sim \mathcal{R}_\mu^{s,a}, \qquad s_{t+1} \mid s_t = s, a_t = a \sim \mathcal{T}_\mu^{s,a}, \qquad (1.1)$$

where $s_t \in \mathcal{S}$ and $a_t \in \mathcal{A}$ are the state of the MDP, and the action taken by the agent at time t, respectively. The environment is controlled via a *policy* $\pi \in \mathcal{P}$. This defines a conditional probability measure on the set of actions, such that $\mathbb{P}_\pi(a_t \in A \mid s^t, a^{t-1}) = \pi(A \mid s^t, a^{t-1})$ is the probability of the action taken at time t being in A, where we use $\mathbb{P}$, with appropriate subscripts, to denote probabilities of events and $s^t \triangleq s_1, \ldots, s_t$ and $a^{t-1} \triangleq a_1, \ldots, a_{t-1}$ denotes sequences of states and actions respectively. We use $\mathcal{P}_k$ to denote the set of k-order Markov policies. Important special cases are the set of *blind* policies $\mathcal{P}_0$ and the set of *memoryless* policies $\mathcal{P}_1$. A policy in $\pi \in \bar{\mathcal{P}}_k \subset \mathcal{P}_k$ is *stationary*, when $\pi(A \mid s_{t-k+1}^t, a_{t-k+1}^{t-1}) = \pi(A \mid s^k, a^{k-1})$ for all t.

S. Sanner and M. Hutter (Eds.): EWRL 2011, LNCS 7188, pp. 177–188, 2012.

The expected utility, conditioned on the policy, states and actions is used to define a *value function* for the MDP μ and a stationary policy π, at stage t:

$$Q_{\mu,t}^{\pi}(s,a) \triangleq \mathbb{E}_{\mu,\pi}(U_t \mid s_t = s, a_t = a), \qquad V_{\mu,t}^{\pi}(s) \triangleq \mathbb{E}_{\mu,\pi}(U_t \mid s_t = s), \qquad (1.2)$$

where the expectation is taken with respect to the process defined jointly by μ, π on the set of all state-action-reward sequences $(\mathcal{S}, \mathcal{A}, \mathbb{R})^*$. The *optimal* value function is denoted by $Q_{\mu,t}^* \triangleq \sup_{\pi} Q_{\mu,t}^{\pi}$ and $V_{\mu,t}^* \triangleq \sup_{\pi} V_{\mu,t}^{\pi}$. We denote the optimal policy[1] for μ by π_{μ}^*. Then $Q_{\mu,t}^* = Q_{\mu,t}^{\pi_{\mu}^*}$ and $V_{\mu,t}^* = V_{\mu,t}^{\pi_{\mu}^*}$.

There are two ways to handle the case when the true MDP is unknown. The first is to consider a set of MDPs such that the probability of the true MDP lying outside this set is bounded from above [e.g. 20, 21, 4, 19, 28, 27]. The second is to use a Bayesian framework, whereby a full distribution over possible MDPs is maintained, representing our subjective belief, such that MDPs which we consider more likely have higher probability [e.g. 14, 10, 31, 2, 12]. Hybrid approaches are relatively rare [16]. In this paper, we derive a method for efficiently calculating near-optimal, robust, policies in a Bayesian setting.

1.1   Bayes-Optimal Policies

In the Bayesian setting, our uncertainty about the Markov decision process (MDP) is formalised as a probability distribution on the class of allowed MDPs. More precisely, assume a probability measure ξ over a set of possible MDPs $\mathcal{M}$, representing our belief. The expected utility of a policy π with respect to the belief ξ is:

$$\mathbb{E}_{\xi,\pi} U_t = \int_{\mathcal{M}} \mathbb{E}_{\mu,\pi}(U_t) \, d\xi(\mu). \qquad (1.3)$$

Without loss of generality, we may assume that all MDPs in $\mathcal{M}$ share the same state and action space. For compactness, and with minor abuse of notation, we define the following value functions with respect to the belief:

$$Q_{\xi,t}^{\pi}(s,a) \triangleq \mathbb{E}_{\xi,\pi}(U_t \mid s_t = s, a_t = a), \qquad V_{\xi,t}^{\pi}(s) \triangleq \mathbb{E}_{\xi,\pi}(U_t \mid s_t = s), \qquad (1.4)$$

which represent the expected utility under the belief ξ, at stage t, of policy π, conditioned on the current state and action.

Definition 1 (Bayes-optimal policy). *A Bayes-optimal policy π_{ξ}^* with respect to a belief ξ is a policy maximising (1.3). Similarly to the known MDP case, we use $Q_{\xi,t}^*, V_{\xi,t}^*$ to denote the value functions of the Bayes-optimal policy.*

Finding the Bayes-optimal policy is generally intractable [11, 14, 18]. It is important to note that a Bayes-optimal policy is not necessarily the same as the optimal policy for the true MDP. Rather, it is the optimal policy given that the true MDP was drawn at the start of the experiment from the distribution ξ. All the theoretical development in this paper is with respect to ξ.

---

[1] We assume that there exists at least one optimal policy. If there are multiple optimal policies, we choose arbitrarily among them.

1.2   Related Work and Main Contribution

Since computation of the Bayes-optimal policy is intractable in the general case, in this work we provide a simple algorithm for finding near-optimal *memoryless* policies in polynomial time. By definition, for any belief ξ, the expected utility under that belief of any policy π is a lower bound on that of the optimal policy π_ξ^*. Consequently, the near-optimal memoryless policy gives us a tight lower bound on the subjective utility.

A similar idea was used in [12], where the stationary policy that is optimal on the *expected MDP* is used to obtain a lower bound. This is later refined through a stochastic branch-and-bound technique that employs a similar upper bound. In a similar vein, [17] uses approximate Bayesian inference to obtain a stationary policy for the current belief. More specifically, they consider two families of expectation maximisation algorithms. The first uses a variational approximation to the reward-weighted posterior of the transition distribution, while the second performs expectation propagation on the first two moments. However, none of the above approaches return the optimal stationary policy.

It is worthwhile to mention the very interesting point-based BEETLE algorithm of Poupart et al. [23], which discretised the belief space by sampling a set of future beliefs (rather than MDPs). Using the convexity of the utility with respect to the belief, they constructed a lower bound via a piecewise-linear approximation of the complete utility from these samples. The approach results in an approximation to the optimal non-stationary policy. Although the algorithm is based on an optimal construction reported in the same paper, sufficient conditions for its optimality are not known.

In this paper, we obtain a tight lower bound for the *current* belief by calculating a nearly optimal *memoryless* policy. The procedure is computationally efficient, and we show that it results in a much tighter bound than the value of the expected-MDP-optimal policy. We also show that it can be used in practice to perform robust Bayesian exploration in unknown MDPs. This is achieved by computing a new memoryless policy once our belief has changed significantly, a technique also employed by other approaches [19, 3, 2, 29, 31]. It can be seen as a principled generalisation of the sampling approach suggested in [29] from a single MDP sample to multiple samples from the posterior. The crucial difference is that, while previous work uses some form of *optimistic* policy, we instead employ a more conservative policy in each stationary interval. This can be significantly better than the policy which is optimal for the expected MDP.

The first problem we tackle is how to compute this policy given a belief over a finite number of MDPs. For this, we provide a simple algorithm based on backwards induction [see 11, for example]. In order to extend this approach to an arbitrary MDP set, we employ Monte Carlo sampling from the current posterior. Unlike other Bayesian sampling approaches [10, 29, 2, 6, 30, 12, 31], we use these samples to estimate a policy that is nearly optimal (within the restricted set of memoryless policies) with respect to the distribution these samples were drawn from. Finally, we provide theoretical and experimental analyses of the proposed algorithms.

2   MMBI: Multi-MDP Backwards Induction

Even when our belief ξ is a probability measure over a finite set of MDPs $\mathcal{M}$, the finding an optimal policy is intractable. For that reason, we restrict ourselves to memoryless polices $\pi \in \mathcal{P}_1$. We can approximate the optimal *memoryless* policy with respect to ξ, by setting the posterior measure given knowledge of the policy so far and the current state, to equal the initial belief, i.e. $\xi(\mu \mid s_t = s, \pi) = \xi(\mu)$ (we do not condition on the complete history, since the policies are memoryless). The approximation is in practice quite good, since the difference between the two measures tends to be small. The policy π_{MMBI} can then be obtained via the following backwards induction. By definition:

$$Q_{\xi,t}^\pi(s,a) = \mathbb{E}_{\xi,\pi}(r_t \mid s_t = s, a_t = a) + \gamma \, \mathbb{E}_{\xi,\pi}(U_{t+1} \mid s_t = s, a_t = a), \qquad (2.1)$$

where the expected reward term can be written as

$$\mathbb{E}_{\xi,\pi}(r_t \mid s_t = s, a_t = a) = \int_{\mathcal{M}} \mathbb{E}_\mu(r_t \mid s_t = s, a_t = a) \, d\xi(\mu), \qquad (2.2a)$$

$$\mathbb{E}_\mu(r_t \mid s_t = s, a_t = a) = \int_{-\infty}^\infty r \, d\mathcal{R}_\mu^{s,a}(r). \qquad (2.2b)$$

The next-step utility can be written as:

$$\mathbb{E}_{\xi,\pi}(U_{t+1} \mid s_t = s, a_t = a) = \int_{\mathcal{M}} \mathbb{E}_{\mu,\pi}(U_{t+1} \mid s_t = s, a_t = a) \, d\xi(\mu), \qquad (2.3a)$$

$$\mathbb{E}_{\mu,\pi}(U_{t+1} \mid s_t = s, a_t = a) = \int_{\mathcal{S}} V_{\mu,t+1}^\pi(s') \, d\mathcal{T}_\mu^{s,a}(s'). \qquad (2.3b)$$

Putting those steps together, we obtain Algorithm 1, which greedily calculates a memoryless policy for a T-stage problem and returns its expected utility.

Algorithm 1. MMBI - Backwards induction on multiple MDPs.

1: **procedure** MMBI($\mathcal{M}, \xi, \gamma, T$)
2:     Set $V_{\mu,T+1}(s) = 0$ for all $s \in \mathcal{S}$.
3:     **for** $t = T, T-1, \ldots, 0$ **do**
4:         **for** $s \in \mathcal{S}, a \in \mathcal{A}$ **do**
5:             Calculate $Q_{\xi,t}(s,a)$ from (2.1) using $\{V_{\mu,t+1}\}$.
6:         **end for**
7:         **for** $s \in \mathcal{S}$ **do**
8:             $a_{\xi,t}^*(s) = \arg\max \{Q_{\xi,t}(s,a) \mid a \in \mathcal{A}\}$.
9:             **for** $\mu \in \mathcal{M}$ **do**
10:                 $V_{\mu,t}(s) = Q_{\mu,t}(s, a_{\xi,t}^*(s))$.
11:             **end for**
12:         **end for**
13:     **end for**
14: **end procedure**

The calculation is greedy, since optimising over π implies that at any step $t + k$, we must condition the belief on past policy steps $\xi(\mu \mid s_{t+k} = s, \pi_t, \ldots, \pi_{t+k-1})$ to calculate the expected utility correctly. Thus, the optimal π_{t+k} depends on both future and past selections. Nevertheless, it is easy to see that Alg. 1 returns the correct expected utility for time step t. Theorem 1 bounds the gap between this and the Bayes-optimal value function when the difference between the current and future beliefs is small.

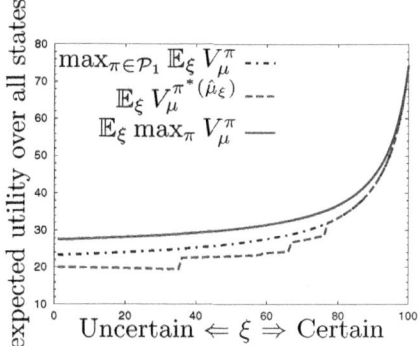

Fig. 1. Value function bounds

Theorem 1. *For any $k \in [t, T]$, let $\xi_k \triangleq \xi(\cdot \mid s^k, a^k)$ be the posterior after k observations. Let λ be a dominating measure on $\mathcal{M}$ and $\|f\|_{\lambda,1} = \int_{\mathcal{M}} |f(\mu)| \, d\lambda(\mu)$, for any λ-measurable function f. If $\|\xi_t - \xi_k\|_{\lambda,1} \leq \epsilon$, for all k, then the policy π_{MMBI} found by MMBI is within $r_{max}(1 - \gamma)^{-2}\epsilon$ of the Bayes-optimal policy π_ξ^*.*

Proof. The error at every stage $k > t$, is bounded as follows:

$$|V_{\xi,k}(s) - \mathbb{E}_\xi(U_k \mid s^k, a^k)| = \left| \int_{\mathcal{M}} [\xi_t(\mu) - \xi_k(\mu)(s)] V_{\mu,k}(s) \, d\lambda(\mu) \right|$$

$$\leq \frac{r_{\max}}{1 - \gamma} \int_{\mathcal{M}} |\xi_t(\mu) - \xi_k(\mu)(s)| \, d\lambda(\mu) \leq \frac{r_{\max}}{1 - \gamma} \epsilon.$$

The final result is obtained via the geometric series. $\qquad\square$

We can similarly bound the gap between the MMBI policy and the ξ-optimal memoryless policy, by bounding $\sup_{k,s,\pi} \|\xi_t(\cdot) - \xi_t(\cdot \mid s_k = s, \pi)\|_{\lambda,1}$.

The ξ-optimal memoryless policy is generally *different* from the policy which is optimal with respect to the expected MDP $\hat{\mu}_\xi \triangleq \mathbb{E}_\xi \mu$, as can be seen via counterexample where $\mathbb{E}_\xi V_\mu^\pi \neq V_{\hat{\mu}_\xi}^\pi$, or even where $\mathbb{E}_\xi \mu \notin \mathcal{M}$. MMBI can be used to obtain a much tighter value function bound than the $\hat{\mu}_\xi$-optimal policy, as shown in Fig. 1, where the MMBI bound is compared to the $\hat{\mu}_\xi$-optimal policy bound and the simple upper bound: $V_\xi^*(s) \leq \mathbb{E}_\xi \max_\pi V_\mu^\pi(s)$. The figure shows how the bounds change as our belief over 8 MDPs changes. When we are more uncertain, MMBI is much tighter than $\hat{\mu}_\xi$-optimal. However, when most of the probability mass is around a single MDP, both lower bounds coincide. In further experiments on online reinforcement learning, described in Sec. 3, near-optimal memoryless policies are compared against the $\hat{\mu}_\xi$-optimal policy.

2.1 Computational Complexity

When $\mathcal{M}$ is finite and $T < \infty$, MMBI (Alg. 1) returns a greedily-optimised policy π_{MMBI} and its value function. When $T \to \infty$, MMBI can be used to calculate an ϵ-optimal approximation by truncating the horizon, as shown below.

Lemma 1. *The complexity of Alg. 1 for bounding the value function error by ϵ, is $\mathcal{O}\left([|\mathcal{M}||\mathcal{S}|^2(|\mathcal{A}|+1) + (1+|\mathcal{M}|)|\mathcal{S}||\mathcal{A}|]\log_\gamma \frac{\epsilon(1-\gamma)}{r_{max}}\right)$, assuming $r_t \in [0, r_{max}]$,*

Proof. Since $r_t \in [0, r_{\max}]$, if we look up to some horizon T, our value function error is bounded by $\gamma^T c$, where $c = H r_{\max}$ and $H = \frac{1}{1-\gamma}$ is the effective horizon. Consequently, we need $T \geq \log_\gamma(\epsilon/c)$ to bound the error by ϵ. For each t, step 5 is performed $|\mathcal{S}||\mathcal{A}|$ times. Each step takes $O(|\mathcal{M}|)$ operations for the expected reward and $O(|\mathcal{S}||\mathcal{M}|)$ operations for the next-step expected utility. The second loop is $\mathcal{O}(|\mathcal{S}|(|\mathcal{A}|+|\mathcal{M}||\mathcal{S}|))$, since it is performed $|\mathcal{S}|$ times, with the max operators taking $|\mathcal{A}|$ operations, while inner loop is performed $|\mathcal{M}|$ times with each local MDP update step 10 takes $|\mathcal{S}|$ operations.   □

Algorithm 2. MSBI: Multi-Sample Backwards Induction

1: **procedure** MSBI(ξ, γ, ϵ)
2:      $n = \left(\frac{3r_{\max}}{\epsilon(1-\gamma)}\right)^3$.
3:      $\mathcal{M} = \{\mu_1, \ldots, \mu_n\}$, $\mu_i \sim \xi$.
4:      MMBI$(\mathcal{M}, p, \gamma, \log_\gamma \frac{\epsilon(1-\gamma)}{r_{\max}})$, with $p(\mu_i) = 1/n$ for all i.
5: **end procedure**

It is easy to see that the most significant term is $\mathcal{O}(|\mathcal{M}||\mathcal{S}|^2|\mathcal{A}|)$, so the algorithmic complexity scales linearly with the number of MDPs. Consequently, when $\mathcal{M}$ is not finite, exact computation is not possible. However, we can use high probability bounds to bound the expected loss of a policy calculated stochastically through MSBI (Alg.2).

MSBI simply takes a sufficient number of samples of MDPs from ξ, so that in ξ-expectation, the loss relative to the MMBI policy is bounded according to the following lemma.

Lemma 2. *The expected loss of* MSBI *relative to* MMBI *is bounded by ϵ.*

Proof. Let $\hat{\mathbb{E}}^n U = \frac{1}{n}\sum_{i=1}^n \mathbb{E}_{\mu_i} U$ denote the empirical expected utility over the sample of n MDPs, where the policy subscript π is omitted for simplicity. Since $\mathbb{E}_\xi \hat{\mathbb{E}}^n U = \mathbb{E}_\xi U$, we can use the Hoeffding inequality to obtain:

$$\xi\left(\left\{\mu^n \,\middle|\, \hat{\mathbb{E}}^n U \geq \mathbb{E}_\xi U + \epsilon\right\}\right) \leq e^{-2n\epsilon^2/c^2}.$$

This implies the following bound:

$$\mathbb{E}_\xi(\hat{\mathbb{E}}^n U - \mathbb{E}_\xi U) \le c\delta + c\sqrt{\frac{\ln(1/\delta)}{2n}} \le c(8n)^{-1/3} + c\sqrt{\frac{(8n)^{1/3}}{2n}} = 3cn^{-1/3}.$$

Let $\mathcal{P}_1$ be the set of memoryless policies. Since the bound holds uniformly (for any $\pi \in \mathcal{P}$), the policy $\hat{\pi}^* \in \bar{\mathcal{P}}_1$ maximising $\hat{\mathbb{E}}^n$ is within $3cn^{-1/3}$ of the ξ-optimal policy in $\mathcal{P}_1$. $\square$

Finally, we can combine the above results to bound the complexity of achieving a small approximation error for MSBI, with respect to expected loss:

Theorem 2. MSBI *(Alg. 2) requires* $\mathcal{O}\left(\left(\frac{6r_{max}}{\epsilon(1-\gamma)}\right)^3 |\mathcal{S}|^2 |\mathcal{A}| \log_\gamma \frac{\epsilon(1-\gamma)}{2r_{max}}\right)$ *operations to be ϵ-close to the best MMBI policy.*

Proof. From Lem. 2, we can set $n = (6c/\epsilon)^3$ to bound the regret by $\epsilon/2$. Using the same value in Lem. 1, and setting $|\mathcal{M}| = n$, we obtain the required result. $\square$

2.2   Application to Robust Bayesian Reinforcement Learning

While MSBI can be used to obtain a memoryless policy which is in expectation close to both the optimal memoryless policy and the Bayes-optimal policy for a given belief, the question is how to extend the procedure to online reinforcement learning. The simplest possible approach is to simply recalculate the stationary policy after some interval $B > 0$. This is the approach followed by MCBRL (Alg. 3), shown below.

Algorithm 3. MCBRL: Monte-Carlo Bayesian Reinforcement Learning

---
1: **procedure** MCBRL($\xi_0, \gamma, \epsilon, B$)
2:     Calculate $\xi_t(\cdot) = \xi_0(\cdot \mid s^t, a^{t-1})$.
3:     Call MSBI(ξ_t, γ, ϵ) and run returned policy for B steps.
4: **end procedure**

---

3   Experiments in Reinforcement Learning Problems

Selecting the number of samples n according to ϵ for MCBRL is computationally prohibitive. In practice, instead of setting n via ϵ, we simply consider increasing values of n. For a single sample ($n = 1$), MCBRL is equivalent to the sampling method in [29], which at every new stage, samples a single MDP from the current posterior and then uses the policy that is optimal for the sampled MDP. In addition, for this particular experiment, rather than using the memoryless policy found, we apply the stationary policy derived by using the first step of the memoryless policy. This incurs a small additional loss. We also compared MCBRL against the common heuristic of acting according to the policy that is optimal with respect to the *expected* MDP $\hat{\mu}_\xi \triangleq \mathbb{E}_\xi \mu$. The algorithm, referred

Algorithm 4. EXPLOIT: Expected MDP exploitation [23]

1: **procedure** EXPLOIT(ξ_0, γ)
2:     **for** $t = 1, \ldots$ **do**
3:         Calculate $\xi_t(\cdot) = \xi_0(\cdot \mid s^t, a^{t-1})$.
4:         Estimate $\hat{\mu}_{\xi_t} \triangleq \mathbb{E}_{\xi_t}\mu$.
5:         Calculate $Q^*_{\hat{\mu}_{\xi_t}}(s, a)$ using discount parameter γ.
6:         Select $a_t = \arg\max_a Q^*_{\hat{\mu}_{\xi_t}}(s, a)$
7:     **end for**
8: **end procedure**

to as the EXPLOIT heuristic in [23], is shown in detail in Alg. 4. At every step, this calculates the expected MDP by obtaining the expected transition kernel and reward function under the current belief. It then acts according to the optimal policy with respect to $\hat{\mu}_\xi$. This policy may be much worse than the optimal policy, even within the class of stationary policies $\mathcal{P}_1$.

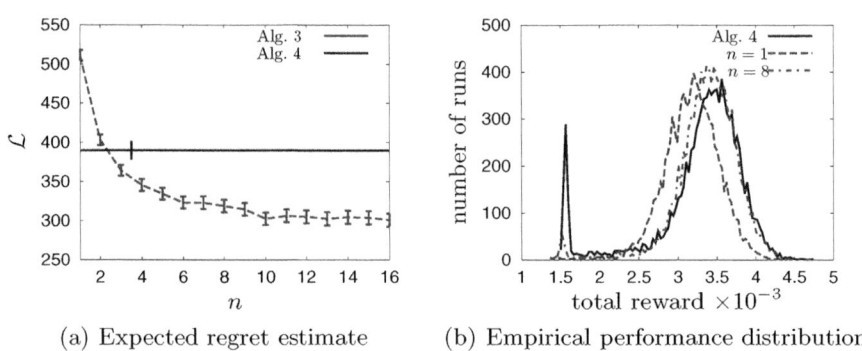

(a) Expected regret estimate   (b) Empirical performance distribution

Fig. 2. Performance on the chain task, for the first 10^3 steps, over 10^4 runs. (a): Expected regret relative to the optimal (oracle) policy. The *sampling* curve shows the regret of Alg. 3, as the number of samples increases, with 95% confidence interval calculated via a 10^4-boostrap. The *expected* curve shows the performance of an algorithm acting greedily with respect to the expected MDP. (b): Empirical distribution of total rewards for: the *expected* MDP approach and MCBRL with $n = 1$ and $n = 8$ samples.

We compared the algorithms on the Chain task [9], commonly used to evaluate exploration in reinforcement learning problems. Traditionally, the task has a horizon of 10^3 steps, a discount factor $\gamma = 0.95$, and the expected total reward $\mathbb{E}_{\mu,\pi} \sum_{t=1}^{T} r_t$ is compared. We also report the expected utility $\mathbb{E}_{\mu,\pi} U_t$, which depends on the discount factor. All quantities are estimated over 10^4 runs with appropriately seeded random number generators to reduce variance.[2] The initial

---

[2] In both cases this expectation is with respect to the distribution induced by the actual MDP μ and policy π followed, rather than with respect to the belief ξ.

Table 1. Comparative results on the chain task. The 80% percentile interval is such that no more than 10% of the runs were above the maximum or below the minimum value. The confidence interval on the accuracy of the mean estimate, is the 95% bootstrap interval. The results for BEETLE and the EM algorithms were obtained from the cited papers, with and the interval based on the reported standard deviation.

Model	$\sum_{t=1}^{1000} r_t$ ($\mathbb{E}\,U$)	80% percentile	confidence interval
Alg. 4	3287 (26.64)	2518 – 3842	3275 – 3299
$n = 1$	3166 (28.50)	2748 – 3582	3159 – 3173
$n = 8$	3358 (29.65)	2932 – 3800	3350 – 3366
$n = 16$	3376 (29.95)	2946 – 3830	3368 – 3384

Model	$\sum_{t=1}^{1000} r_t$	Standard interval
BEETLE [23]	1754	1712–1796
AMP-EM [17]	2180	2108–2254
SEM [17]	2052	2000 –2111

belief about the state transition distribution was set to be a product-Dirichlet prior [see 11] with all parameters equal to $|\mathcal{S}|^{-1}$, while a product-Beta prior with parameters $(1, 1)$ was used for the rewards.

Figure 2 summarises the results in terms of total reward. The left hand side (2(a)) shows the expected difference in total reward between the optimal policy π^* and the used policy π, over T steps, i.e. the regret: $\mathcal{L} = \mathbb{E}_{\mu,\pi} \sum_{t=1}^{T} r_t - \mathbb{E}_{\mu,\pi} \sum_{t=1}^{T} r_t$. The error bars denote 95% confidence intervals obtained via a 10^4-bootstrap [15]. For $n = 1$, MCBRL performs worse than the expected MDP approach, in terms of *total reward*. On the other hand, as the number of samples increase, its performance monotonically improves.

Some more detail on the behaviour of the algorithms is given in Figure 2(b), which shows the empirical performance distribution in terms of total reward. The expected MDP approach has a high probability of getting stuck in a sub-optimal regime. On the contrary, MCBRL, for $n = 1$, results in significant over-exploration of the environment. However, as n increases, MCBRL explores significantly less, while the number of runs where we are stuck in the sub-optimal regime remains small ($< 1\%$ of the runs). Table 1 presents comparative results on the chain task for Alg. 4 and for MCBRL for $n \in \{1, 8, 16\}$ in terms of the total reward received in 10^3 steps. This enables us to compare against the results reported in [23, 17]. While the performance of Alg. 4 may seem surprisingly good, it is actually in line with the results reported in [23]. Therein, BEETLE only outperformed Alg. 4 in the *Chain* task when stronger priors were used. In addition, we would like to note that while the case $n = 1$ is worse than Alg. 4 for the total reward metric, this no longer holds when we examine the expected utility, where an improvement can already be seen for $n = 1$.

4   Discussion

We introduced MMBI, a simple backwards induction procedure, to obtain a near-optimal memoryless policy with respect to a belief over a finite number of MDPs. This was generalised to MSBI, a stochastic procedure, whose loss is close in expectation to MMBI, with a gap that depends polynomially on the number of samples, for a belief on arbitrary set of MDPs. It is shown that MMBI results in a much tighter lower bound on the value function that the value of the $\hat{\mu}_\xi$-optimal policy. In addition, we prove a bound on the gap between the value of the MMBI policy and the Bayes-optimal policy. Our results are then applied to reinforcement learning problems, by using the MCBRL algorithm to sample a number of MDPs at regular intervals. This can be seen as a principled generalisation of [29], which only draws one sample at each such interval. Then MSBI is used to calculate a near-optimal memoryless policy within each interval. We show experimentally that this performs significantly better than following the $\hat{\mu}_\xi$-optimal policy. It is also shown that the performance increases as we make the bound tighter by increasing the number of samples taken.

Compared to results reported for other Bayesian reinforcement learning approaches on the *Chain* task, this rather simple method performs surprisingly well. This can be attributed to the fact that at each stage, the algorithm selects actions according to a nearly-optimal stationary policy.

In addition, MSBI itself could be particularly useful for *inverse* reinforcement learning problems (see for example [1, 22]) where the underlying dynamics are unknown, or to multi-task problems[26]. Then it would be possible to obtain good stationary policies that take into account the uncertainty over the dynamics, which should be better than using the expected MDP heuristic. More specifically, in future work, MMBI will be used to generalise the Bayesian methods developed in [13, 25] for the case of unknown dynamics.

In terms of direct application to reinforcement learning, MSBI could be used in the inner loop of some more sophisticated method than MCBRL. For example, it could be employed to obtain tight lower bounds for the leaf nodes of a planning tree such as[12]. By tight integration with such methods, we hope to obtain improved performance, since we would be considering wider policy classes. In a related direction, it would be interesting to see examine better upper bounds [8, 7, 24] and in particular whether the information relaxations discussed by Brown et al. [5] could be extended to the Bayes-optimal case.

Acknowledgements. Many thanks to Matthijs Snel and Shimon Whiteson for extensive discussions on the optimality of the MMBI algorithm, and for helping to discover an error. In addition, I would like to thank Nikos Vlassis and the anonymous reviewers for helpful comments. This work was partially supported by the EU-Project IM-CLeVeR, FP7-ICT-IP-231722, and the Marie Curie Project ESDEMUU, Grant Number 237816.

References

[1] Abbeel, P., Ng, A.Y.: Apprenticeship learning via inverse reinforcement learning. In: Proceedings of the 21st International Conference on Machine Learning (ICML 2004) (2004)

[2] Asmuth, J., Li, L., Littman, M.L., Nouri, A., Wingate, D.: A Bayesian sampling approach to exploration in reinforcement learning. In: UAI 2009 (2009)

[3] Auer, P., Jaksch, T., Ortner, R.: Near-optimal regret bounds for reinforcement learning. In: Proceedings of NIPS 2008 (2008)

[4] Brafman, R.I., Tennenholtz, M.: R-max-a general polynomial time algorithm for near-optimal reinforcement learning. The Journal of Machine Learning Research 3, 213–231 (2003)

[5] Brown, D.B., Smith, J.E., Sun, P.: Information relaxations and duality in stochastic dynamic programs. Operations Research 58(4), 785–801 (2010)

[6] Castro, P.S., Precup, D.: Smarter Sampling in Model-Based Bayesian Reinforcement Learning. In: Balcázar, J.L., Bonchi, F., Gionis, A., Sebag, M. (eds.) ECML PKDD 2010. LNCS, vol. 6321, pp. 200–214. Springer, Heidelberg (2010)

[7] de Farias, D.P., Van Roy, B.: The linear programming approach to approximate dynamic programming. Operations Research 51(6), 850–865 (2003)

[8] de Farias, D.P., Van Roy, B.: On constraint sampling in the linear programming approach to approximate dynamic programming. Mathematics of Operations Research 293(3), 462–478 (2004)

[9] Dearden, R., Friedman, N., Russell, S.J.: Bayesian Q-learning. In: AAAI/IAAI, pp. 761–768 (1998)

[10] Dearden, R., Friedman, N., Andre, D.: Model based Bayesian exploration. In: Laskey, K.B., Prade, H. (eds.) Proceedings of the 15th Conference on Uncertainty in Artificial Intelligence (UAI 1999), July 30-August 1, pp. 150–159. Morgan Kaufmann, San Francisco (1999)

[11] DeGroot, M.H.: Optimal Statistical Decisions. John Wiley & Sons (1970)

[12] Dimitrakakis, C.: Complexity of stochastic branch and bound methods for belief tree search in Bayesian reinforcement learning. In: 2nd International Conference on Agents and Artificial Intelligence (ICAART 2010), Valencia, Spain, pp. 259–264. ISNTICC, Springer (2009)

[13] Dimitrakakis, C., Rothkopf, C.A.: Bayesian multitask inverse reinforcement learning. In: European Workshop on Reinforcement Learning, EWRL 2011 (2011)

[14] Duff, M.O.: Optimal Learning Computational Procedures for Bayes-adaptive Markov Decision Processes. PhD thesis, University of Massachusetts at Amherst (2002)

[15] Efron, B., Tibshirani, R.J.: An Introduction to the Bootstrap. Monographs on Statistics & Applied Probability, vol. 57. Chapmann & Hall, ISBN 0412042312 (November 1993)

[16] Fard, M.M., Pineau, J.: PAC-Bayesian model selection for reinforcement learning. In: NIPS 2010 (2010)

[17] Furmston, T., Barber, D.: Variational methods for reinforcement learning. In: Teh, Y.W., Titterington, M. (eds.) Proceedings of the 13th International Conference on Artificial Intelligence and Statistics (AISTATS). JMLR: W&CP, vol. 9, pp. 241–248

[18] Gittins, C.J.: Multi-armed Bandit Allocation Indices. John Wiley & Sons, New Jersey (1989)

[19] Jacksh, T., Ortner, R., Auer, P.: Near-optimal regret bounds for reinforcement learning. Journal of Machine Learning Research 11, 1563–1600 (2010)

[20] Kaelbling, L.P.: Learning in Embedded Systems. PhD thesis, ept of Computer Science, Stanford (1990)

[21] Kearns, M., Singh, S.: Near-optimal reinforcement learning in polynomial time. In: Proc. 15th International Conf. on Machine Learning, pp. 260–268. Morgan Kaufmann, San Francisco (1998)

[22] Ng, A.Y., Russell, S.: Algorithms for inverse reinforcement learning. In: Proc. 17th International Conf. on Machine Learning, pp. 663–670. Morgan Kaufmann (2000)

[23] Poupart, P., Vlassis, N., Hoey, J., Regan, K.: An analytic solution to discrete Bayesian reinforcement learning. In: ICML 2006, pp. 697–704. ACM Press, New York (2006)

[24] Rogers, L.C.G.: Pathwise stochastic optimal control. SIAM Journal on Control and Optimization 46(3), 1116–1132 (2008)

[25] Rothkopf, C.A., Dimitrakakis, C.: Preference Elicitation and Inverse Reinforcement Learning. In: Gunopulos, D., Hofmann, T., Malerba, D., Vazirgiannis, M. (eds.) ECML PKDD 2011. LNCS, vol. 6913, pp. 34–48. Springer, Heidelberg (2011)

[26] Snel, M., Whiteson, S.: Multi-Task Reinforcement Learning: Shaping and Feature Selection. In: EWRL 2011 (2011)

[27] Strehl, A.L., Littman, M.L.: An analysis of model-based interval estimation for Markov decision processes. Journal of Computer and System Sciences 74(8), 1309–1331 (2008)

[28] Strehl, A.L., Li, L., Littman, M.L.: Reinforcement learning in finite MDPs: PAC analysis. The Journal of Machine Learning Research 10, 2413–2444 (2009)

[29] Strens, M.: A bayesian framework for reinforcement learning. In: ICML 2000, pp. 943–950. Citeseer (2000)

[30] Wang, T., Lizotte, D., Bowling, M., Schuurmans, D.: Bayesian sparse sampling for on-line reward optimization. In: ICML 2005, pp. 956–963. ACM, New York (2005)

[31] Wyatt, J.: Exploration control in reinforcement learning using optimistic model selection. In: Danyluk, A., Brodley, C. (eds.) Proceedings of the Eighteenth International Conference on Machine Learning (2001)

Optimized Look-ahead Tree Search Policies

Francis Maes, Louis Wehenkel, and Damien Ernst

University of Liège
Dept. of Electrical Engineering and Computer Science
Institut Montefiore, B28, B-4000, Liège - Belgium

Abstract. We consider in this paper look-ahead tree techniques for
the discrete-time control of a deterministic dynamical system so as to
maximize a sum of discounted rewards over an infinite time horizon.
Given the current system state x_t at time t, these techniques explore the
look-ahead tree representing possible evolutions of the system states and
rewards conditioned on subsequent actions u_t, u_{t+1}, ... When the computing budget is exhausted, they output the action u_t that led to the
best found sequence of discounted rewards. In this context, we are interested in computing good strategies for exploring the look-ahead tree. We
propose a generic approach that looks for such strategies by solving an
optimization problem whose objective is to compute a (budget compliant) tree-exploration strategy yielding a control policy maximizing the
average return over a postulated set of initial states.

This generic approach is fully specified to the case where the space of
candidate tree-exploration strategies are "best-first" strategies parameterized by a linear combination of look-ahead path features – some of
them having been advocated in the literature before – and where the optimization problem is solved by using an EDA-algorithm based on Gaussian distributions. Numerical experiments carried out on a model of the
treatment of the HIV infection show that the optimized tree-exploration
strategy is orders of magnitudes better than the previously advocated
ones.

Keywords: Real-time Control, Look-ahead Tree Search, Estimation of
Distribution Algorithms.

1 Introduction

Many interesting problems in the field of engineering and robotics can be casted
as optimal control problems for which one seeks to find policies optimising a
sum of discounted rewards over an infinite time horizon. Among the various
approaches that have been proposed in the literature to solve these problems
when both the system dynamics and the reward function are assumed to be
known, a quite recent one has caught our attention. In this approach, which
was first published in [4], a new type of policy relying on look-ahead trees was
proposed. The key idea in this paper is to explore, at any control opportunity t,
in an optimistic way the look-ahead tree starting from the current state x_t and
whose branchings are determined by the possible sequences of control actions

S. Sanner and M. Hutter (Eds.): EWRL 2011, LNCS 7188, pp. 189–200, 2012.

$u_t, u_{t+1}, \ldots$. More specifically, this strategy expands at every iteration a leaf of the tree for which the discounted sum of rewards collected from the root to this leaf (u-value) plus an optimistic estimation of the future discounted rewards is maximal, and when the computing budget is exhausted it returns the action u_t (or the sequence of actions $u_t, u_{t+1}, \ldots$) leading to the leaf with the best u-value. The results reported by the authors are impressing: with rather small trees they managed to control in an efficient way quite complex systems.

We tested this approach on some even more complex benchmark problems than those considered in [4], and the results were still excellent. This raised the following question in our mind: while this *problem independent* tree-exploration strategy yields excellent results, would it be possible to obtain better results by computing automatically a *problem dependent* look-ahead tree-exploration strategy? From there came the idea of looking within a broader family of tree-exploration strategies for a best one for a given optimal control problem and a given real-time computing budget. This led us to the main contribution presented in this paper, namely the casting of the search for an optimal tree-exploration strategy as an optimization task exploiting prior knowledge about the optimal control problem and computing budget available for real-time control.

The present work has also been influenced by other authors which have sought to identify good tree/graph exploration strategies for other decision problems than the one tackled here but for which the solution could also be identified in a computationally efficient way by exploring in a clever way a tree/graph structure. One of the most seminal works in this field is the A^* algorithm [3] that uses a best-first search to find a shortest path towards a goal state. In A^*, the function used for evaluating the nodes is the sum of two terms: the length of the so far shortest path towards the current node and an optimistic estimate of the shortest path from this node to a goal state (a so-called "admissible" heuristic). Note the close connection that exists between the optimistic planning algorithm of [4] and the A^* algorithm even if they apply to different types of planning problems. Several authors have sought to learn good admissible heuristics for the A^* algorithm. For example, we can mention the LRTA* algorithm [5] which is a variant of A^* which is able to learn over multiple trials an optimal admissible heuristic. It has been shown later on, in the work on real-time dynamic programming from Barto and al. [1], that LRTA* has a behaviour similar to the asynchronous value iteration algorithm. More recent works for learning strategies to efficiently explore graphs have focused on the use of supervised regression techniques using various approximation structures to solve this problem (e.g., linear regression, neural networks, k-nearest neighbours) [7,8,10]). To position the contribution of this paper with respect to this large body of work, we may say that, to the best of our knowledge, this paper is the first one where the search for a tree-exploration based control strategy is explicitly casted as an optimization problem which is afterwards solved using a derivative-free optimisation algorithm.

The rest of this paper is organized as follows. In Section 2, we specify the class of control problems that we consider and present the look-ahead tree-exploration approach in this context. Section 3 presents our framework for optimizing look-ahead tree-exploration strategies and fully specifies an algorithm for solving this problem in a near-optimal way. This algorithm is compared in Section 4 with several other tree-exploration methods. Finally, Section 5 concludes and presents further research directions.

2   Problem Formulation

We start this section by describing the type of optimal control problem that will be considered throughout this paper. Afterwards, we describe look-ahead tree exploration based control policies as well as the specific problem we address.

2.1   Optimal Control Problem

We consider the general class of deterministic time-invariant dynamical systems whose dynamics is described (in discrete-time) by:

$$x_{t+1} = f(x_t, u_t)   t = 0, 1, \ldots \tag{1}$$

where for all t, the state x_t is an element of the state space X and the action u_t is an element of the action space U. We assume in this paper that U is finite (and not too large) but we do not restrict X.

To the transition from t to $t+1$ is associated an instantaneous reward $\rho(x_t, u_t) \in \mathbb{R}^+$ and, for every initial state x_0, we define the (infinite horizon, discounted) return of a (stationary) control policy $\mu : X \to U$ as:

$$J^\mu(x_0) = \lim_{T \to \infty} \sum_{t=0}^{T} \gamma^t \rho(x_t, u_t) \tag{2}$$

subject to $u_t = \mu(x_t)$ and $x_{t+1} = f(x_t, u_t)$ $\forall t > 0$, and where $\gamma \in [0, 1[$ is a discount factor. Within the set $\mathcal{U}$ of stationary policies, we define optimal policies $\mu^* \in \mathcal{U}$ such that, for every initial state $x_0 \in X$, we have:

$$J^{\mu^*}(x_0) \geq J^\mu(x_0), \forall \mu \in \mathcal{U}. \tag{3}$$

In this paper, we are interested in finding a good approximation of an optimal policy in the sense that its return is close to J^{μ^*}.

2.2   Look-ahead Tree Exploration Based Control Policies

The control policies considered in this paper are the combination of two components: an *exploration* component that exploits the knowledge of f and ρ to generate a look-ahead tree representing the evolution of the system states and the rewards observed given various sequences of actions $u_t, \ldots, u_{t+n}$, and a *decision* component that exploits this information to select the action u_t. The concept of look-head tree is represented on Figure 1.

In the following, we denote look-ahead trees by τ. A look-ahead tree is composed of nodes $n \in \tau$ that are triplets (u_{t-1}, r_{t-1}, x_t) where $u_{t-1} \in U$ is an action whose application on the parent's node state x_{t-1} leads to the reward

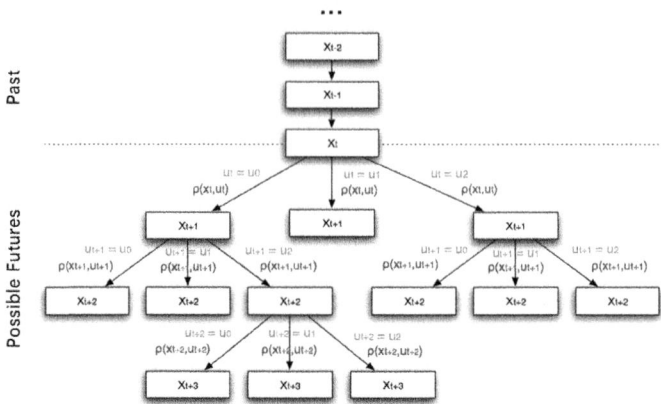

Fig. 1. Example of look-ahead tree in a system with three possible control actions. Each node of the tree is associated to an action u, a reward value ρ and a state x.

$r_{t-1} = \rho(x_{t-1}, u_{t-1})$ and the successor state $x_t = f(x_{t-1}, u_{t-1})$. We denote by $parent(n)$ the parent node of $n \in \tau$. When clear from the context, we use $r(n)$ (*resp.* $x(n)$) to denote the reward (*resp.* state) attached to node n. Internal nodes of a look-ahead tree are said to be *expanded (or closed)* while its leaves are *open*. An expansion of an open node consists in adding to this node a number $|U|$ of children corresponding to the different actions u that can be applied to the state x of this node. We use $path(n)$ to denote the complete sequence of nodes from the root node n_0 to a node n at depth $d(n)$. Such a path corresponds to a trajectory $(x_t, u_t, r_t, x_{t+1}, \ldots, u_{t+d(n)-1}, r_{t+d(n)-1}, x_{t+d(n)})$ of the system.

A typical example of look-ahead tree search algorithm is the "uniform tree search" (or breadth-first) algorithm, that first develops a tree representing the evolutions of the system for all sequences of actions of length d, and then selects as action u_t the first action of a path of the tree along which the highest sum of discounted cumulated rewards ($\sum_{t'=t}^{t+d-1} \gamma^{t'} r_{t'}$) is collected. The computing budget necessary for developing a uniform tree of depth d is $\mathcal{O}(|U|^d)$, which may often lead to the case where for a given computational budget the tree cannot be developed up to a depth which is sufficient to obtain a well performing policy. This is often true in the context of real-time control of a system and is a reason for looking for other types of look-ahead tree search policies.

2.3   Budget Constrained Path-Scoring Based Tree Exploration

From now on, we will mainly focus on a particular subclass of look-ahead tree search policies which is easy to implement, works well in an anytime setting and is sufficiently rich. Every policy of this subclass is fully defined by a scoring function $h(\cdot)$ (with values $\in \mathbb{R}$) that assigns scores to look-ahead search tree paths. These policies develop the tree incrementally in such a way that they always fully expand the leaf whose corresponding path has the largest score.

Input: the current state of the system x_t

Input: path scoring function $h(\cdot) \in \mathbb{R}$

Output: an action u_t

Set $n_0 = (\emptyset, \emptyset, x_t)$ *(root node)*

Set $\tau = \{n_0\}$ *(initial tree)*

Set $v(n_0) = 0$

repeat

    Select a terminal node $n = \operatorname{argmax} v(n)$

    foreach $u \in U$ **do**

        Add child node $n' = \big(u, \rho(x(n), u), f(x(n), u)\big)$ to n

        Set $v(n') = h(path(n'))$

    end

until *computational budget is exhausted*

return *the first action of a path towads an open node n and that maximizes the discounted cumulated rewards $\sum_{t'=t}^{t+d(n)-1} \gamma^{t'-t} r_{t'}$ (u-value)*

Algorithm 1. Generic budget constrained tree exploration based control

They stop the tree development when the computational budget is exhausted, in which case they output as action u_t, the first action taken along a path in the tree along which the highest sum of discounted rewards (u-value) is observed. A generic tabular representation of this path-scoring look-ahead tree control policy is given by Algorithm 1, where $v(n)$ denotes the score associated to node n using the scoring function $h(\cdot)$.

Note that the uniform tree search algorithm discussed previously is a particular case of Algorithm 1, using heuristic $h^{mindepth}(path(n)) = -d(n)$. The approach proposed by [4] for problems with bounded rewards is also a particular case of path-scoring based look-ahead tree search with heuristic $h^{optimistic}(path(n)) = \sum_{t'=t}^{t+d(n)-1} \gamma^{t'-t} r_{t'}(n) + B_r \frac{\gamma^d}{1-\gamma}$, where $\{r_{t'}(n)\}_{t'=t}^{t+d(n)-1}$ are the rewards collected along the path leading to n.

3   Optimized Look-ahead Tree Exploration Based Control

In this section we present a generic approach for constructing optimal budget constrained look-ahead tree exploration based control policies, and we then describe the fully specified instance of this approach that is tested in Section 4.

3.1   Generic Optimized Look-ahead Tree Exploration Algorithm

Let $\mu^h(x)$ denote the look-ahead control policy described by Algorithm 1 with a given scoring function $h \in \mathcal{H}$. For identifying a good scoring function $h(\cdot)$, we propose to proceed as follows. First, we define a rich enough set of candidate scoring functions $\mathcal{H}$. Second, we select a subset X_0 of X in a way that it "covers well" the area from which the system is likely to be controlled. Finally we pose the following optimization problem:

$$\operatorname*{argmax}_{h \in \mathcal{H}} R_{X_0}(\mu^h)$$

where $R_{X_0}(\mu^h)$ is the mean return of policy μ^h when starting from states in X_0, i.e. $R_{X_0}(\mu^h) = \frac{1}{|X_0|} \sum_{x_0 \in X_0} J^{\mu^h}(x_0)$. An algorithm solving this problem yields optimized look-ahead tree policies adjusted to the problem at hand (dynamics, rewards, and budget constraints).

3.2   A Particular Instance

We now instantiate our generic approach for the particular case of linear functions of path features optimized by an estimation of distribution algorithm.

Set of candidate path scoring functions. In our study, we consider a generic set of candidate path scoring functions $\mathcal{H}^\phi$ that are using a vector function $\phi(\cdot) \in \mathbb{R}^p$ of *path features*. Such features are chosen beforehand and they may describe any aspect of a path such as cumulated rewards, depth or states. In the next section, we give an example of a generic $\phi(\cdot)$ that can be applied to any optimal control problem with continuous state variables.

Scoring functions $h_\theta^\phi \in \mathcal{H}^\phi$ are defined as parameterized linear functions: $h_\theta^\phi(path(n)) = \langle \theta, \phi(path(n)) \rangle$, where $\theta \in \mathbb{R}^p$ is a vector of parameters and $\langle \cdot, \cdot \rangle$ is the dot-product operator. Each value of θ defines a candidate tree-exploration strategy h_θ^ϕ; we thus have $\mathcal{H}^\phi = \{h_\theta^\phi\}_{\theta \in \mathbb{R}^p}$.

Optimization problem. Given our candidate set of tree-exploration strategies, the optimization problem can be reformulated as follows:

$$\operatorname*{argmax}_{\theta \in \mathbb{R}^p} R_{X_0}(\theta) \text{ with } R_{X_0}(\theta) = \frac{1}{|X_0|} \sum_{x_0 \in X_0} J^{\mu^{h_\theta^\phi}}(x_0) \tag{4}$$

Since $J^{\mu^{h_\theta^\phi}}(x_0)$ involves an infinite sum, it has to be approximated. This can be done with arbitrary precision by truncating the sum with a sufficiently large *horizon* limit. Given a large enough horizon H, we hence approximate it by:

$$J^\mu(x_0) \approx \sum_{t=0}^{H-1} \gamma^t \rho(x_t, u_t) | u_t = \mu(x_t), x_{t+1} = f(x_t, u_t). \tag{5}$$

Optimization algorithm. To solve Equation 4, we suggest the use of derivative-free global optimization algorithms, such as those provided by meta-heuristics. In this work, we used a powerful, yet simple, class of metaheuristics known as *Estimation of Distribution Algorithms* (EDA) [6]. EDAs rely on a probabilistic model to describe promising regions of the search space and to sample good candidate solutions. This is performed by repeating iterations that first *sample* a population of N candidates using the *current* probabilistic model and then *fit* a *new* probabilistic model given the $b < N$ best candidates. Any kind of probabilistic model may be used inside an EDA. The simplest form of EDAs uses one marginal distribution per variable to optimize and is known as the *univariate marginal distribution algorithm* [9]. We have adopted this approach that, although simple, proves to be quite effective to solve Equation 4.

Our EDA algorithm proceeds as follows. There is one Normal distribution per parameter $f \in \{1, 2, \ldots, p\}$. At first iteration, there distributions are initialized as standard Normal distributions. At each iteration, N candidates are sampled using the current distributions and evaluated. At the end of the iteration, the p distributions are re-estimated using the $b < N$ best candidates of current iteration. The policy that is returned corresponds to the θ parameters that led to the highest observed value of $R_{X_0}(\theta)$ after a number i_{max} of EDA iterations.

4   Experiments

We describe numerical experiments comparing optimized look-ahead tree policies against previously proposed look-ahead tree policies on a toy problem and on a much more challenging one that is of some interest to the medical community.

4.1   Path Features Function

Our optimization approach relies on a path feature function $\phi(\cdot) \in \mathbb{R}^p$, that, given a path in the look-ahead search tree, outputs a vector of features describing this path. The primary goal of the $\phi(\cdot)$ function is to project paths of varying length into a fixed-dimension description. Both global properties (e.g. depth, cumulated rewards, actions histogram) and local properties (e.g. state contained in the terminal node, last actions) may be used inside $\phi(\cdot)$.

We propose below a particular $\phi(\cdot)$ function that can be used for any control problem with continuous state variables, *i.e.* for which $X \subset \mathbb{R}^m$. Despite its simplicity, this function lead to optimized look-ahead tree policies performing remarkably well in our test-beds.

Our feature function is motivated by the fact that two highly relevant variables to look-ahead tree search are *depth* and *reward*. In a sense, penalizing or favoring *depth* enables to trade-off between breadth-first and depth-first search and acting on *reward* enables to trade-off between greedy search and randomized search. Since the best way to perform search may crucially depend on the current state variables, we propose to combine *depth* and *reward* with *state* variables. This is performed in the following way:

$$\phi^{simple}(n_0, \ldots, n_d) = \big(x_1(n_d), \ldots, x_m(n_d),$$
$$dx_1(n_d), \ldots, dx_m(n_d),$$
$$r(n_d)x_1(n_d) \ldots, r(n_d)x_m(n_d)\big)$$

where d is the depth, *i.e.* the length of path $n_0, \ldots, n_d$, and $r(n_d)$ is the last reward perceived on the path $n_0, \ldots, n_d$. If the dimension of state space is m, ϕ^{simple} computes vectors of $p = 3m$ features. Note that these features mostly depend on the leaf n_d and only capture global information of the path through the depth d. Of course, several other feature functions that better exploit global information of the path could be used, but ϕ^{simple} already provides a basis to construct very well performing policies as shown below.

4.2   Baselines and Parameters

In the following, we consider the look-ahead tree policies defined by the following path scoring tree-exploration strategies:

$$h^{mindepth}(n_0, \ldots, n_d) = -d \qquad h^{optimistic}(n_0, \ldots, n_d) = \sum_{i=0}^{d} \gamma^i r(n_i) + B_r \frac{\gamma^d}{1 - \gamma}$$

$$h^{greedy1}(n_0, \ldots, n_d) = r(n_d) \qquad h^{greedy2}(n_0, \ldots, n_d) = \gamma^d r(n_d)$$

$$h^{optimized}(n_0, \ldots, n_d) = \left\langle \theta^*, \phi^{simple}(n_0, \ldots, n_d) \right\rangle$$

$h^{mindepth}$ corresponds to *uniform tree search* (a.k.a. breadth-first search). $h^{optimistic}$ is the heuristic proposed by [4]. $h^{greedy1}$ and $h^{greedy2}$ are greedy look-ahead tree exploration strategies *w.r.t.* immediate rewards.

To find the θ^* parameter, we use the EDA algorithm with $i_{max} = 50$ iterations, $N = 100$ candidates per iteration and $b = 10$ best candidates; one full optimization thus involves 5000 look-ahead tree exploration strategy evaluations.

4.3   Synthetic Problem

To compare optimized look-ahead tree policies with previously proposed ones, we adopt the synthetic optimal control problem also used in [4]. In this problem, a state is composed of a position y and a velocity v and there are two possible control actions: $U = \{-1, +1\}$. The dynamics are as follows:

$$(y_{t+1}, v_{t+1}) = (y_t, v_t) + (v_t, u_t)\Delta t \qquad \rho((y_t, v_t), u_t) = \max(1 - y_{t+1}^2, 0)$$

where $\Delta t = 0.1$ is the time discretization step. In this problem, the reward is obviously upper bounded by $B_r = 1$, and this bound is therefore used in the optimistic heuristic in our simulations, as in [4].

The set of initial states X_0 used for optimizing the exploration policy is composed of 10 initial states randomly sampled from the domain $[-1, 1] \times [-2, 2]$, and the discount factor is $\gamma = 0.9$. The evaluation of the average return by the EDA algorithm is truncated at horizon $H = 50$.

We are interested in the mean return values $R_{X_0}(\mu^h)$ for different amounts of computational resources. The *budget* is the number of node expansions allowed to take one decision. For a given budget B, the policy expands B nodes and computes $B|U|$ transitions and values for the nodes. As in [4], we consider values $B = 2^{d+1} - 1$ with $d \in \{1, \ldots, 18\}$, corresponding to complete trees of depth d. We optimize one tree-exploration strategy per possible value of budget B, *i.e.* tree-exploration strategies are specific to a given computational budget.

The results are reported on the left part of Figure 2. It can be seen that the optimized tree-exploration strategy performs significantly better on X_0 than all other tree-exploration strategies, in the whole range of budget values. In particular, it reaches a mean discounted return of 7.198 with a budget $B = 63$, which is still better than the return of 7.179 of the best-performing tree-exploration strategies with the maximal budget $B = 524287$. In other words,

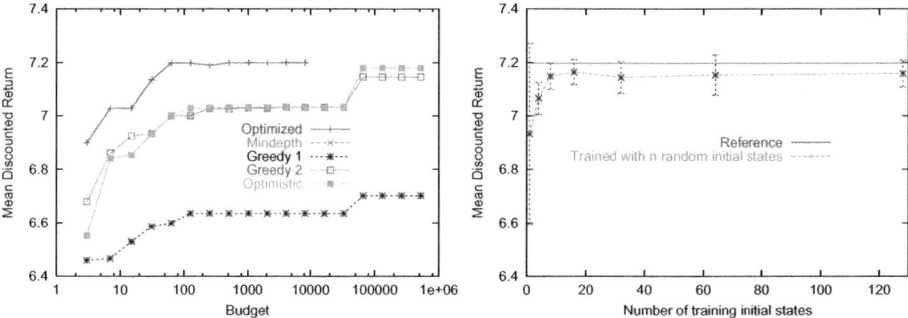

Fig. 2. Left: mean discounted return of five look-ahead tree policies given different amounts of computational resources. Right: Comparison of the tree-exploration strategy optimized on X_0 with the one optimized on randomly sampled initial states.

with a budget $B = 63$, the optimized tree-exploration strategy outperforms exhaustive breadth-first search with a maximal depth of 18.

As an example, the strategy optimized for a budget $B = 63$ is equal to:

$$h^{optimized-63}(n_0, \ldots, n_d) = x_1(n_d)(0.3249 + 1.3368r(n_d) - 2.9695d)$$
$$+ x_2(n_d)(0.9078 - 0.2566r(n_d) + 0.4561d),$$

where $x_1(n_d)$ is the current position and $x_2(n_d)$ is the current velocity.

Notice that for the moment, we evaluated the optimized tree-exploration strategy on the same sample X_0 that was used for optimization. These results are thus "in-sample" results for our method. In order to assess out-of sample behavior and robustness of our algorithm, we performed an additional set of experiments reported on the right of Figure 2. In these experiments, we optimized our strategy with training samples X_0' of growing sizes n and generated independently from the sample X_0 used for the purpose of evaluating the average return (the same as the one in the left part of the figure). We carried out these evaluations by averaging the results over 10 runs for a computing budget $B = 63$ and report the mean and standard deviation of the average returns. When optimized with samples X_0' and evaluated on X_0, the optimized strategy is slightly inferior to the one obtained when optimizing on X_0, but it still outperforms all the other tree-exploration strategies (for all the runs when $n \geq 4$).

4.4   HIV Infection Control

We now consider the challenging problem described in [2]. The aim is to control the treatment of a simulated HIV infection. Prevalent HIV treatment strategies involve two types of drugs that will generically be called here "drug 1" and "drug 2". The negative side effects of these drugs in the long term motivate the investigation of optimal strategies for their use. The problem is represented by a six-dimensional nonlinear model and has four actions: "no drugs", "drug 1", "drug 2", "both drugs". The system is controlled with a sampling time of 5 days and we seek for an optimal strategy over a horizon of a few years.

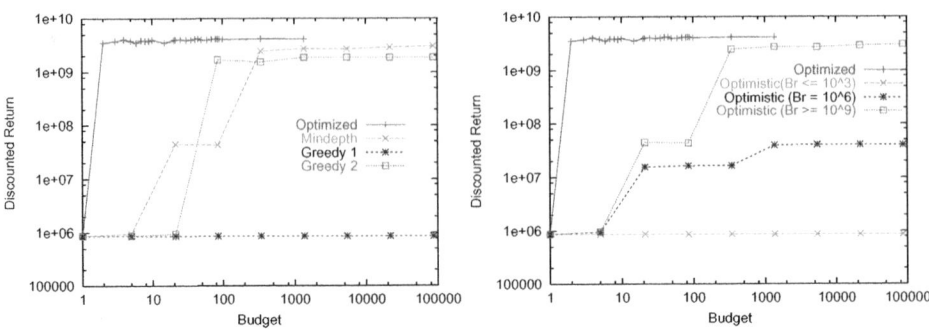

Fig. 3. Left: Discounted returns when starting from the *unhealthy* initial state given different amounts of computational resources. Right: Comparison of the optimized tree-exploration strategy with various optimistic heuristics.

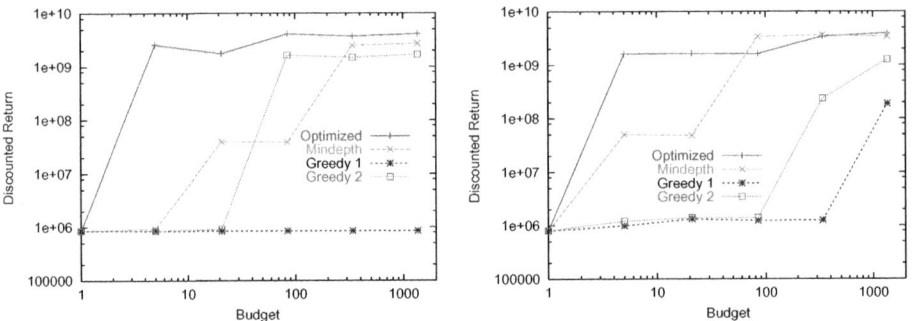

Fig. 4. Discounted returns when starting from two different initial states

We consider a single initial state $x_0 = (163573, 5, 11945, 46, 63919, 24)$ which corresponds to an *unhealthy* stable equilibrium and which represents the state of a patient with a very low immune response. The maximum horizon is $H = 300$ and the discount factor is $\gamma = 0.98$.

We use the same heuristics as previously except $h^{optimistic}$ which cannot be applied directly since no exact bound on the reward is known for this problem. Since state and reward variables may have very large numerical values (these variables typically range from 10^0 to 10^6), we slightly modified ϕ^{simple} by applying the $log(\cdot)$ function on state variables $x_d(n_d)$ and on rewards $r(n_d)$. Except for this difference, we used the same parameters as in the previous example.

Figure 3 reports the results for growing budget values B. Since optimized look-ahead tree policies worked very well for small values of B, we performed a more fine sampling of its performance for small values of B. The gain of using optimized look-ahead tree policies is striking on this domain: with a budget of only $B = 2$, the optimized tree-exploration strategy outperforms the best heuristic (including uniform tree search) with $B = 87381$ (3.48e9 against 3.02e9). Note that, the root node (corresponding to current state) being always expanded first, the policy with $B = 2$ only chooses which node to expand at the second step, based on the successor states $f(x_t, u_t)$ and on the immediate rewards $\rho(x_t, u_t)$. Therefore, in this case, the depth d is not relevant in the feature function.

Besides the particular case where $B = 2$, the optimized tree-exploration strategy significantly outperforms all other strategies on the whole range of budget values. With $B = 85$, the optimized policy gives similar results to the best results reported by [2] with a reinforcement learning approach (4.13e9 *v.s.* 4.16e9).

To analyze the behavior of an "optimistic" approach on this problem, since B_r is not known exactly, we have tried the $h^{optimistic}$ heuristic with all values $B_r = 10^k$ with $k \in \{-9, -6, -3, \ldots, 12, 15, 18\}$. The scores of these heuristics are reported on the right of Figure 3. We observed three different behaviors. When $B_r <= 10^3$, the optimistic term is very small and the heuristic acts as $h^{greedy1}$. When $B_r >= 10^9$, the optimistic term takes all the importance and the heuristic acts as $h^{mindepth}$. When $B_r = 10^6$, we obtain an intermediate heuristic.

Since we optimized the tree-exploration strategy from a single initial state, it is of interest to assess to what extent it generalizes well when starting from other initial states. The left part of Figure 4 reports the returns when starting from an initial state which is close to the one used during optimization: $(163500, 4, 12000, 50, 63000, 20)$. The results here are quite similar to those of Figure 3, which shows that small differences in the initial state do not significantly degrade the quality of the optimized strategy on this problem. To experiment the robustness of our approach, we then tried with a completely different initial state that corresponds to a healthy patient that just got infected by the virus: $(10^6, 3198, 0, 0, 1, 10)$. Although this initial state is totally different from the one that was used during optimization, our optimized strategy still works reasonably well and nearly always outperforms the other tree-exploration strategies.

5   Conclusion and Further Work

In this paper, we have proposed a new approach to design look-ahead tree search policies for optimally controlling a system over an infinite horizon. It departs from the existing techniques by the fact that it optimizes the tree exploration technique in an off-line stage, by tuning it to the system dynamics, the reward function, and to the on-line computing budget. The resulting optimized look-ahead tree policy has been tested on two-benchmark problems and simulation results show that for similar or better performances it requires significantly less on-line computational resources than other look-ahead policies.

While the optimized look-ahead tree exploration strategies tested in this paper where performing very well, we believe that there is still room for improving their performances. This could be done for example by using other search spaces for candidate tree-exploration algorithms as well as other optimization techniques for looking for the best one. In particular, we conjecture that using incremental tree-exploration algorithms which do not score anymore a terminal node based only on the information contained in the path connecting this node to the top one, may be a more relevant technical choice.

We have assumed in this paper that we were dealing with optimal control problems having a finite, and not too large, action space. However, many interesting problems have very large or even continuous action spaces for which it is not possible for look-ahead tree policies to develop a node so that it has successors for every possible action. One way to extend our approach to such a setting would be to use a set of candidate tree-exploration algorithms that do not only point to the nodes that should be preferably developed but also to a subset of control actions that should be used to expand these nodes.

While we have in this paper considered a deterministic and fully observable setting, related types of policies have also been proposed in the literature for stochastic, adversarial and/or partially observable settings many of them belonging to the class of Monte-Carlo tree search techniques. A key issue for these techniques to work well is to have good tree-exploration strategies. Investigating whether the systematic approach proposed in this paper for designing such strategies could be used in such settings would be very relevant.

Acknowledgements. Damien Ernst acknowledges the financial support of the Belgian National Fund of Scientific Research (FNRS) of which he is a Research Associate. This paper presents research results of the European excellence network PASCAL2 and of the Belgian Network BIOMAGNET, funded by the Interuniversity Attraction Poles Programme, initiated by the Belgian State, Science Policy Office.

References

1. Barto, A.G., Bradtke, S.J., Singh, S.P.: Learning to act using real-time dynamic programming. Artificial Intelligence 72, 81–138 (1995)
2. Ernst, D., Stan, G., Goncalves, J., Wehenkel, L.: Clinical data based optimal STI strategies for HIV; a reinforcement learning approach. In: Proceedings of the 45th IEEE Conference on Decision and Control (2006)
3. Hart, P., Nilsson, N., Raphael, B.: A Formal Basis for the Heuristic Determination of Minimum Cost Paths. IEEE Transactions on Systems Science and Cybernetics 4(2), 100–107 (1968)
4. Hren, J.-F., Munos, R.: Optimistic Planning of Deterministic Systems. In: Girgin, S., Loth, M., Munos, R., Preux, P., Ryabko, D. (eds.) EWRL 2008. LNCS (LNAI), vol. 5323, pp. 151–164. Springer, Heidelberg (2008)
5. Korf, R.E.: Real-time heuristic search. Artificial Intelligence 42(2-3), 189–211 (1990)
6. Lozano, J.A., Larra Naga, P.: Estimation of Distribution Algorithms. A New Tool for Evolutionary Computation, pp. 99–124. Kluwer Academic Publishers (2002)
7. Maes, F.: Learning in Markov Decision Processes for Structured Prediction. PhD thesis, Pierre and Marie Curie University, Computer Science Laboratory of Paris 6 (LIP6) (October 2009)
8. Minton, S.: Machine Learning Methods for Planning. Morgan Kaufmann Publishers Inc., San Francisco (1994)
9. Pelikan, M., Mühlenbein, H.: Marginal distributions in evolutionary algorithms. In: Proceedings of the International Conference on Genetic Algorithms Mendel 1998, Brno, Czech Republic, pp. 90–95 (1998)
10. Yoon, S.W., Fern, A., Givan, R.: Learning heuristic functions from relaxed plans. In: International Conference on Automated Planning and Scheduling (ICAPS 2006), pp. 162–171 (2006)

A Framework for Computing Bounds
for the Return of a Policy

Cosmin Păduraru, Doina Precup, and Joelle Pineau

McGill University, School of Computer Science, Montreal, Canada

Abstract. We present a framework for computing bounds for the return of a policy in finite-horizon, continuous-state Markov Decision Processes with bounded state transitions. The state transition bounds can be based on either prior knowledge alone, or on a combination of prior knowledge and data. Our framework uses a piecewise-constant representation of the return bounds and a backwards iteration process. We instantiate this framework for a previously investigated type of prior knowledge – namely, Lipschitz continuity of the transition function. In this context, we show that the existing bounds of Fonteneau et al. (2009, 2010) can be expressed as a particular instantiation of our framework, by bounding the immediate rewards using Lipschitz continuity and choosing a particular form for the regions in the piecewise-constant representation. We also show how different instantiations of our framework can improve upon their bounds.

1 Introduction

We are interested in reinforcement learning (RL) problems that arise in domains where guaranteeing a certain minimum level of performance is important. In domains such as medicine, engineering, or finance, there are often large material or ethical costs associated with executing a policy that performs poorly. In order for RL to have a positive impact on these areas, it is therefore necessary to find ways to guarantee that the policy produced by an RL algorithm will achieve a certain level of performance.

The performance of a policy is measured via its *return*, the (possibly discounted) sum of rewards obtained when executing the policy. The return is a random variable, and its value can change from one execution to the next depending on the stochasticity in the environment. In order to establish guarantees for the performance of a policy, many researchers (e.g., Brunskill et al. (2008) or Kakade et al., (2003)) have developed bounds on the *value function*, which is the expected value of the return.

While it is undoubtedly useful to have information about a policy's value function, there may well be problems where we want to know more than the average performance – for instance, problems where the policy will only be executed a small number of times. In such problems, it can be useful to have a worst-case analysis for the return that the policy may obtain during its execution. This paper presents a framework for performing such an analysis.

Our framework transforms guarantees about the effect of individual transitions into guarantees about the return. For this, we require a human expert to either provide information about the range of each transition's outcomes, or to provide some regularity conditions that can then be used together with transition data to infer these ranges. For

S. Sanner and M. Hutter (Eds.): EWRL 2011, LNCS 7188, pp. 201–212, 2012.

infinite state spaces, this prior knowledge would obviously have to be expressed in a compact form. We also require information about the range of immediate rewards that can be obtained from states in a certain region. Our framework transforms the information about the transition and reward function into bounds on the policy's return.

The framework can handle either discrete or continuous state spaces. In order to handle continuous state spaces, we use a piecewise-constant representation for the bounds. Other representations could have been chosen, but the piecewise-constant representation has the advantage that the required prior knowledge can be expressed in a relatively simple manner - an important feature when working with human experts.

We show how the framework can be instantiated for a particular type of prior knowledge (Lipschitz continuity of the deterministic transition model) that has already been investigated by Fonteneau et al. (2009, 2010). The fact that transition and reward bounds can simply be plugged into our framework, rather than relying on a less flexible closed-form solution, allows us to identify algorithmic choices that can perform better in practice than the bounds of Fonteneau et al. We also show empirically that, given accurate knowledge of the one-step transition and reward bounds, our framework can be used to produce informative return bounds in a stochastic domain.

2   Framework Description

We are concerned with decision-making problems where $S \subset \mathbb{R}^d$ is the set of states, $A \subset \mathbb{R}$ is the set of actions, $f : S \times A \to S$ is the transition function and $r : S \times A \to \mathbb{R}$ is the (deterministic) reward function. The transition function can be probabilistic, in which case $f(s,a)$ is a random variable. Our goal is to compute lower bounds[1] for the return obtained by starting in some state x_0 and following the deterministic[2] policy $\pi : S \to A$ for K steps, where the return is defined as

$$R_K^\pi(x_0) = \sum_{k=0}^{K-1} \gamma^k r(x_k, \pi(x_k)) \tag{1}$$

s.t. $\forall k > 0, x_k \sim f(x_{k-1}, \pi(x_{k-1}))$, where $\gamma \in (0,1]$ is a discount factor.

Our algorithmic framework involves discretizing the state space and maintaining a lower bound for each of the resulting regions. The bounds are computed by iterating backwards. At each stage $k \in \{1, \ldots K\}$ we update $\hat{R}_k^\pi(\omega)$, the bound on the k-step return for region ω, based on the sum of two quantities: a lower bound on the rewards in ω, and the smallest of all the bounds for the regions that states in ω can transition to when π is taken. We allow for different discretizations for different horizons, and denote by Ω_k the set of regions for horizon k.

---

[1] We could easily extend our framework to compute upper bounds as well; while this may also have interesting applications, such as informing exploration strategies, we focus on lower bounds for now.

[2] Although we assume deterministic rewards and policies, these are not major restrictions. We can always add a state variable for probabilistic rewards, and make the reward function a deterministic function of that variable. In addition, the policies produced by many reinforcement learning methods are deterministic, and policies of interest in the target domains we envision are also usually deterministic.

Before the formal presentation of the algorithm, we note that there are two assumptions that are central to this work. The **first assumption** is that we have a method for computing, given any state $s \in S$, a region $f^\pi(s)$ such that either $f(s, \pi(s)) \in f^\pi(s)$, if f is deterministic, or the support of $f(s, \pi(s))$ is included in $f^\pi(s)$, if f is stochastic. Any choice of f^π that respects this condition can be used, but the smaller the region produced, the tighter the final bounds will be. We will discuss possible methods for computing f^π further in the paper.

Our **second assumption** is that we have a way to lower bound the immediate reward $r(x, \pi(x))$ and the next-state bound $\inf_{\{\omega' \in \Omega_{k-1} | \omega' \cap f^\pi(x) \neq 0\}} \hat{R}_k^\pi(\omega')$ over $x \in \omega$, for any k and any $\omega \subset S$. This is a general assumption, which encompasses many types of domain knowledge. For instance, if the reward function is differentiable with a finite number of local minima, and all the zeros of the derivative can be computed (as in the puddle world experiments), it can be minimized analytically. In contrast, Section 3 discusses how these functions are minimized when the dynamics are Lipschitz continuous and a batch of transitions are available. We will discuss this assumption further in Section 5.

We are now ready to describe our algorithmic framework. We will use the following recursive algorithm for computing a lower bound for R_K^π:

Algorithm 1. *Recursive lower bound for the K-step return*

1. For each $\omega \in \Omega_0$, let $\hat{R}_0^\pi(\omega) = \inf_{x \in \omega} r(x, \pi(x))$
2. Loop through $k = 1, \ldots K$
   For each $\omega \in \Omega_k$, compute the lower bound $\hat{R}_k^\pi(\omega)$ using

$$\hat{R}_k^\pi(\omega) = \inf_{x \in \omega} r(x, \pi(x)) + \gamma \inf_{x \in \omega} \left[\inf_{\{\omega' \in \Omega_{k-1} | \omega' \cap f^\pi(x) \neq 0\}} \hat{R}_{k-1}^\pi(\omega') \right]$$

The following result shows that Algorithm 1 produces valid lower bounds for the K-step return.

Theorem 1. *For any $k = 0, \ldots K$, any $\omega \in \Omega_k$, and any $s \in \omega$ we have that*

$$\hat{R}_k^\pi(\omega) \leq R_k^\pi(s)$$

Proof. By induction over k.

For $k = 0$, the proof is immediate given the construction of $\hat{R}_0^\pi$.
For $k \to k+1$, let $\omega \in \Omega_k$ and $s \in \omega$. We have that

$$R_{k+1}^\pi(s) \geq r(s, \pi(s)) + \gamma \inf_{s' \sim f(s, \pi(s))} R_k^\pi(s')$$

Because $s \in \omega$ we have

$$r(s, \pi(s)) \geq \inf_{x \in \omega} r(x, \pi(x)),$$

and, using $f(s, \pi(s)) \subset \{\omega' \in \Omega_{k-1} | \omega' \cap f^\pi(s) \neq 0\}$ and the inductive assumption,

$$\inf_{s' \sim f(s, \pi(s))} R_k^\pi(s') \geq \inf_{x \in \omega} \left[\inf_{\{\omega' \in \Omega_{k-1} | \omega' \cap f^\pi(x) \neq 0\}} \hat{R}_k^\pi(\omega') \right].$$

These equations, together with the update equation in Step 3 of the algorithm, lead to the conclusion that $\hat{R}_{k+1}^\pi(\omega) \leq R_{k+1}^\pi(s)$.

3   Implementation for Lipschitz Continuity

In this section, we borrow ideas from Fonteneau et al. (2009, 2010) for computing f^π and the bounds in our second assumption by incorporating prior knowledge in the form of Lipschitz continuity of the transition model. Since their work did not consider discounting, we will only consider non-discounted problems ($\gamma = 1$) in this section, in order to directly compare to their bounds.

3.1   Notation and Assumptions

Let S, A, and $S \times A$ be normed spaces, and denote the distance metric on all of them by d (it should be obvious from the context which one is being used). Define the distance between state-action pairs to be (Fonteneau et al.,2009):

$$d\left((s^1,a^1),(s^2,a^2)\right) = d\left(s^1,s^2\right) + d\left(a^1,a^2\right). \tag{2}$$

The transition function, reward and policy are said to be Lipschitz continuous if there exist constants L_f, L_r and L_π such that

$$d(f(s^1,a^1),f(s^2,a^2)) \leq L_f d\left((s^1,a^1),(s^2,a^2)\right), \tag{3}$$

$$d(r(s^1,a^1),r(s^2,a^2)) \leq L_r d\left((s^1,a^1),(s^2,a^2)\right) \tag{4}$$

and

$$d(\pi(s^1),\pi(s^2)) \leq L_\pi d\left(s^1,s^2\right) \tag{5}$$

respectively, for all $s^1,s^2 \in S$ and $a^1,a^2 \in A$. Similarly to the work of Fonteneau et al., the transition function is assumed to be deterministic.

In addition to Lipschitz continuity, we assume that there exists a set of n observed transitions of the form $(s_i,a_i,r(s_i,a_i),f(s_i,a_i))$, collected using an arbitrary and potentially unknown policy. Note that the actions in the batch do not have to match the actions taken by π, so this is an off-policy evaluation problem. Having this set of observed transitions allows them to bound the outcome $f(s,a)$ for any state-action pair (s,a), by taking for some $i \in \{1,\dots n\}$ a ball of center $f(s_i,a_i)$ and radius $L_f d((s,a),(s_i,a_i))$, as illustrated in Figure 1.

3.2   Previous Work

In their first paper, Fonteneau et al. (2009) provide a lower bound for finite-horizon value functions in Lipschitz-continuous deterministic systems with continuous state and action spaces. Given K observed transitions (one for each step of the horizon), their bound has the form

$$Q_K^\pi(x_0,\pi(x_0)) \geq \sum_{k=1}^{K} \left[r(s_k,a_k) - L_Q^{K-k}\left(d(s_k,f(s_{k-1},a_{k-1})) + d(a_k,\pi(f(s_{k-1},a_{k-1})))\right)\right], \tag{6}$$

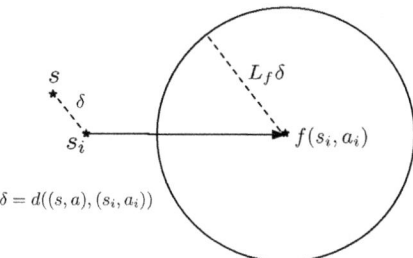

Fig. 1. Illustration of using Lipschitz continuity to bound $f(s,a)$

where L_Q is defined as

$$L_Q^{K-k} = L_r \left(\sum_{i=0}^{K-k-1} [L_f(1+L_\pi)]^i \right),$$

and is shown to be a valid Lipschitz constant for the value function.

In a second paper, Fonteneau et al. (2010) compute a similar bound for open-loop policies over discrete actions. This new bound can be derived from the one above if we replace L_π and the distances between actions with zero. They use this bound to show how it can inform an algorithm for computing "cautious" policies that avoid unexplored regions of the state space. Since the main focus of this work is analyzing the tightness of bounds, we will not be concerned with their cautious generalization algorithm in this paper.

3.3   Framework Instantiation

We first discuss the construction of f^π. In order to maintain simplicity of computation and exposition, we construct $f^\pi(x)$ to have the same value for all $x \in \omega$, and we denote this value by $f^\pi(\omega)$. We construct $f^\pi(\omega)$ as a ball of center $f(s_\omega, a_\omega)$ and radius

$$L_f \left[d(c_\omega, s_\omega) + d(\pi(c_\omega), a_\omega) + \rho_\omega + \sup_{x \in \omega} d(\pi(x), \pi(c_\omega)) \right],$$

where (s_ω, a_ω) is a state-action pair from the batch that we associate with partition ω, and c_ω is the center and ρ_ω the radius of a hypersphere containing ω. One heuristic way of doing this association is by taking the closest state-action pair to $(c_\omega, \pi(c_\omega))$. Another way, proposed by Fonteneau et al. (2009), is to search for the sequence of states that minimizes the K-step bound[3]. The Lipschitz property, together with the triangle inequality, guarantee that $\forall x \in \omega$ we have $f(x, \pi(x)) \in f^\pi(\omega)$.

Depending on the other choices made as part of Algorithm 1, we can either replicate the bounds of Fonteneau et al., or produce new bounds, as we will now illustrate.

---

[3] In our experiments, we found that both methods performed similarly.

Replicating the bounds of Fonteneau et al. Although Fonteneau et al. arrived at their results via a different approach, their bound can be exactly replicated within our framework if we use the following choices in Algorithm 1:

- Have **a single ball in each** Ω_k. As discussed earlier, and illustrated in Figure 1, bounding the outcomes of state transitions using Lipschitz continuity produces spherical bounds. Therefore, it is natural to consider each Ω_k to be composed of a single ball, with $\Omega_K = \{\omega_0\} = \{x_0\}$ and $\Omega_{k-1} = \{\omega_{k-1}\} = \{f^\pi(\omega_k)\}$, $\forall k < K$.
- **Bound policy changes using Lipschitz continuity.** Assume that π is Lipschitz continuous; this allows the following upper bound to be constructed:

$$\sup_{x \in \omega} d(\pi(x), \pi(c_\omega)) \le \sup_{x \in \omega} L_\pi d(x, c_\omega)) = L_\pi \rho_\omega$$

- **Bound rewards using Lipschitz continuity.** Assume that the reward is Lipschitz continuous; together with the triangle inequality and the Lipschitz continuity of the policy, this can be used to show that

$$\inf_{x \in \omega_k} r(x, \pi(x)) \ge r(s_{\omega_k}, a_{\omega_k}) - L_r \left[\sup_{x \in \omega_k} [d(s_{\omega_k}, x) + d(a_{\omega_k}, \pi(x))] \right]$$

$$\ge r(s_{\omega_k}, a_{\omega_k}) - L_r \left[d(s_{\omega_k}, c_{\omega_k}) + d(a_{\omega_k}, \pi(c_{\omega_k})) + \rho_{\omega_k} + \sup_{x \in \omega_k} d(\pi(x), \pi(c_{\omega_k})) \right]$$

$$\ge r(s_{\omega_k}, a_{\omega_k}) - L_r \left[d(k) + (1 + L_\pi) L_f (d(k+1) + (1 + L_\pi) \rho_{\omega_{k+1}}) \right],$$

where $d(k) = d(s_{\omega_k}, f(s_{\omega_{k+1}}, a_{\omega_{k+1}})) + d(a_{\omega_k}, \pi(f(s_{\omega_{k+1}}, a_{\omega_{k+1}})))$ and $d(K) = 0$.
- **Bound** $\inf_{\{\omega' \in \Omega_{K-k+1} | \omega' \cap f^\pi(x) \neq 0\}} \hat{R}_{k-1}^\pi(\omega')$ **using Lipschitz continuity.** Because of the construction of f^π and Ω_k, $f^\pi(x) \subset \omega_{k-1}$ for all $x \in \omega_k$. Therefore, we can replace $\inf_{x \in \omega_k} \left[\inf_{\{\omega' \in \Omega_{k-1} | \omega' \cap f^\pi(x) \neq 0\}} \hat{R}_{k-1}^\pi(\omega') \right]$ with $\hat{R}_{k-1}^\pi(\omega_{k-1})$.

Given the above, the update equation in Step 2 of Algorithm 1 takes the form

$$\hat{R}_k^\pi(\omega_k) = \hat{R}_{k-1}^\pi(\omega_{k-1}) + r(s_{\omega_k}, a_{\omega_k})$$
$$- L_r \left[d(k) + (1 + L_\pi) L_f (d(k+1) + (1 + L_\pi) \rho_{\omega_{k+1}}) \right].$$

By writing out this recursive sum and re-arranging the terms, we obtain the bound of Fonteneau et al. (2009).

Note that the Fonteneau et al. (2010) paper considers blind policies (policies where the action at each stage does not depend on the state), so the distance between actions simply disappears in that bound.

Proposed alternatives. We now consider two alternative algorithm instantiations that can lead to tighter bounds than previous results.

The first instantiation, which we will call **single-ball-rmin**, is similar to the Fonteneau et al. approach, except that it performs **direct reward minimization** instead of using Lipschitz continuity to bound the rewards. More exactly, in the experiments we use constraint optimization to compute $\inf_{x \in \omega} r(x, \pi(x))$. This can provide tighter bounds than Lipschitz-based bounds for high values of the Lipschitz constant. Note that such replacement of single-step bounds is particularly easy to perform within our framework.

The second instantiation will be called **multi-partition-rmin**, because it partitions each Ω_k into multiple regions. Using a single region for each k, as in the single-ball algorithm, has the disadvantage that it does not account for how π may take different actions in different parts of Ω_k. Therefore, in this algorithm we partition each Ω_k; for computational simplicity, we will take the partition boundaries to be axis-parallel. The multi-partition algorithm also performs direct minimization for bounding the rewards.

For this paper, we consider the two alternatives, single-ball-rmin and multi-partition-rmin. However, we note that other interesting instantiations are also possible, an obvious one being to perform direct minimization of $\sup_{x \in \omega} d(\pi(x), \pi(c_\omega))$ instead of using Lipschitz continuity. We leave such extensions for future work.

3.4 Discussion of Bounds Based on Lipschitz Continuity

The tightness of the bounds produced by all the methods described above will depend on the value of the Lipschitz constants L_f, L_r and L_π, which is related to the prior knowledge we have about the system and the characteristics of the domain. These values will influence all methods in a similar way. Another factor that influences all methods will be the distances $d(c_\omega, s_\omega) + d(\pi(c_\omega), a_\omega)$. As these distances get smaller, the bounds get tighter. The component $d(\pi(c_\omega), a_\omega)$, in particular, will depend on the ability to find transitions in the batch that take actions similar to those that π would take at the centers of the ω regions. So, to sum up, the more transitions there are in the batch, and the closer the actions of those transitions are to the policy we are trying to evaluate, the tighter the bounds will be.

For the multi-partition method to be effective, we need the union of the $f^\pi(\omega)$ for $\omega \in \Omega_k$ to be small relative to the size of $f^\pi(\omega)$ that would be computed by the single-ball algorithm. The size of the union relative to a single projection will depend on π, the policy we are evaluating. This influence of the nature of π is difficult to express quantitatively, because π can affect different parts of the state space in different ways. This difficulty in including more information about π in a closed-form bound makes it difficult to predict what type of bound will perform better, but is also one of the main reasons for proposing the multi-partition method, since it is the only method that attempts to include such information.

Finally, the multi-partition bound will become tighter as the density of partitions in Ω increases. Because of this, the multi-partition method suffers from the curse of dimensionality when applied in its most basic form: the number of regions required to effectively cover the state space is exponential in the dimensionality of S, and the computational complexity of computing $f^\pi(\omega)$ for $\omega \in \Omega_k$, the most expensive part of the multi-partition algorithm, is $O(n * |\Omega_k| * |\Omega_{k-1}|)$. However, it should be the case in most problems that the policy π that we are evaluating only visits a small region of the state space. In such problems, the forward projections computed using the multi-partition method, and therefore the Ω_k, are small relative to the size of the state space. In either case, the multi-partition method will be more computationally expensive than the single-ball method.

4   Empirical results

4.1   Deterministic Problems with Unknown Model and Lipschitz Continuous Dynamics

Puddle World. For our first set of experiments, we consider the deterministic puddle world domain, as implemented by Fonteneau et al. (2010). The puddle world is a two-dimensional problem in which an autonomous agent has four discrete actions available, taking it in each of the cardinal directions. There are two regions (the puddles) where the agent receives negative rewards, and a goal region where the agent receives positive rewards. We use the same starting state as Fonteneau et al., and the same mechanism for generating an initial batch of transitions (we place them uniformly throughout the state space). Complete details for this implementation of puddle world can be found in Fonteneau et al. (2010). The policy π that we evaluate is an open-loop policy that takes the agent around the puddle and to the goal region by going down for five steps, right for six steps, up for eight steps and right once again for the final step.

Fonteneau et al. (2010) use the bounds they compute as an intermediate part in their algorithm for computing cautiously generalizing policies. However, in this paper we are primarily concerned with the tightness of different types of bounds, so we will not carry out their experiment in full. Instead, we compare the bounds that are computed for π by the method of Fonteneau et al. (2010) to the single-ball-rmin and multi-partition-rmin algorithms. For both of these algorithms, $\inf_{x \in \omega} r(x, \pi(x))$ was computed using the Matlab `fmincon` function, which performs constrained optimization for smooth differentiable functions.

The results, presented in Figure 2(a), show the tightness of the different bounds for two different sets of stored transitions. The starting states of the transitions were placed uniformly throughout the state space, at distances of 0.05 along each axis for the first set of results and 0.02 along each axis for the second set of results. The multi-partition results shown in this figure are for 200,000 partitions in each Ω_k. We also include the trivial bound obtained by multiplying the horizon by the smallest immediate reward (denoted by $K * R_{\min}$ on the graph).

The first thing to note is that, in this domain, the single-ball-rmin algorithm produces bounds that are the same order of magnitude as the actual return - its values for the two samples sizes are -98.09 and -81.74, respectively, compared to a correct value of -54.08. This happens because the domain is quite well-behaved, with $L_f = 1$ and $L_\pi = 0$ (meaning that the effect of the policy can be ignored, since the policy is open-loop). The errors that accumulate over repeated projections are therefore solely due to the distances between the observed transitions and the centers of the projected regions.

The multi-partition-rmin algorithm is not performing as well as single-ball-rmin in this domain. Because the policy does not depend on the state, there is no advantage to incorporating information about how the policy changes in different parts of Ω_k. Instead, the multi-partition algorithm produces looser bounds because of the extra approximations it needs to use in order to handle state-dependent policies.

Finally, because of the large value of L_r in this domain ($1.3742 * 10^6$), methods that lower bound the reward using Lipschitz continuity perform poorly. For instance, the bound computed by the Fonteneau et al. method using a sample size of 80656 transitions

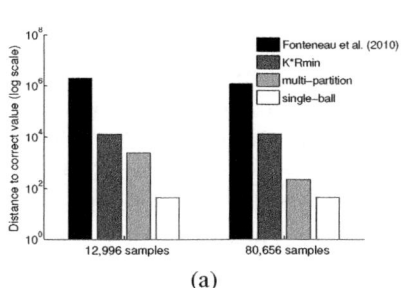
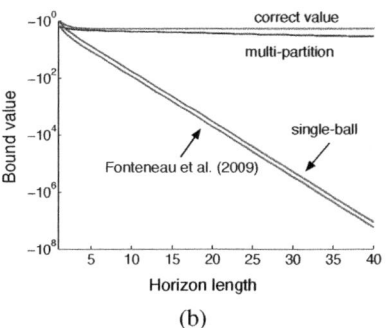

(a)   (b)

Fig. 2. Bound values for (a) the puddle world domain, and (b) the one-dimensional problem. Smaller values indicate tighter bounds.

is equal to $-55.94 * 10^4$. In comparison, the trivial bound obtained by multiplying the smallest reward by the horizon is actually significantly tighter, having the value $-1.3 * 10^4$.

The main conclusions that we draw from this experiment is that the additional step of analytically minimizing the immediate reward can offer significant improvements with respect to Lipschitz-continuity-based bounds, and that the multi-partition method is unlikely to be useful if the policy does not depend on the state.

1D Problem with Stabilizing Policy. In this section, we consider a simple one-dimensional problem to illustrate that the multi-partition approach can lead to tighter bounds when the policy depends on the state.

The domain we are using has $S = \mathbb{R}$, $A = \mathbb{R}$, $f(s,a) = s + a$ and $r(s,a) = -|s|$. We compute lower bounds for a policy that always takes the agent halfway to state 0, that is, $\pi(s) = -s/2$. We assume that we have a batch of observed transitions that was collected using a different policy, one for which the action at state s is uniformly distributed in $[-s, 0]$. The starting states for the observed transitions are generated uniformly randomly in the $[-1, 1]$ interval, and the starting state for evaluating π was -1. The Lipschitz constants for the reward and transition function are $L_f = 1$ and $L_r = 1$, while $L_\pi = 0.5$, so $L_f(1 + L_\pi) = 1.5$.

We compare the lower bound computed using the single-ball-rmin and multi-partition-rmin algorithms to the analytic bound of Fonteneau et al. (2009). The number of partitions in Ω for the multi-partition method was set to 1000. The results are presented in Figure 2(b). The size of the batch was 1000 transitions. The results were averaged over 50 trials, with all trials exhibiting the same pattern.

Note the logarithmic scale on the y-axis of the figure. The Fonteneau et al. (2009) single-ball bounds scale exponentially in the horizon K. This is due to the fact that a factor of $L_f(1 + L_\pi)^K$ appears in their Lipschitz constant for the value function L_Q, which then gets multiplied by the distance between the end point of a transition and the start of another. A small improvement in performance can be observed when using the single-ball-rmin algorithm instead of the method of Fonteneau et al. (2009). However, in this problem the effect of multiplying by an exponential function of $L_f(1 + L_\pi) =$

1.5 is still dominant with respect to the effect of bounding the reward using Lipschitz continuity.

The main point of this experiment is to illustrate how not accounting for state-dependent policy effects can produce bounds that scale up exponentially in the horizon, even for very simple problems, and that the multi-partition algorithm can alleviate this issue. While we acknowledge that the experiment was crafted specifically to highlight this issue, we believe that it is illustrative of settings that may arise in practical applications. For instance, in dynamical system control, the controller is often designed so that it adjusts the system towards a particular desired state, or range of states.

4.2   Stochastic Problem with known Model

Our framework can also be used in problems with noise in the transition model, as long as this noise has bounded support. In this section, we illustrate this on a stochastic problem where we make use of the known model to compute f^π.

Our testbed is the halibut fishery problem described by Ermon et al. (2010). In this problem, the state and action space are continuous and one-dimensional: the state variable represents the fish stock at the beginning of the year (in millions of pounds), and the action is the amount of fish to be harvested in a particular year (also expressed in millions of pounds). Denoting the state by s and the action by a, the stochastic transition function has the form:

$$f_w(s,a) = (1-m)(s-a) + w\frac{r_0(s-a)}{1+(s-a)/M},$$

where $w \sim U[0.89, 1.06]$ is random noise in the natural population replenishment, and $m = 0.15, M = 196.3923$, and $r_0 = 0.543365$ are constants. Since the noise is multiplied by a positive quantity and then added to the rest of the transition function, we construct $f^\pi(s)$ to be equal to $[f_{0.89}(s, \pi(s)), f_{1.06}(s, \pi(s))]$.

The reward function is defined as a sum of three components, the first being the price obtained for the harvested fish, the second being a variable labor and energy cost, and the third being a fixed capital cost that is incurred every year the action is greater than zero. The exact form of the reward function is

$$r(s,a) = pa - \int_{s-a}^{s} \frac{c}{qy^b}dy - K,$$

where $p = 4.3 \cdot 10^6$, $c = 200,000$, $q = 9.07979 \cdot 10^{-7}$, $b = 2.55465$, and $K = 5 \cdot 10^6$. The discount factor γ was equal to $\frac{1}{1.05}$, the initial state was $s_0 = 90.989$ million pounds, and the horizon was $T = 33$.

We computed bounds for two policies. The first policy was a version of the risk-adverse policy reported in Ermon et al. (2010). Their policy relies on two values, S^* and s^*, and at each time step t (time steps correspond to years) selects the action $a_t = s_t - s^*$ if $s_t > S^*$, and $a_t = 0$ otherwise (there is no harvest at time t if $s_t < S^*$). The policy used by Ermon et al. is non-stationary, in that the optimal values of S^* and s^* change depending on the time step; however, since they only report the values for the first year, we used the first-year values ($S^* = 176.5$ and $s^* = 133$) for all time steps.

Our bound, computed using 1000 partitions, was equal to $84.88 \cdot 10^7$, compared to a value of $86.70 \cdot 10^7$ if the noise is such that there is minimum fish stock replenishment every year. Note that our bound, unlike the min-replenishment bound, does not have knowledge of what is the worst outcome of each individual transition. We also computed an approximation of the bound via Monte Carlo sampling, by generating one million returns and taking the smallest value. This produced a bound equal to $95.75 \cdot 10^7$, which appears to be fairly optimistic.

The second policy is to simply harvest an amount equal to $a_t = 16$ million pounds every year. With this policy, if minimum fish stock replenishment occurs every year, then the population is exhausted in 21 years. Our bound informs us (by looking at the sequence of worst states at each step) that in the worst case the fish population may be exhausted in 19 years, providing a pessimistic bound but only being off by two years. On the other hand, the worst-case scenario is difficult to identify with the Monte Carlo approach: out of one million simulations, there is none that predicts exhaustion in less than 33 years.

Our results show that our bounds can provide accurate information about the worst-case behavior of stochastic systems when the stochastic transitions have bounded support and f^π provides accurate information. In this particular example, we had the best possible f^π, by computing it using the known model. A topic for further research is investigating under what conditions similar results can be achieved when full model information is replaced by a batch of transitions.

5   Discussion and Future Work

We proposed a framework that transforms prior knowledge about the one-step dynamics (transition and reward function) into bounds on the cumulative return. The type of the prior knowledge required is described by the two assumptions in Section 2. The first assumption, that we can compute f^π, is concerned with bounding the effect of the transition function. We can obtain useful instantiations of f^π using different forms of prior knowledge – in this paper, we computed f^π based either on Lipschitz continuity of the transition function, or on full knowledge of the stochastic transition model. The second assumption, that we can lower-bound the immediate reward and the next-state bound, can also be satisfied by different forms of prior knowledge. Some examples of functions that can be lower-bounded include Lipschitz continuous functions, differentiable functions for which the zeros of the derivative can be analytically computed (such as the reward function in the puddle world example), monotonic one-dimensional functions (the reward function in the 1D problem and the fisheries domain), or piecewise-constant functions with few discontinuities (such as the next-state bound in the fisheries domain). In the future, we hope to work together with practitioners in applied fields, and identify what type of prior knowledge is available for the systems in which they are interested.

An interesting area of investigation is analyzing the interplay between the observed transitions and the policy π that is evaluated. As we discussed in Section 3.4, the multi-partition bounds become tighter the more transitions there are with starting states close to the regions in Ω, the more the actions for those transitions are similar to those that

π would take, and also depending on the nature of π. While these effects are difficult to quantify theoretically, a comprehensive empirical analysis would further the understanding of the multi-partition method's scope of applicability. We also discussed in Section 3.4 how the multi-partition method may suffer from the curse of dimensionality; we would like to further investigate possibilities for addressing this issue.

In this paper we focused solely on producing tight bounds. In the future, we plan to also look at how the bounds produced can be used for selecting what policy to implement. The problem of computing policies that maximize a lower bound on the value function has received a fair amount of attention. Nilim and El Ghaoui (2005) and Delage and Mannor (2009) address this problem in finite state spaces from a game theoretic perspective. In infinite state spaces, Fonteneau et al. (2010) compute lower-bound-maximizing policies by using lower bounds computed using the Lipschitz continuity assumptions (as described in Section 3), while Ermon et al. (2010) address a similar problem in the context of halibut fishery management under convexity assumptions on the reward and the value function.

In some applications we may be interested in exploratory rather than cautious policies. In such domains, we can use upper bounds computed similarly to the lower bounds to derive exploratory policies that maximize the upper bounds – an idea dating back at least to the work of Kaelbling (1993).

Acknowledgments. The authors would like to thank Raphael Fonteneau for providing his implementation of the algorithm described in Fonteneau et al. (2010), and Arthur Guez for helpful discussions regarding this work. Funding was provided by the National Institutes of Health (grant R21 DA019800) and the NSERC Discovery Grant program.

References

1. Brunskill, E., Leffler, B., Li, L., Littman, M., Roy, N.: CORL: A continuous-state offset-dynamics reinforcement learner. In: Proceedings of the International Conference on Uncertainty in Artificial Intelligence, pp. 53–61 (2008)
2. Delage, E., Mannor, S.: Percentile Optimization for Markov Decision Processes with Parameter Uncertainty. Operations Research 58(1), 203–213 (2009)
3. Ermon, S., Conrad, J., Gomes, C., Selman, B.: Playing games against nature: optimal policies for renewable resource allocation. In: Proceedings of The 26th Conference on Uncertainty in Artificial Intelligence (2010)
4. Fonteneau, R., Murphy, S., Wehenkel, L., Ernst, D.: Inferring bounds on the performance of a control policy from a sample of trajectories. In: IEEE Symposium on Adaptive Dynamic Programming and Reinforcement Learning (ADPRL), pp. 117–123 (2009)
5. Fonteneau, R., Murphy, S.A., Wehenkel, L., Ernst, D.: Towards Min Max Generalization in Reinforcement Learning. In: Filipe, J., Fred, A., Sharp, B. (eds.) ICAART 2010. CCIS, vol. 129, pp. 61–77. Springer, Heidelberg (2011)
6. Kaelbling, L.P.: Learning in embedded systems. MIT Press (1993)
7. Kakade, S., Kearns, M., Langford, J.: Exploration in Metric State Spaces. In: International Conference on Machine Learning, vol. 20, p. 306 (2003)
8. Nilim, A., El Ghaoui, L.: Robust Control of Markov Decision Processes with Uncertain Transition Matrices. Operations Research 53(5), 780–798 (2005)

Transferring Evolved Reservoir Features in Reinforcement Learning Tasks

Kyriakos C. Chatzidimitriou[1,2], Ioannis Partalas[3], Pericles A. Mitkas[1,2], and Ioannis Vlahavas[3]

[1] Dept. of Electrical & Computer Engineering,
Aristotle University of Thessaloniki, Greece
[2] Informatics and Telematics Institute, Centre for Research and Technology Hellas
kyrcha@issel.ee.auth.gr, mitkas@eng.auth.gr
[3] Dept. of Informatics, Aristotle University of Thessaloniki, Greece
{partalas,vlahavas}@csd.auth.gr

Abstract. The major goal of transfer learning is to transfer knowledge acquired on a source task in order to facilitate learning on another, different, but usually related, target task. In this paper, we are using neuroevolution to evolve echo state networks on the source task and transfer the best performing reservoirs to be used as initial population on the target task. The idea is that any non-linear, temporal features, represented by the neurons of the reservoir and evolved on the source task, along with reservoir properties, will be a good starting point for a stochastic search on the target task. In a step towards full autonomy and by taking advantage of the random and fully connected nature of echo state networks, we examine a transfer method that renders any inter-task mappings of states and actions unnecessary. We tested our approach and that of inter-task mappings in two RL testbeds: the mountain car and the server job scheduling domains. Under various setups the results we obtained in both cases are promising.

1 Introduction

Reinforcement learning (RL) [9] deals with the problem of how an agent, situated in an environment and interacting with it, can learn a policy, a mapping of states to actions, in order to maximize the total amount of reward it receives over time, by inferring on the immediate feedback returned in the form of scalar rewards as a consequence of its actions. RL has enjoyed increased popularity due to its ability to deal with complex and limited feedback problems, while it is believed to be an appropriate paradigm for creating fully autonomous agents in the future [8]. Despite the suitability to solve such problems, RL algorithms often require a considerable amount of training time especially for complex problems. A solution to speed up the learning procedure is through *transfer learning* (TL).

TL refers to the process of using knowledge that has been acquired in a previous learned task, the *source task*, in order to enhance the learning procedure in a new and more complex task, the *target task*. The tasks can belong to either

S. Sanner and M. Hutter (Eds.): EWRL 2011, LNCS 7188, pp. 213–224, 2012.

the same domain, for example two mazes with different structure, or to different domains, for example checkers and chess board games. The more similar those two tasks are, the easier it is to transfer knowledge between them. TL is to play a crucial role in the development of fully autonomous agents since it is believed that it would be a core component in lifelong learning agents that persist over time [12].

For agents to perform well under a RL regime in real world problems, there is the need for *function approximators* (FAs) to model the policy and be able to generalize well to unseen environment states. Fully autonomous agents will need FAs that will adapt to the environment at hand with little, if any, human intervention [8]. Following this trend, we selected the *echo state network* (ESN) to be our FA of choice. Its recurrent neural network nature makes it appropriate for use as FA in agents dealing with sequential decision making problems, because it enables temporal computations and can process non-linear and non-Markovian state signals (Section 2).

In order to augment the capabilities of the adaptive FA approach, in this work, we tested methods of transferring reservoir topologies (or alternatively reservoir features). These topologies were adapted in a source RL task to be used as templates for the initial population of the neuroevolution procedure on a target RL task. Besides testing network transfer under the standard way of using mappings between the source and target, state variables and actions, we evaluated a transfer method agnostic of any mappings, taking advantage of the random, fully connected nature of ESNs. This procedure is performed at the microscopic level, that is in the level of the topology and the weights of the reservoir (Section 3).

Our methodology is evaluated empirically on two RL test-beds: a) the *mountain car* problem from the area of control and b) the *server job scheduling* problem from the area of autonomic computing (Section 4). The results of the experiments from both testbeds (Section 5) are promising under several different metrics with respect to the base search approach. We discuss related work in the area and how our approach differs from it (Section 6), closing with the conclusions of our research and plans for future work (Section 7).

2   Background

2.1   Echo State Networks

The idea behind *reservoir computing* (RC) and in particular ESNs [4] is that a random *recurrent neural network* (RNN), created under certain algebraic constraints, could be driven by an input signal to create a rich set of dynamics in its reservoir of neurons, forming non-linear response signals. These signals, along with the input signals, could be combined to form the so-called *read-out function*, a linear combination of features, $y = \mathbf{w}^T \cdot \phi(\mathbf{x})$, which constitutes the prediction of the desired output signal, given that the weights, w, are trained accordingly.

The reservoir consists of a layer of K input units, connected to N reservoir units through a $N \times K$ weighted connection matrix W^{in}. The connection matrix of the reservoir, W, is a $N \times N$ matrix. Optionally a backprojection matrix W^{back} could be present, with dimensions $N \times L$, where L is the number of output units, connecting the outputs back to the reservoir neurons. The weights from input units (linear features) and reservoir units (non-linear features) to the output are collected into a $L \times (K + N)$ matrix, W^{out}. For this work, the reservoir units use $f(x) = tanh(x)$ as an activation function, while the output units use either $g(x) = tanh(x)$ or the identity function, $g(x) = x$.

Best practices for generating ESNs, that is procedures for generating the random connection matrices W^{in}, W and W^{back}, can be found in [4]. Briefly, these are: (i) W should be sparse, (ii) the mean value of weights should be around zero, (iii) N should be large enough to introduce more features for better prediction performance, (iv) the spectral radius, ρ, of W should be less than 1 to practically (and not theoretically) ensure that the network will be able to function as an ESN. Finally, a weak uniform white noise term can be added to the features for stability reasons.

In this work, we consider discrete time models and ESNs without backprojection connections. As a first step, we scale and shift the input signal, $\mathbf{u} \in \mathbb{R}^K$, depending on whether we want the network to work in the linear or the non-linear part of the sigmoid function. The reservoir feature vector, $\mathbf{x} \in \mathbb{R}^N$, is given by Equation 1:

$$\mathbf{x}(t + 1) = \mathbf{f}(\mathbf{W}^{in}\mathbf{u}(t + 1) + \mathbf{W}\mathbf{x}(t) + \mathbf{v}(t + 1)) \tag{1}$$

where $\mathbf{f}$ is the element-wise application of the reservoir activation function and $\mathbf{v}$ is a uniform white noise vector. The output, $\mathbf{y} \in \mathbb{R}^L$, is then given by Equation 2:

$$\mathbf{y}(t + 1) = \mathbf{g}(\mathbf{W}^{out}[\mathbf{u}(t + 1)|\mathbf{x}(t + 1)]) \tag{2}$$

with $\mathbf{g}$, the element-wise application of the output activation function and $|$, the aggregation of vectors.

For RL tasks with K continuous states and L discrete actions, we can use an ESN to model a Q-value function, where each network output unit l, can be mapped to an action $a_l, l = 1 \ldots L$, with the network output value y_l denoting the long-term discounted value, $Q(\mathbf{s}, a_l)$ of performing action a_l, when the agent is at state $\mathbf{s}$. Given $g(x) = x$, this Q-value can be represented by an ESN as:

$$y_l = Q(\mathbf{s}, a_l) = \sum_{i=1}^{K} w_{li}^{out}s_i + \sum_{i=K+1}^{K+N} w_{li}^{out}x_{i-K}, l = 1, \ldots, L \tag{3}$$

while actions can be chosen under the ϵ-greedy policy [9].

Linear Gradient Descent (GD) SARSA TD-learning can be used to adapt weights [9,10], where the update equations take the form of:

$$\delta = r + \gamma Q(\mathbf{s}, a') - Q(\mathbf{s}, a_l) \tag{4}$$

$$\mathbf{w}_l^{out'} = \mathbf{w}_l^{out} + \alpha\delta[\mathbf{s}|\mathbf{x}] \tag{5}$$

with a' the next action to be selected and α the learning rate.

2.2   NeuroEvolution of Augmented Reservoirs

NeuroEvolution of Augmented Topologies (NEAT) [7] is a topology and weight evolution of artificial neural networks algorithm, constructed on four principles that made it a reference algorithm in the area of NE. First of all, the network, i.e. the phenotype, is encoded as a linear genome (genotype), making it memory efficient with respect to algorithms that work with full weight connection matrices. Secondly, using the notion of *historical markings*, newly created connections are annotated with innovation numbers. NEAT during crossover aligns parent genomes by matching the innovation numbers and performs crossover on these matching genes (connections). The third principle is to protect innovation through *speciation*, by clustering organisms into species in order for them to have time to optimize by competing only in their own niche. Last but not least, NEAT starts with minimal networks, that is networks with no hidden units, in order (a) to initially start with a minimal search space and (b) to justify every complexification made in terms of fitness. NEAT complexifies networks through the application of structural mutations, by adding nodes and connections, and further adapts the networks through weight mutation by perturbing or restarting weight values. The above successful ideas could be used in other NE settings in the form of a meta-search evolutionary procedure. In our case, we follow these ideas to achieve an efficient search in the space of ESNs.

NeuroEvolution of Augmented Reservoirs (NEAR) [2] utilizes NEAT as a meta-search algorithm and adapts its four principles to the ESN model of neural networks. The structure of the evolutionary search algorithm is exactly the same as in NEAT with adaptations being made mainly with respect to gene representation, crossover with historical markings, clustering and including some additional evolutionary operators related to ESNs. An important difference from NEAT is that both evolution and learning are used in order to adapt networks to the problem at hand. NEAR, to its advantage, incorporates TD learning in order to make a local gradient descent search on the output matrix, W^{out}, of the ESN and to locate good solutions that reside nearby instead of performing just evolutionary search. In this work we use NEAR with the Lamarckian type of evolution, where learned weights, W^{out}, are transferred from generation to generation instead of being reset to zero before each generation.

3   Transfer of Reservoir Topologies

The idea behind our work is that certain parts of high performing ESNs in the source task could be reused as templates of the networks that make up the initial population in the target task. These certain parts should be ESN properties that distinguish one network from the other. In our case we have selected:

- the reservoir, denoted by matrix W, and along with it, the number N of neurons in the reservoir, the graph of the topology and the connection weights,
- the density D of the reservoir, and
- the spectral radius ρ, a factor used to dampen signals in the reservoir.

Our goal is to *alleviate completely* the problem of mappings in state and action variables between tasks by just transferring the reservoir connection matrix W and its particularities, leaving the other matrices W^{in} and W^{out} to be handled by the NEAR method. In particular, like in the standard way of generating ESN, W^{in} is randomly initialized and later adapted through NEAR, while W^{out} is also initialized randomly and under an evolutionary weight mutation operator that uses perturbation, appropriate weights are derived. This makes the transfer, agnostic of any state or action mappings, since these are randomly created through the matrices W^{in} and W^{out}, only to be later adapted through NEAR to the problem at hand.

We have targeted our approach on transferring evolved reservoir repositories using the following methodologies. The prime symbol is used to denote properties of the target task ESN.

1. **Reservoir-Transfer**: The mapping agnostic method, that transfers the reservoir matrix $W' = W$ and the spectral radius $\rho' = \rho$ and randomly initializes W^{out} and W^{in} matrices, as discussed in the paragraph above (Figure 1).

2. **Mapping**: Use mappings that relate state and action variables from the source to the target task in the same way as that presented in [15] for the NEAT algorithm. The matrix W is transferred as is. The state and action mappings indicate which connection weight from matrices W^{in} and W^{out} of the final network of the source task, will be set to which position in matrices W'^{in} and W'^{out} of the initial networks of the target task. Such mappings are provided by a domain expert (Figure 2). This specific setup was chosen in order to test the mapping agnostic approach against a reference inter-task mapping methodology using neural networks.

3. **Mapping+Doubling**: Use mappings to account for W^{in} and W^{out} weights, but also increase the reservoir neurons (i.e. the hidden neurons) in order to account for the increased task complexity. The increase is directly proportional to the number of state and action variables growth, from the source to the target task. We have used doubling because in our testbeds we have a doubling in the number of state and action variables. Thus, we have created the new reservoir matrix W' to contain the matrix W in its upper left ($1 \leq i' \leq N$, $1 \leq j' \leq N$) and lower right ($N+1 \leq i' \leq N'$, $N+1 \leq j' \leq N'$) blocks, with the rest of the matrix elements set to 0 (Figure 3). This method was chosen in order to survey whether, in the presence of a higher dimensional target task, the computational units of the network need to be augmented as well. In fact, we wanted to experimentally test if the two reservoirs could handle the same number of variables each as in the source task and let the neuroevolution algorithm grow connections between them.

In all the above cases the basic properties of an ESN, N, D and ρ, are transferred implicitly in the first two cases (through matrix W) and explicitly in the case of ρ (by setting it initially in all target task genomes).

We have focused our attention on source-target task pairings that diverge from each other due to an increase in the dimensional complexity of the problem,

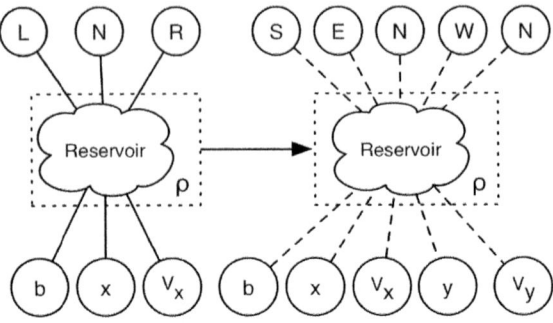

Fig. 1. In this setting only the reservoir, implicitly including the N and the D proper-ties, is transferred along with spectral radius ρ. W^{in} and W^{out} matrices are randomly initialized and adapted through the NEAR process. The figure is an example of reser-voir transfer in the mountain car domain discussed in the next section.

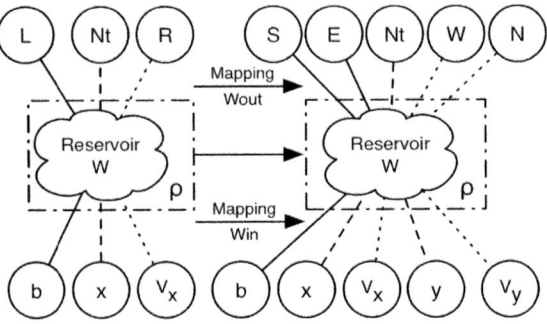

Fig. 2. Besides reservoir transfer, weights found in the source task for W^{in} and W^{out} matrices are transferred to the target task using inter-task mappings as described in [15]

but belong to the same domain. For example, situating the agent from a two dimensional (2D) problem to a three dimensional (3D) version or increasing the number of things it has to control, leading to an increase in the number of sensors (state variables) and actuators (actions). The main objectives of transfer learning are: (a) increased *asymptotic performance* of the transfer enabled agent over the basic one and (b) improvement in adaptation time to reach pre-specified thresholds of performance, a metric known as `time-to-threshold`.

4   Domains

4.1   Mountain Car

For the mountain car (MC) domain we use the version by [6]. In the stan-dard 2D task an underpowered car must be driven up to a hill. The state of the

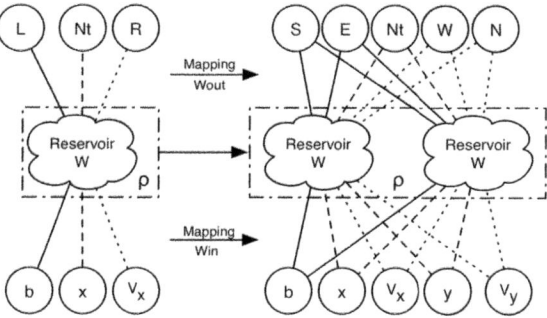

Fig. 3. Going a step beyond, and exploiting the doubling with respect to the number of state and action variables, the reservoir is doubled along with inter-task mappings

environment is described by two continuous variables: horizontal position $x \in [-1.2, 0.6]$ and velocity $v_x \in [-0.007, 0.007]$. The actions are {Neutral, Left and Right} which modify the velocity by $-0.001, 0$ and 0.001, respectively. At each time step, an amount of $-0.0025 * \cos 3x$ is added to the velocity, in order to represent the gravity.

In our case, each episode starts with the car at a random position in the valley, having a random speed, and ends when x becomes greater than 0.5. At each time step, the agent selects among the three available actions and receives a reward of -1. The objective is to move the car to the goal state as fast as possible.

The 3D MC extends the 2D task by adding an extra spatial dimension. The 3D task was originally proposed in [14]. The state is composed by four continuous variables: the coordinates in space x, and $y \in [-1.2, 0.6]$, as well as the velocities v_x and $v_y \in [-0.07, 0.07]$. The available actions are {Neutral, West, East, South, North}.

Along with the version discussed above, one more MC version was tested, which we will call the non-Markovian (NM) one, since the speed variable or variables in the 3D case, are kept from the agent, and only its position or positions in the 3D case are given, making the task more challenging. Each episode lasts 2500 time steps in all four MC versions after which another episode begins.

In this work we use the mountain car software[1] that is based on version 3.0 of the RL-Glue library[2] [11].

4.2 Server Job Scheduling

Server job scheduling (SJS) [17] is a domain that belongs to the realm of autonomic computing. It was previously used for performing transfer learning with the NEAT algorithm in [15]. Certain types of jobs are waiting in the job queue of a server to be processed as new jobs arrive in the queue. Each job type has a

---

[1] Available at http://library.rl-community.org/
[2] Available at http://glue.rl-community.org/

utility function that changes over time and represents the user anticipation over having the job scheduled. Certain users want quicker response from the server, while others are not as eager. The goal of the scheduler is to pick a job (action) based on the status of the queue (state), receiving as utility the value of the function of the scheduled job type at the timestep of execution (immediate reward). The performance is calculated as the sum of utilities (long term reward) when the queue empties. Each task begins with the scheduler finding 100 jobs in the queue, while at each time step a new job is added to the queue for the first 100 ticks. Each episode lasts 200 timesteps. For the state variable and action setup we used the modeling found in [17]. Job types 1 and 3 were used for the source tasks and all four jobs for the target task. The state variable mapping was to match the inputs concerned with job type 1 with the inputs of job type 2 and the inputs about job type 3 with inputs of job type 4.

5   Experiments

Each one of the experiments was conducted 10 times. The population was initialized with 100 individual networks and was evolved over 50 generations. In each generation, every individual was allowed to learn over 100 episodes with randomly initializing the starting state variables both in the MC and the SJS domains. In each generation, the champion network was evaluated for 1000 additional episodes with random restarts and its performance was recorded as the average fitness over those episodes. In the presence of transfer learning, before starting the evolution of the networks in the target task, the initial population was initialized as discussed in Section 3 with the seed network being the champion network produced by a neuroevolution procedure using the NEAR methodology in the source task.

Table 1 presents the results of the four approaches for all testbeds. The results illustrate the asymptotic performance of the agent for each task, i.e. the total reward received over time under the policy produced by the champion ESN over the 100 randomly initialized episodes. Our first observation is that, as the testbed difficulty increases, the performance gains of transferring reservoir features is even more evident.

The SJS domain can be considered a much more difficult domain than the MC, at least in terms of the number of state variables and actions the agent has to handle. In this specific testbed, as far as the average asymptotic performance is concerned, all transfer methods outperform evolution from scratch with a difference that is statistically significant at the 95% level under t-test. Moreover a smaller variation in the asymptotic behavior of the algorithms is observed under the transfer regimes. This is of particular value, when we are dealing with autonomous agents, which have to build such mechanisms without human supervision.

Last but not least, even though the mapping method has better performance overall, it does not statistically dominate over the mapping agnostic method. In fact, this could indicate that the reservoir along with its properties captures

Table 1. The average and standard deviation of the generalization performance for all the domains and algorithms under examination

Domain	Mean	Sd
MC-M-Scratch	-100.04	3.51
MC-M-Reservoir	-96.47	4.48
MC-M-Mapping	-97.67	4.40
MC-M-Map+Double	-97.95	4.32
MC-NM-Scratch	-293.52	47.86
MC-NM-Reservoir	-288.52	62.31
MC-NM-Mapping	-261.92	59.30
MC-NM-Map+Double	-279.71	33.67
SJS-Scratch	-5243.08	82.20
SJS-Reservoir	-5187.02	31.24
SJS-Mapping	-5176.42	16.74
SJS-Map+Double	-5187.17	30.46

crucial information of the domain in this inter-task transfer. This kind of information could be specific sub-graphs existing in the reservoir, which together with specific weight values in the connections, calculate temporal features needed to produce efficient policies faster than when starting from scratch.

In both testbeds, the reservoir transfer methods win most of the races towards the performance threshold targets than the baseline version. This can be seen in the Figures 4, 5 and 6. The reservoir transfer method wins the race 7 times, followed by the mapping transfer method with 6 times and the mapping with reservoir doubling with 1 times, indicating a dominance of the transfer methods versus evolving a network from scratch with respect to this criterion. We also, note that in the Markovian variation of the MC domain, the agnostic method is the best algorithm as it preserves a slightly better performances against its rivals.

6   Related Work

This section presents related work in TL and contrasts it with the proposed approach. For a broader review of TL in reinforcement learning domains the reader can refer to a comprehensive survey of the field [12].

The transfer of neural networks has been studied in [15] and [1]. More specifically, Taylor et al. [15] proposed a method, named *Transfer via inter-task mappings for Policy Search Reinforcement Learning* (TVITM-PS), that initializes the weights of the target task, using the learned weights of the source task, by utilizing mapping functions. In particular, in [15], they proposed a TL method for policy search algorithms where the policies are represented as neural networks. The internal structure of the source networks along with the weights are copied to the target task using predefined or learned mapping functions. We have evaluated this method with respect to ESNs and the NEAR algorithm, which is

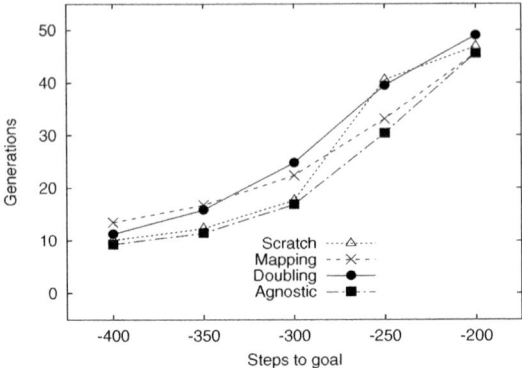

Fig. 4. The number of generations to reach pre-specified performance thresholds in the Markovian MC problem

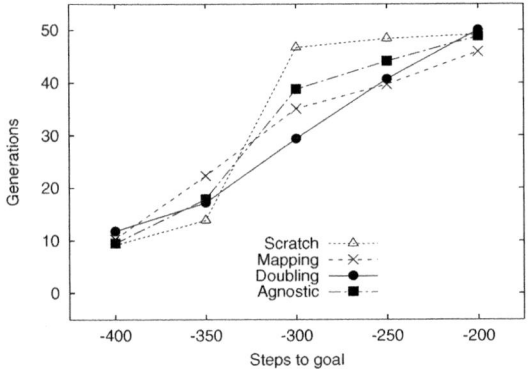

Fig. 5. The number of generations to reach pre-specified performance thresholds in the non-Markovian MC testbed

the first time to our knowledge. Such a test has value since ESNs have inherently more connections than NEAT derived networks. Additionally, we have tested a method that tries to take advantage of the fully, random, connectivity of ESNs and does not require the use of mappings, with promising results.

Finally, in [1] Bahceci and Miikkulainen introduce a method that transfers pattern-based heuristic in games. A population of evolved neural networks (which represent the patterns) to a target task as the starting population. In contrast our method allows different action and state spaces between the source and target tasks, and also is agnostic of any mappings.

Several other approaches have been proposed in the past for transfer learning. More specifically, [3] describes an algorithm which reuses policies from previously

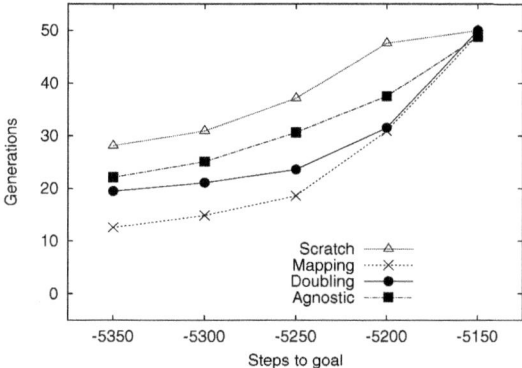

Fig. 6. The number of generations to reach pre-specified utility thresholds for the SJS domain

learned source tasks, while methods that use rules extracted from experience previously gained in a source task have been proposed in [5,16]. Both our testbeds have been used before in TL research. The 2D to 3D MC source-target pair has been used in [13], while the SJS in [15].

7   Conclusions and Future Work

In this paper we presented a method in the field of transfer learning and more specifically a method of transferring reservoir topologies and weights from a source task in order to facilitate adaptation of ESNs in a more complex target task. The main contribution of this paper is a method that renders any inter-task mappings of states and actions unnecessary. We believe that this is a step towards the fully automation of the transfer learning procedure.

We have tested the approach in three different problems from two different domains. Results indicated that as the difficulty of the task increases in terms of complexity in the states and actions, the gains from transferring knowledge are more evident. We hypothesize that as problems become increasingly difficult, the population initialization of NE methods will play an even more crucial role. The mapping approach is the method that outperformed any other, since it is difficult to beat expert knowledge with randomly assigned weights. On the other hand, transferring just reservoir and its properties seems to be a viable solution when the problem's dimensionality increases and mappings are becoming obscure. The NEAR methodology as is, is capable of finding solutions quite rapidly in the above problems, so the statistically significant improvement in the asymptotic performance on the server job scheduling task, is a promising result for further gains through reservoir transfer in much more demanding environments.

We plan to investigate the reservoir transfer approaches in more difficult testbeds like the keepaway domain. Additionally, we plan to work on cross-domain reservoir transfer, where the source and the target tasks belong to different domains. Finally, another issue that deserves further research is the sensitivity of the transfer learning method discussed in this paper with the presence of noise for both the source and the target tasks.

References

1. Bahceci, E., Miikkulainen, R.: Transfer of evolved pattern-based heuristics in games. In: IEEE Symposium on Computational Intelligence and Games (2008)
2. Chatzidimitriou, K.C., Mitkas, P.A.: A neat way for evolving echo state networks. In: European Conference on Artificial Intelligence, pp. 909–914 (2010)
3. Fernández, F., Veloso, M.: Probabilistic policy reuse in a reinforcement learning agent. In: 5th International Joint Conference on Autonomous Agents and Multiagent Systems, pp. 720–727 (2006)
4. Jaeger, H.: Tutorial on training recurrent neural networks, covering BPTT, RTRL, EKF and the "echo state network" approach. Tech. Rep. GMD Report 159, German National Research Center for Information Technology (2002)
5. Madden, M.G., Howley, T.: Transfer of experience between reinforcement learning environments with progressive difficulty. Artif. Intell. Rev. 21(3-4), 375–398 (2004)
6. Singh, S.P., Sutton, R.S.: Reinforcement learning with replacing eligibility traces. Machine Learning 22(1-3), 123–158 (1996)
7. Stanley, K.O., Miikkulainen, R.: Evolving neural networks through augmenting topologies. Evolutionary Computation 10(2), 99–127 (2002)
8. Stone, P.: Learning and multiagent reasoning for autonomous agents. In: Proceedings of the 20th International Joint Conference on Artificial Intelligence, pp. 13–30 (January 2007), http://www.ijcai-07.org/
9. Sutton, R.S., Barto, A.G.: Reinforcement Learning: An Introduction. MIT Press, Cambridge (1998)
10. Szita, I., Gyenes, V., Lőrincz, A.: Reinforcement Learning with Echo State Networks. In: Kollias, S.D., Stafylopatis, A., Duch, W., Oja, E. (eds.) ICANN 2006. LNCS, vol. 4131, pp. 830–839. Springer, Heidelberg (2006)
11. Tanner, B., White, A.: Rl-glue: Language-independent software for reinforcement-learning experiments. Journal of Machine Learning Research 10, 2133–2136 (2010)
12. Taylor, M., Stone, P.: Transfer learning for reinforcement learning domains: A survey. Journal of Machine Learning Research 10, 1633–1685 (2009)
13. Taylor, M.E., Jong, N.K., Stone, P.: Transferring Instances for Model-Based Reinforcement Learning. In: Daelemans, W., Goethals, B., Morik, K. (eds.) ECML PKDD 2008, Part II. LNCS (LNAI), vol. 5212, pp. 488–505. Springer, Heidelberg (2008)
14. Taylor, M.E., Kuhlmann, G., Stone, P.: Autonomous transfer for reinforcement learning. In: AAMAS 2008: Proceedings of the 7th International Joint Conference on Autonomous Agents and Multiagent Systems, pp. 283–290 (2008)
15. Taylor, M.E., Whiteson, S., Stone, P.: Transfer via inter-task mappings in policy search reinforcement learning. In: 6th International Joint Conference on Autonomous Agents and Multiagent Systems, pp. 1–8 (2007)
16. Torrey, L., Shavlik, J., Walker, T., Maclin, R.: Skill Acquisition Via Transfer Learning and Advice Taking. In: Fürnkranz, J., Scheffer, T., Spiliopoulou, M. (eds.) ECML 2006. LNCS (LNAI), vol. 4212, pp. 425–436. Springer, Heidelberg (2006)
17. Whiteson, S., Stone, P.: Evolutionary function approximation for reinforcement learning. Journal of Machine Learning Research 7, 877–917 (2006)

Transfer Learning via Multiple Inter-task Mappings

Anestis Fachantidis[1], Ioannis Partalas[1], Matthew E. Taylor[2],
and Ioannis Vlahavas[1]

[1] Department of Informatics, Aristotle University of Thessaloniki
{afa,partalas,vlahavas}@csd.auth.gr
[2] Department of Computer Science, Lafayette College
taylorm@cs.lafayette.edu

Abstract. In this paper we investigate using multiple mappings for transfer learning in reinforcement learning tasks. We propose two different transfer learning algorithms that are able to manipulate multiple inter-task mappings for both model-learning and model-free reinforcement learning algorithms. Both algorithms incorporate mechanisms to select the appropriate mappings, helping to avoid the phenomenon of negative transfer. The proposed algorithms are evaluated in the Mountain Car and Keepaway domains. Experimental results show that the use of multiple inter-task mappings can significantly boost the performance of transfer learning methodologies, relative to using a single mapping or learning without transfer.

1 Introduction

In recent years, a wealth of *Transfer Learning* (TL) methods have been developed in the context of *reinforcement learning* (RL) tasks. Typically, when an RL agent leverages TL, it uses knowledge acquired in one or more (*source*) tasks to speed up its learning in a more complex (*target*) task.

Although the majority of the work in this field presumes that the source task is connected in an obvious or natural way with the target task, this may not the case in many real life applications where RL transfer could be used. These tasks may have different state and action spaces, or even different reward and transition functions. One way to tackle this problem is to use functions that map the state and action variables of the source task to state and action variables of the target task. These functions are called *inter-task mappings* [12].

While inter-task mappings have indeed been used with successfully in several settings, we identify several shortcomings. First, an agent typically uses a hand-coded mapping, requiring the knowledge of a domain expert. If human intuition cannot be applied to the problem, selecting an inter-task mapping may be done randomly, requiring extensive experimentation and time not typically available in complex domains or in real applications. On the other hand, even if a correct mapping is used, it is fixed and applied to the entire state-action space, ignoring

S. Sanner and M. Hutter (Eds.): EWRL 2011, LNCS 7188, pp. 225–236, 2012.

the important possibility that different mappings may be better for different regions of the target task.

This paper examines the potential impact of using multiple mappings in TL. More specifically, we propose two different transfer learning algorithms that are able to manipulate multiple inter-task mappings for both model-learning and model-free reinforcement learning algorithms. Both algorithms incorporate mechanisms for the selection of the appropriate mappings in order to avoid the *negative transfer* phenomenon, in which TL can actually decrease learning performance. The proposed algorithms are evaluated in the domains of Mountain Car and Keepaway and experimental results show that using multiple inter-task mappings can improve the performance of transfer learning.

2   Transfer via Multiple Inter-task Mappings

In order to enable a transfer learning procedure across tasks that have different state variables and action sets, one must define how these tasks are related to each other. One way to represent this relation is to use a pair (X_S, X_A) of inter-task mappings [12,8], where the function X_S maps a target state variable to a source state variable and X_A maps an action in the target task to an action in the source task.

Inter-task mappings have been used for transferring knowledge in several settings like TD-learning [12], policy search and model-based algorithms [13,9]. All these approaches use only one pair of inter-task mappings which is usually dictated by a human expert.

Defining multiple mappings of the target task to the source task is a domain-dependent process and requires the involvement of a domain expert. Methods that automatically construct the mapping functions have been recently proposed [10], but are currently very computationally expensive and outside the scope of this paper. Additionally, one could generate *all* feasible mappings and use them. A problem of this approach is that in domains with many features, generating the mappings becomes prohibitive in terms of both computational and memory requirements.

2.1   Transferring with Multiple Inter-task Mappings in Model Based Learners

Model-based Reinforcement Learning algorithms, such as Fitted R-max [1] and PEGASUS [3], use their experiences to learn a model of the transition and reward functions of a task. These models are used to either produce new mental experiences for direct learning (e.g., Q-Learning) or with dynamic programming methods and planning in order to construct a policy. Contrary to direct learning agents, model based RL agents have some added options for transfer learning as they can also transfer the transition and/or reward function of a source task. Models of these functions can have many representations, such as a batch of observed instances or a neural network that approximates its transition and/or reward function.

Transferring instances in model-based RL has been proposed in the TIM-BREL algorithm [9], in which a single-mapping TL algorithm showed significant improvement over the no-transfer case. Our proposed method builds upon the TIMBREL algorithm, but can autonomously use multiple inter-task mappings.

In addition to empirically demonstrating the effectiveness of using an instance-based RL algorithm with multiple mappings, we will also argue the existence of a theoretical connection between the described setting and another setting more extensively studied: multi-task, single-mapping transfer. In a multi-task transfer learning problem, an agent is transferring knowledge from more than one source task into a target task that is more complex, but still related to the source tasks. In order to avoid negative transfer, the agent must decide what information to transfer from which source task, and when to transfer this information [2].

In this case, recently proposed methods can assist the agent when selecting the appropriate source task, as well as when and what knowledge to transfer. Lazaric et al. [2] defines two probabilistic measures, *compliance* and *relevance*. These measures can assist an agent to determine the most similar source tasks to the target task (compliance) and also to select which instances to transfer from that source task (relevance). Results showed that this is a feasible and efficient method for multi-task trasfer—we will extend this method to use multiple inter-task mappings, rather than multiple source tasks.

We consider each inter-task mapping function as a hypothesis, proposed to match the geometry and dynamics between the source and target task. By considering a source task through this hypotheses, we directly differentiate the source task's outcomes for a fixed input. Mapping states and actions from a target task to a source task not only transforms the way we view and use the source task, but also the way it behaves and responds to a fixed target task's state-action query (before these are mapped). Thus, every mapping X_i can be considered as a constructor of a new *virtual source task* S_{X_i}. This naturally **re-formulates the problem of finding the best mapping as a problem of finding the most compliant virtual source task**. Additionally, this re-formulation transforms the problem of finding out when to sample from a certain mapping to a problem of sample relevance.

Based on previously defined notation [2], we define the compliance of a target task transition τ_i and a mapping X_i as:[1]

$$\lambda_{X_i} = \frac{1}{Z^P Z^R} (\sum_{j=1}^{m} \lambda_{ij}^P)(\sum_{j=1}^{m} \lambda_{ij}^R)$$

The compliance between the target task (not only one instance of it) and the virtual source task S_{X_i} generated by the mapping X_i is defined as:

$$\Lambda = \frac{1}{t} \sum_{i=1}^{t} \lambda_i P(s)$$

---

[1] P and R refer respectively to the transition and reward functions of the target task. Z^P and Z^R are normalization constants detailed elsewhere [2].

To implement this new definition of compliance we design a novel multiple-mappings TL algorithm, *COmpliance aware transfer for Model Based REinforcement Learning* (COMBREL), which is able to select mappings based on their compliance as defined above. The equally important notion of relevance is left for future work.

First, COMBREL implements a simple and straightforward segmentation of the state-action space, which we call *mapping regions*. These regions allow for different compliance values—and thus different best mappings—for different regions of the state-action space. This adds the flexibility that each state-action region has its own best mapping instead of computing one best mapping for the whole target task. Additionally, every mapping region can be considered as a different target task (just like each mapping may be considered a different source task). A compliance value is calculated for each mapping region, averaging the compliance of every state-action instance in the region. It is important to note that mapping regions are created based on a fixed resolution. In this paper, a low resolution of four mapping regions, per state variable, was selected assisting on the efficiency of the method and on its actual usefulness since, as (intuitively) the higher the resolution, the less the information we can obtain for each new (thinner) segment of the state space. Its important to note that setting the resolution of the mapping region's segmentation is, for the time being, an experimental choice and not an informed decision.

For COMBREL's underlying model-based algorithm we use Fitted R-Max, building upon the single-mapping transfer algorithm TIMBREL [9]. TIMBREL cooperates with a model-based RL algorithm like Fitted R-Max by assisting its model approximation, by transferring source task instances near the points that need to be approximated. In the absence of such source task instances, no transfer takes place. However, the specific use of Fitted R-Max should not be considered as a limitation of the method as COMBREL is easily adaptable to other instance-based RL algorithms.

The flow of our proposed method and algorithm, COMBREL (see Algorithm 1), is as follows. On lines 1-5, the algorithm records source task transitions and then starts training in the target task. A set of regions is defined, segmenting the state-action space so each can have a different best mapping. Lines 5-8 compute the compliance of the last target task transition with each of the mappings, X_i. This value is then added to the regional compliance estimate. On lines 9 and 10, if the agent's model-based algorithm is unable to approximate a state with its current data, it starts transferring source task data. To select the best mapping for the current agent's state, on lines 11 to 15 the agent decides on the average compliance of the current region for each mapping (i.e., virtual source task). When found, it translates samples from it and combines them with target task samples to create a model. Then, the algorithm iterates and continues learning.

2.2   Multiple Inter-task Mappings in TD Learners

Having discussed using multiple inter-task mappings in model-based RL methods, we now turn our attention to model-free, temporal difference agents.

Algorithm 1. COMBREL - Multiple Mappings in Model Based Learners

---

1: Learn in the source task, recording m (s, a, r, s') transitions.
2: **while** training in the target task **do**
3:     **for** (e epidodes) **do**
4:         Define D_i regions of the state-action space in the target task
5:         Record Target Task transitions $\langle x_T, a_T \rangle$
6:         Find Region membership D for each $\langle x_T, a_T \rangle$
7:         For each Mapping X_i define a virtual source task S_i
8:         Update region's compliance estimate, adding compliance λ_i of $\langle x_T, a_T \rangle$ with each virtual source task S_i
9:         **if** the model-based RL algorithm is unable to accurately estimate $T_T(x_T, a_T)$ or $R_T(x_T, a_T)$ **then**
10:            **while** $T_T(x_T, a_T)$ or $R_T(x_T, a_T)$ does not have sufficient data **do**   ▷ Use the most compliant mapping to save instances near $\langle x_T, a_T \rangle$
11:                Find $|X|$ pairs (X_S, X_A) of inter-task mappings
12:                Find Region membership of $\langle x_T, a_T \rangle$
13:                **for** ($|X|$ iterations) **do**
14:                    Find X^{best} based on current region's compl. estimate Λ_{D_i} with X_i
15:                    $X^{best} \leftarrow X_{\max(\Lambda_{D_i})}$ ▷ The most compliant
16:                Translate samples from S^{best} and put them in $\hat{T}$
17:                Locate 1 or more instances in $\hat{T}$ near the $\langle x_T, a_T \rangle$ to be estimated.

We assume that an RL agent has been trained in the source task and that it has access to a function $Q_{source}(s, a')$, which returns an estimation of the Q value for a state s' and action a' of the source task. The agent is currently being trained in the target task, learning a function $Q_{target}(s, a)$ and senses the state s. In order to transfer the knowledge from the source task, we find the best mapping and we add the corresponding values from the source task via the selected mapping. Algorithm 2 shows the pseudocode of the transferring procedure.

On lines 3–10, the algorithm finds the best mapping from instances that are recognized in the target task. More specifically, we define the best mapping as the one with the maximum mean sum of the values for each action in the source task (lines 6–8). There are alternate ways to select the best mapping, such as finding the maximum action value, but we leave such explorations to future work.

After the best mapping is found, the algorithm adds to the Q-values from the source task to the Q-values from the target task (lines 11-12). We use a mapping function, g_A, which is an inverse function of f_A and maps a source action to its equivalent target action. Note that if a target action is not mapped to a source action, the algorithm does not add an extra value. Depending on the impact of the source task's Q-values, the agent will now be biased to select actions from the set $A_a = \{g_A^{bestMapping}(a') | a' \in A_{source}^{s'}\}$. The impact of the source task depends strongly on how much time was spent learning it. Specifically, if the source task was trained for a small number of episodes then the Q-values will be also small and afterwards the agent will probably select actions from A_a for a small period (assuming pessimistic initialization). As learning proceeds, the values of the target Q-function will increase and the initial bias will be overridden.

Algorithm 2. Value-Addition procedure: Multiple Mappings in TD Learners

---

1: **procedure** VALUE-ADDITION($s, M_S, M_A, Q_{target}, Q_{source}$)
2:     $bestMeanQValue \leftarrow 0$; $bestMapping \leftarrow 0$
3:     **for** $i \leftarrow 1 \ldots N$ **do**
4:         $s' \leftarrow M_S^i(s)$; $meanQValue \leftarrow 0$
5:         **for all** $a' \in A_{source}(s')$ **do**
6:             $meanQValue \leftarrow meanQValue + Q_{source}(s', a')$
7:         $meanQValue \leftarrow meanQValue/|A_{source}(s')|$
8:         **if** $meanQValue > bestMeanQValue$ **then**
9:             $bestMeanQValue \leftarrow meanQValue$; $bestMapping \leftarrow i$
10:     $s' \leftarrow M_S^{bestMapping}(s)$
11:     **for** $a' \in A_{source}(s')$ **do**
12:         $Q_{target}(s, g_A^{bestMapping}(a')) \leftarrow Q_{target}(s, g_A^{bestMapping}(a')) + Q_{source}(s', a')$

---

3   Domains

3.1   Mountain Car

For the mountain car domain we use the version proposed by [4]. In the standard 2D task an underpowered car must be driven up to a hill. The state of the environment is described by two continuous variables, the horizontal position $x \in [-1.2, 0.6]$, and velocity $v_x \in [-0.007, 0.007]$. The actions are {Neutral, Left and Right} move the car to the goal state, as quickly as possible. The 4D mountain car extends the 2D task by adding an extra spatial dimension [9]. The state is composed by four continuous variables (4D), the coordinates in space x, and $y \in [-1.2, 0.6]$, as well as the velocities v_x and $v_y \in [-0.07.0.07]$. The available actions are {Neutral, West, East, South, North}.

3.2   Keepaway

Keepaway [6], is a subset of the RoboCup robot soccer domain, where K keepers try to hold the ball for as long as possible, while T takers (usually $T = K - 1$) try to intercept the ball. The agents are placed within a fixed region at the start of each episode, which ends when the ball leaves this region or the takers intercept it.[2]

The task becomes harder as extra keepers and takers are added to the fixed-sized field, due to the increased number of state variables on one hand, and the increased probability of ball interception in an increasingly crowded region on the other.

4   Experiments and Results

4.1   Transferring with COMBREL in Mountain Car 4D

To test the efficiency of the model-based proposed approach we conducted a series of experiments in the Mountain Car 4D domain. First, a standard Q-learning

---

[2] For more information please refer to the original paper [6].

agent learned Mountain Car 2D while recording instances from its experiences. These instances, in the form of tuples $\langle s, a, r, s' \rangle$, formed a dataset of 100,000 instances, covering the state-space. The number of experiences gathered from Mountain Car 2D allowed us to create multiple datasets of varying size. These datasets will be used as the source task knowledge.

The first experiment was conducted using a fixed sample size (2,000 instances). We tested COMBREL and TIMBREL with Multiple Mappings (TIM-BERL+MM) (no selection mechanism) against eight versions of single-mapping TIMBREL. Each of these versions used a different state-action mapping. This experiment revealed the importance of using multiple mappings when we have limited experimentation time and no knowledge or intuition of the best mappings that should be used. COMBREL outperformed single mapping TIMBREL for any of its versions, in their first run.

TIMBREL with its best mapping, identified in the previous experiment, is tested against the multiple mappings versions in three more experiments, using 1,000, 5,000, and 10,000 source task instances . This experiment compares the performance of COMBREL against TIMBREL and TIMBREL+MM, independently of the sample size. Furthermore this experiment allows us to analyse the sample efficiency of the three algorithms while varying the source task sample size.

About the set of parameters used for Fitted R -Max and TIMBREL, these were the same as [9]. In our experiments COMBREL, TIMBREL+MM and TIMBREL used a model breadth parameter $b = 0.4$ and a minimum fraction parameter of 10%

We ran 20 trials for each of the compared methods and for each of the source task sample sizes. Figure 1(a) shows the performance of the compared algorithms when transferring 1,000 source task instances. There is a clear performance increase with the use of COMBREL, when compared to TIMBREL+MM and the No-Transfer case (the difference was statistically significant at a 95% level). When compared to TIMBREL the gain is still statistically significant, but only for episodes 80 and above. Figure 1(b) shows results from a similar experiment, but with a 2,000 instances from the source task. COMBREL demonstrated an even greater performance increase, but it was more unstable, showing the sensitivity of the multiple mappings methods with the transferred sample size. Experiments with 5,000 and 10,000 source task instances showed a similar and more stable performance gain (but are omitted for space constraints). It is important to note the poor performance of TIMBREL+MM compared to the other algorithms. This shows that using a set of mappings simultaneously, without previous domain knowledge and any selection mechanism, can result to negative transfer. Table 1 shows the four state mappings used by COMBREL in our experiments, as well as the number of times they were selected. We can observe that the more "intuitively correct" mappings were found and selected most often. Mappings which seem irrelevant were also used some times, but likely in state space regions where they were useful.

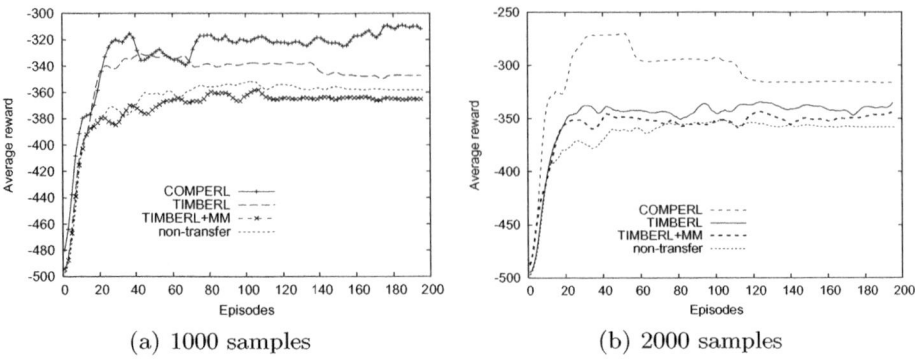

(a) 1000 samples   (b) 2000 samples

Fig. 1. Average reward in a period of 200 episodes. The curves are smoothed with a window of 10 episodes

Table 1. Four state space mappings in Mountain Car 4D and the number of times they were selected by COMBREL, in an indicative episode

4D state var.	Mapping 1	Mapping 2	Mapping 3	Mapping 4
x	x	x	v_x	x
y	x	v_x	v_x	x
v_x	v_x	x	x	v_x
v_y	v_x	v_x	x	v_x
Times used	269.992	52.075	29.966	322.411

4.2   Transferring with Value-Addition in Keepaway

This section evaluates the proposed approach in Keepaway. The dimensions of the keepaway region is set to 25m × 25m and remains fixed for all source and target tasks. The algorithm that is used to train the keepers is the SMDP variation of Sarsa(λ) [7]. Additionally, we use linear tile-coding for function approximation with settings shown to work well in the Keepaway domain [5].[3]

To evaluate the performance of the proposed approach we use the *time-to-threshold* metric [12], which measures the time required to achieve a predefined performance threshold in the target task. Typically, the threshold is set empirically, after preliminary experimentation in the target task. In our case this threshold corresponds to a number of seconds that keepers maintain ball possession. In order to conclude that the keepers have learned the task successfully, the average performance of 1,000 consecutive episodes must be greater than the threshold. We compare (1) the time-to-threshold without transfer learning with (2) the time-to-threshold with transfer learning plus the training time in the source task.

---

[3] We use 32 tilings are used for each variable. The width of each tile is set to 3 meters for the distance variables and 10 degrees for the angle variables. We set the learning rate, α, to 0.125, ϵ to 0.01 and λ to 0. These values remain fixed for all experiments.

Finally, an important aspect of the experiments concerns the way that the mapping will be produced. For the Keepaway domain, the production of the mappings can semi-automatic. More specifically, any K^t vs. T^t task can be mapped to a K^s vs. T^s task, where $K^s < K^t$ and $T^s < T^t$, simply by deleting $K^t - K^s$ teammates and $T^t - T^s$ opponents of the keeper with ball possession. The actual number of different mappings is:

$$\binom{K^t - 1}{K^s - 1}\binom{T^t}{T^s} = \frac{(K^t - 1)!T^t!}{(K^s - 1)!(K^t - K^s)!T^s!(T^t - T^s)!}$$

Transfer into 4 vs. 3 from 3 vs. 2. This subsection evaluates the performance of the proposed approach on the 4 vs. 3 target task using a threshold of 9 simulated seconds. We use the 3 vs. 2 task as the source and experiment with different number of training episodes, ranging from 0 (no transfer) to 3,200.

Table 2 shows the training time and average performance (in seconds) in the source task, as well as time-to-threshold and total time in the target task for different amount of training episodes in the source task. The results are averaged over 10 independent runs and the last column displays the standard deviation. The time-to-threshold without transfer learning is about 13.38 simulated hours.

Table 2. Training time and average performance in the source task, as well as time-to-threshold and total time in the target task. The best time-to-threshold and total time are in **bold**.

	3 vs. 2		4 vs. 3		
#episodes	train time	performance	time-to-thr.	total time	st. dev.
0	0	–	13.38	13.38	2.02
100	0.11	4.38	13.19	13.30	1.77
200	0.23	4.67	12.59	12.82	2.10
400	0.72	6.71	12.08	12.80	1.70
800	1.73	8.52	10.28	12.01	0.97
1600	4.73	12.20	3.48	**8.21**	1.16
2500	8.42	16.02	4.16	12.44	0.60
3200	12.17	16.84	**2.76**	14.95	0.28

We first notice that the proposed approach leads to lower time-to-threshold in the target task compared to the standard algorithm that does not use transfer learning. This is due to the fact that the more the training episodes in the source task the better the Q-function that is learned. Note that for 800 episodes of the 3 vs. 2 task, the keepers are able to hold the ball for an average of 8.5 seconds, while for 1600 episodes their performance increases to 12.2 seconds. As the number of the training episodes in the source task increase the time that is required to reach the threshold decreases, showing that our method successfully improves performance in the target task.

The total time of the proposed approach in the target task is also less than the time-to-threshold without transfer learning in many cases. The best performance

Table 3. Average training times (in hours) for 5 vs. 4. The results are averaged over 10 independent trials. The best time-to-threshold and total time are in **bold**.

task	source #episodes	5 vs. 4 tr. time	time-to-thr.	total time	st. dev.
-	0	0	26.30	26.30	2.85
3 vs. 2	1600	4.73	15.54	20.27	3.24
3 vs. 2	2500	8.42	8.88	17.31	3.23
3 vs. 2	3200	12.17	**3.43**	**15.60**	1.24
4 vs. 3	4000	7.52	10.07	17.59	1.49
4 vs. 3	6000	12.32	9.26	21.58	1.92

is 8.21 hours, which corresponds to a reduction of 38.6% of the time-to-threshold without transfer learning. This performance is achieved when training the agents for 1600 training episodes in the source task. This result shows that rather than directly learning on the target task, it is actually faster to learn on the source task, use our transfer method, and only then learn on the target task.

In order to detect significant difference among the performances of the algorithms we use the paired t-test with 95% confidence. We perform seven paired t-tests, one for each pair of the algorithm without transfer learning and the cases with transfer learning. The test shows that the proposed approach is significantly better when it is trained with 800 and 1600 episodes.

Scaling up to 5 vs. 4. We also test the performance of the proposed in the 5 vs. 4 target task. The 5 vs. 4 threshold is set to 8.5 seconds. The 3 vs. 2 task with 1,600, 2,500 and 3,200 training episodes and the 4 vs. 3 task with 4,000 and 6,000 training episodes are used as source tasks. Table 3 shows the training times, time-to-threshold and their sum for the different source tasks and number of episodes averaged over 10 independent runs along with the standard deviation.

Firstly, we note that in all cases the proposed approach outperforms learning without transfer learning. It is interesting to note that the best time-to-threshold is achieved when using 3 vs. 2 as a source task is achieved with fewer episodes than when using 4 vs. 3 as a source task. This means that a relatively simple source task may provide *more* benefit than a more complex source task. In addition, the 3 vs. 2 task requires less training time, as it is easier than the 4 vs. 3 task. We perform 5 paired t-tests, one for each case of the proposed approach against learning without transfer. The proposed method achieves statistically significantly higher performance (at the 95% confidence level) in all cases.

5   Related Work

[9] proposes an algorithm that implements Transfer Learning for Model - Based RL agents. It is based on transferring instances from a source task to a target task and using them with the model-based RL algorithm Fitted R-Max to assist it in the construction of a model of the target task. Although it demonstrates

promising results and a significant performance gain it uses only one hand-coded mapping between the source and the target task. Our proposed model-based multiple mappings method is based on TIMBREL but extends it with the use of a multiple mappings mechanism able to autonomously select mappings while learning.

[12] constructs an autonomous mapping selection method (MASTER) which is able to select a mapping based on the similarity between transitions in the source and target task . MASTER learns a model of the action effects in the source task. In the target task, it first selects actions randomly, sampling target task transitions. It then compares this transitions with queries on the source task model to calculates the error (difference). It selects the best mapping to minimize this error. Our model-based method is also based on the similarity of the effects between the two tasks but it makes an off-line calculation of the error thus, not spending agent time with random environment interaction. Also as our method is built up upon TIMBREL, it requires no explicit model learning in the source task as it transfer only instance from it.

In [8], each possible inter-state mapping is considered as an expert and with the use of a multi-armed bandit algorithm the agent decides which mapping to use. This method shows significant improvement but its performance is surpassed in the long run as it continues to explore always taking "advice" from low return experts.

For space limitations reasons, the reader is directed to read more about the various transfer learning methodologies in more comprehensive treatments [11].

6   Conclusions and Future Work

In this paper, we examined the benefit of using multiple-mappings in the transfer learning setting. To avoid negative transfer a multiple-mappings method must be supported by a mechanism able to select the most compliant and relevant mappings.

Results on our proposed methods showed a statistically significant benefit over the single mapping transfer algorithm as also from the multiple mapping method that uses all the mappings simultaneously. Future work includes discovering more sophisticated ways to discretize the state space in COMBREL, avoiding discontinuities in the mapping selection function between nearby states in the state space. Furthermore, whereas COMBREL implements the novel notion of a mappings compliance, this scheme can be extended to another important probabilistic measure, that of relevance [2], where we would not only care about the best mapping but also on when to sample from it and how much.

References

1. Jong, N.K., Stone, P.: Model-Based Exploration in Continuous State Spaces. In: Miguel, I., Ruml, W. (eds.) SARA 2007. LNCS (LNAI), vol. 4612, pp. 258–272. Springer, Heidelberg (2007)

2. Lazaric, A.: Knowledge Transfer in Reinforcement Learning. PhD thesis, Politecnico di Milano (2008)
3. Ng, A., Coates, A., Diel, M., Ganapathi, V., Schulte, J., Tse, B., Berger, E., Liang, E.: Autonomous inverted helicopter flight via reinforcement learning. In: Experimental Robotics, vol. IX, pp. 363–372 (2006)
4. Singh, S.P., Sutton, R.S.: Reinforcement learning with replacing eligibility traces. Machine Learning 22(1-3), 123–158 (1996)
5. Stone, P., Kuhlmann, G., Taylor, M.E., Liu, Y.: Keepaway Soccer: From Machine Learning Testbed to Benchmark. In: Bredenfeld, A., Jacoff, A., Noda, I., Takahashi, Y. (eds.) RoboCup 2005. LNCS (LNAI), vol. 4020, pp. 93–105. Springer, Heidelberg (2006)
6. Stone, P., Sutton, R.S.: Keepaway Soccer: A Machine Learning Testbed. In: Birk, A., Coradeschi, S., Tadokoro, S. (eds.) RoboCup 2001. LNCS (LNAI), vol. 2377, pp. 214–223. Springer, Heidelberg (2002)
7. Sutton, R.S., Barto, A.G.: Reinforcement Learning, An Introduction. MIT Press (1998)
8. Talvitie, E., Singh, S.: An experts algorithm for transfer learning. In: Twentieth International Joint Conference on Artificial Intelligence (IJCAI 2007), pp. 1065–1070 (2007)
9. Taylor, M.E., Jong, N.K., Stone, P.: Transferring Instances for Model-Based Reinforcement Learning. In: Daelemans, W., Goethals, B., Morik, K. (eds.) ECML PKDD 2008, Part II. LNCS (LNAI), vol. 5212, pp. 488–505. Springer, Heidelberg (2008)
10. Taylor, M.E., Kuhlmann, G., Stone, P.: Autonomous transfer for reinforcement learning. In: 7th International Joint Conference on Autonomous Agents and Multiagent Systems, pp. 283–290 (2008)
11. Taylor, M.E., Stone, P.: Transfer learning for reinforcement learning domains: A survey. Journal of Machine Learning Research 10(1), 1633–1685 (2009)
12. Taylor, M.E., Stone, P., Liu, Y.: Transfer learning via inter-task mappings for temporal difference learning. Journal of Machine Learning Research 8, 2125–2167 (2007)
13. Taylor, M.E., Whiteson, S., Stone, P.: Transfer via inter-task mappings in policy search reinforcement learning. In: 6th International Joint Conference on Autonomous Agents and Multiagent Systems, pp. 37:1–37:8 (2007)

Multi-Task Reinforcement Learning: Shaping and Feature Selection

Matthijs Snel and Shimon Whiteson

Intelligent Systems Lab Amsterdam (ISLA),
University of Amsterdam, 1090 GE Amsterdam, Netherlands
m.snel, s.a.whiteson@uva.nl

Abstract. Shaping functions can be used in multi-task reinforcement learning (RL) to incorporate knowledge from previously experienced source tasks to speed up learning on a new target task. Earlier work has not clearly motivated choices for the shaping function. This paper discusses and empirically compares several alternatives, and demonstrates that the most intuive one may not always be the best option. In addition, we extend previous work on identifying good representations for the value and shaping functions, and show that selecting the right representation results in improved generalization over tasks.

1 Introduction

Transfer learning approaches to reinforcement learning (RL) aim to improve performance on some *target tasks* by leveraging experience from some *source tasks*. Clearly, the target tasks must be related to the source tasks in order for transfer to be beneficial. In *multi-task reinforcement learning* (MTRL), this relationship is formalized with a *domain*, a distribution over tasks from which the source and target tasks are independently drawn [18]. For example, consider a domain consisting of a fixed goal location in a room with obstacles, where each obstacle configuration constitutes a different task. In this case, the agent could exploit the fact that the goal is always in the same place to learn new tasks more quickly.

Various approaches to MTRL exist, such as using the source tasks to form a prior for Bayesian RL [7, 21] or transfering state abstractions [19] (see [18] for an overview of transfer learning approaches for RL). In this paper, we consider a different approach in which the source tasks are used to learn a *shaping function* that guides learning in the target tasks. Shaping functions [10] augment a task's base reward function, which is often sparse, with additional artificial rewards. By capturing prior knowledge about the task, manually designed shaping functions have proven successful at speeding learning in single-task RL (e.g. [1, 3, 20]). In MTRL, a shaping function can be learned automatically from the source tasks. In the above example, an agent that observes that the goal location is fixed in the source tasks could learn a shaping function that rewards approaching this location.

S. Sanner and M. Hutter (Eds.): EWRL 2011, LNCS 7188, pp. 237–248, 2012.
© Springer-Verlag Berlin Heidelberg 2012

This paper makes two contributions. First, it investigates several strategies for learning shaping functions. Earlier work learned a shaping function based on an approximation of the optimal value function V^* or Q^* for each source task [14, 6, 17]. While these approaches are intuitive, they have never been explicitly motivated. We discuss and compare several alternatives, and demonstrate that approximating the optimal value function across tasks may not always be the best option.

Second, this paper also considers what representation the shaping function should employ. Konidaris and Barto observed that the best state features to use for shaping can be different from those used to represent the value function for each task. However, their experiments relied on manually selecting these features. Snel and Whiteson proposed a definition of the relevance of a feature for shaping and provided a method that uses this definition to automatically construct a shaping function representation. However, while their experiments validate their definition, they do not demonstrate that automatically finding the right shaping features can improve performance. We extend this work by casting the problem of learning a shaping function as a regression problem to which feature selection methods from supervised learning can be applied. Using this approach, we demonstrate that effective feature selection improves generalization from the source tasks, which in turn improves the shaping function's ability to speed target task learning.

2   Background and Notation

An MDP is a tuple $m = \langle \mathcal{S}_m, \mathcal{A}_m, \mathcal{B}_m, P_m, R_m, \gamma \rangle$ with set of states $\mathcal{S}_m$, set of actions $\mathcal{A}_m$, and set of admissible action pairs $\mathcal{B}_m \subseteq \mathcal{S}_m \times \mathcal{A}_m$. A transition from s to s' given a has probability $P_m^{sas'}$ and results in expected reward $R_m^{sas'}$. The expected γ-discounted return in m when taking a in s under policy π_m is $Q_m^{\pi}(s, a)$, and under the optimal policy π_m^* it is $Q_m^*(s, a)$.

A *domain* d is a distribution $\Pr(m)$, where $m \in \mathcal{M}$, a set of MDPs. The domain has state set $\mathcal{S}_d = \bigcup_m \mathcal{S}_m$, action set $\mathcal{A}_d = \bigcup_m \mathcal{A}_m$ and admissible action pairs $\mathcal{B}_d = \bigcup_m \mathcal{B}_m$, allowing for MDPs with different state and action spaces.

This paper considers *potential-based* shaping functions of the form $F_{s'a'}^{sa} = \gamma \Phi(s', a') - \Phi(s, a)$, where $\Phi : \mathcal{S} \times \mathcal{A} \mapsto \mathbb{R}$ is a *potential function* that assigns a measure of "desirability" of a state-action pair and s' and a' are the next state and action. Ng et al. [10] showed that such shaping functions preserve the optimal policy of the MDP and Wiewiora et al. [20] showed that using them is equivalent to initializing the agent's Q-table to Φ.

3   Approximating the Optimal Shaping Function

In this section, we formulate a definition of an optimal potential-based shaping function and propose four different ways of approximating it. We focus on a transfer learning scenario in which the agent aims to maximize the total reward accrued while learning on a set of target tasks; this measure also implicitly captures the time to reach a good policy. Furthermore, for simplicity we assume a tabular learning algorithm L is employed on each target task. In this setting, an optimal shaping function is one based on a potential function that maximizes expected return across target tasks:

$$\Phi_L^* = \operatorname*{argmax}_{\Phi} \mathrm{E}\left[R_m | \Phi, L\right] = \operatorname*{argmax}_{\Phi} \sum_{m \in \mathcal{M}} \Pr(m) \mathrm{E}\left[\sum_{t=0}^{\infty} \gamma^t r_{t,m} | \Phi, L\right], \qquad (1)$$

where R_m is the return accrued in task m and $r_{t,m}$ is the immediate reward obtained on timestep t in task m. Note that the task should be seen as a hidden variable since the potential function does not take it as input. This means that the best potential function may perform poorly in some tasks; however, on average across tasks, it will do better than any other function.

Since shaping with Φ is equivalent to initializing the Q-table with it, solving (1) is equivalent to finding the best cross-task initialization of the Q-table. Unfortunately, there is no obvious way to compute such a solution efficiently, and search approaches quickly become impractical. Therefore, in the following sections we discuss four strategies for efficiently approximating Φ_L^*.

3.1   Initialization Closest to Q_m^*

Intuitively, a good initialization of the Q-function is the one that is closest in expectation to the optimal value function Q_m^* of the target task m the agent will face. Since choosing Φ is equivalent to choosing such an initialization, one way to approximate Φ_L^* is by minimizing the *expected mean squared error* (EMSE) across tasks:

$$EMSE(\Phi) = \sum_{m \in \mathcal{M}} \Pr(m) \sum_{s,a \in \mathcal{B}_m} \Pr(s,a|m) \Big[Q_m^*(s,a) - \Phi(s,a)\Big]^2, \qquad (2)$$

where $\Pr(s,a|m)$ is a task-dependent weighting over state-action pairs that determines how much each pair contributes to the error.

It is not immediately clear how to select $\Pr(s,a|m)$, though a minimal requirement is that $\Pr(s,a|m) = 0$ for all $(s,a) \notin \mathcal{B}_m$. The simplest option is to set $\Pr(s,a|m) = 1/|\mathcal{B}_m|$ for all $(s,a) \in \mathcal{B}_m$. Since the $EMSE(\Phi)$ averages over Q_m^*, another option is to use the distribution induced by π_m^*. However, this may be problematic as it represents only the distribution over (s,a) *after* learning. There may be state-action pairs that are never visited under π_m^* but that are visited during learning and for which $EMSE(\Phi)$ is thus important. The best choice of $\Pr(s,a|m)$ may thus be the distribution over (s,a) *during* learning. A disadvantage is that this requires applying L to all $m \in \mathcal{M}$.

Given a choice of $\Pr(s,a|m)$, we can find $\hat{\Phi}_m = \operatorname{argmin}_{\Phi} EMSE(\Phi)$ by setting the gradient of (2) to zero and solving, yielding:

$$\hat{\Phi}_*(s,a) = \sum_{m \in \mathcal{M}} \Pr(m|s,a) Q_m^*(s,a) \qquad \forall s,a \in \mathcal{B}_d. \qquad (3)$$

The notation $\hat{\Phi}$ indicates an approximation to Φ_L^*. The subscript indicates that this approximation averages over the optimal solution for each m. For (s, a) in $\mathcal{B}_d$ but not in $\mathcal{B}_m$ for a given m, we define $Q_m^*(s, a) \equiv 0$. In any case, these values are already excluded from the average by the weighting $\Pr(m|s, a)$.

While approaches similar to (3) have been used successfully in previous work [6, 14, 17], they are not guaranteed to be optimal. Since they minimize error with respect to the optimal value function that *results* from learning, they may be too optimistic *during* learning, which the shaping function is meant to guide. Consequently, they may cause the agent to over-explore the target task, to the detriment of total reward. Section 4 will provide experimental support for this claim.

3.2   Initialization Closest to $\tilde{Q}_m$

In some cases (e.g., when using function approximation or an on-policy algorithm with a soft policy), the agent's Q-function on a sournce or target task m will never reach Q_m^*, even after learning. When this occurs, it may be better to base the potential function on an average over $\tilde{Q}_m$, the value function to which the learning algorithm converges. The derivation is the same as for the previous section, yielding:

$$\hat{\Phi}_{\tilde{Q}}(s) = \sum_{m \in \mathcal{M}} \Pr(m|s, a)\tilde{Q}_m(s, a) \qquad \forall s, a \in \mathcal{B}_d. \tag{4}$$

3.3   Best Fixed Cross-Task Policy

While $\hat{\Phi}_{\tilde{Q}}$ is less optimistic than $\hat{\Phi}_*$, it may still be too optimistic since it is also based on minimizing error with respect to a value function *after* learning. To address this issue, we can instead base the potential function on the optimal *cross-task value function*. A cross-task value function describes the expected return for a fixed *cross-task policy*, i.e., one that assigns the same probability to a given state-action pair regardless of what task it is used in.

We define μ^* to be the policy that maximizes expected total reward given the initial state distribution $\Pr(S_0 = s)$:

$$V_d^\mu = \sum_s \Pr(S_0 = s) \sum_a \mu(s, a)Q_d^\mu(s, a) \tag{5}$$

$$Q_d^\mu(s, a) = \sum_{m \in \mathcal{M}} \Pr(m|s, a)Q_m^\mu(s, a), \tag{6}$$

where V_d^μ is the domain-wide value of μ, $\Pr(m|s, a)$ is a weighting term like in section 3.1, and $Q_m^\mu(s, a)$ is the value of pair (s, a) in m under μ. Unfortunately, it seems that no Bellman-like equation can be derived from (6), so we are left with a dependency on $Q_m^\mu(s, a)$. One consequence of this is that we are restricted to policy search methods for trying to find μ^*.

Using these definitions, we can define a potential function based on the optimal cross-task value function:

$$\hat{\varPhi}_{\mu^*}(s, a) = Q_d^{\mu^*}(s, a) \qquad \forall s, a \in \mathcal{B}_d. \tag{7}$$

The key distinction between this potential function and both $\hat{\varPhi}_{\bar{Q}}$ and $\hat{\varPhi}_*(s, a)$ is that it averages expected return of a single fixed policy. In contrast, in $\hat{\varPhi}_{\bar{Q}}(s)$ and $\hat{\varPhi}_*(s, a)$ averages are computed over different policies for each task.

3.4   Averaging MDP

A potential drawback of the previous approaches is that solutions to all tasks must be computed, or a policy search must be done to find μ^*. If the task models are available, then we could take the average of these models according to $\Pr(m)$, and compute the solution to this *averaging MDP*, defined as

$$\bar{R}^{sas'} = \sum_{m \in \mathcal{M}} \Pr(m|s, a) R_m^{sas'} \tag{8}$$

$$\bar{P}^{sas'} = \sum_{m \in \mathcal{M}} \Pr(m|s, a) P_m^{sas'}. \tag{9}$$

It might not be immediately clear what the difference, if any, between the solution to the averaging MDP and for example equation 3 is. As section 4 will show, it turns out that they are not necessarily the same, although there exist tasks for which they are. Furthermore, the solution to the averaging MDP does usually not correspond to μ^*, since the latter may not be deterministic (analogous to a reactive policy for a POMDP).

The potential function equals the solution to the averaging MDP:

$$\hat{\varPhi}_{\mathrm{avg}}(s, a) = \bar{Q}^*(s, a) \qquad \forall s, a \in \mathcal{B}_d. \tag{10}$$

Finally, it should be clear that in this case we should use $\Pr(s, a|m) = 1/|\mathcal{B}_m|$.

4   Shaping Function Evaluation

This section empirically compares the four different potential functions proposed in the previous sections to a baseline agent that does not use shaping. For comparison purposes, we assume we have perfect knowledge of the domain and compute each potential function using all tasks in the domain [1]. Evaluation occurs using a sample of tasks from the same domain. This enables us to compare the potential functions' maximum potential, untainted by sampling error. In the next sections, we discuss and evaluate cases in which only a sample of tasks from the domain is available.

---

[1] With perfect knowledge of the domain, a better approach would be to store the optimal policies for each task and select the appropriate one while interacting with the target task. However, our current approach serves to demonstrate the theoretical advantages of each potential function type.

4.1 Domain

To illustrate a scenario in which $\hat{\Phi}_*$, the average over optimal value functions, is not the optimal potential function, we define a *cliff domain* based on the episodic cliff-walking grid world from Sutton and Barto [16]. One task from this domain is shown in Fig. 1a. The agent starts in **S** and needs to reach the goal location **G**, while avoiding stepping into the cliff represented by the black area.

The domain consists of all permutations with the goal and start state in opposite corners of the same row or column with a cliff between them (8 tasks in total). Each task is a 4x4 grid world with deterministic actions N, E, S, W, states (x, y), and a -1 step penalty. Falling down the cliff results in -1000 reward and teleportation to the start state. The distribution over tasks is uniform.

4.2 Method

We compute each potential function according to the definitions given in section 3. Since we cannot compute the cross-task policy μ^* exactly, we use an evolutionary algorithm (EA) to approximate it.

To illustrate how performance of a given Φ depends on the learning algorithm, we use two standard RL algorithms, Sarsa(0) and Q(0). Since for Q-Learning, $\hat{\Phi}_{\tilde{Q}}(s) = \hat{\Phi}_*(s)$, we use Sarsa's solution for $\hat{\Phi}_{\tilde{Q}}(s)$ for both algorithms. Both algorithms use an ϵ-greedy policy with $\epsilon = 0.1$, $\gamma = 1$, and the learning rate $\alpha = 1$ for Q-Learning and $\alpha = 0.4$ for Sarsa, maximizing the performance of both algorithms for the given ϵ.

We also run an additional set of experiments in which the agent is given a "cliff sensor" that indicates the direction of the cliff (N, E, S, W) if the agent is standing right next to it. Note that the addition of this sensor makes no difference for learning a single task: it does not change the optimal policy, nor does the number of states per task increase. However, the number of states in the domain does increase: one result of adding the sensor is that tasks no longer have identical state spaces.

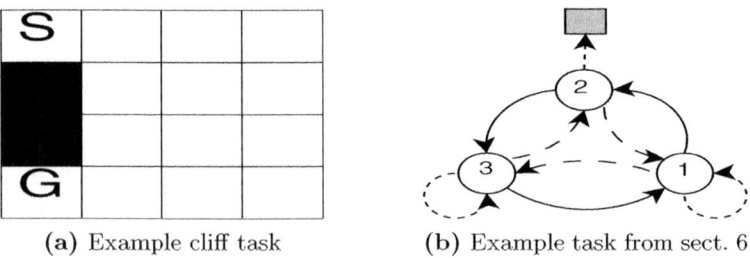

(a) Example cliff task   (b) Example task from sect. 6

Fig. 1. Example tasks from the domains used for evaluation in section 4 (a) and 6 (b). In (b), line types solid, dashed, and dotted represents actions 1, 2, and 3; grey square is terminal.

Table 1. Mean total reward for various shaping configurations and learning algorithms on the cliff domain. All numbers $\times 10^4$. Intervals represent the 95% confidence interval.

<table>
<tr><td colspan="3" align="center">(a) Without sensor</td><td colspan="3" align="center">(b) With sensor</td></tr>
<tr><td></td><td>**Q-Learning**</td><td>**Sarsa**</td><td></td><td>**Q-Learning**</td><td>**Sarsa**</td></tr>
<tr><td>$\hat{\Phi}_{\mu^*}$</td><td>-19.77 ± 2.43</td><td>-512 ± 130</td><td>$\hat{\Phi}_{\mu^*}$</td><td>−</td><td>−</td></tr>
<tr><td>$\hat{\Phi}_{\text{avg}}$</td><td>-5.83 ± 0.13</td><td>-4.48 ± 0.11</td><td>$\hat{\Phi}_{\text{avg}}$</td><td>−</td><td>−</td></tr>
<tr><td>No shaping</td><td>-5.86 ± 0.12</td><td>-3.86 ± 0.10</td><td>No shaping</td><td>-5.85 ± 0.13</td><td>-3.96 ± 0.11</td></tr>
<tr><td>$\hat{\Phi}_*$</td><td>-5.13 ± 0.17</td><td>-3.96 ± 0.11</td><td>$\hat{\Phi}_*$</td><td>-5.44 ± 0.12</td><td>-3.67 ± 0.12</td></tr>
<tr><td>$\hat{\Phi}_{\tilde{Q}}$</td><td>-4.74 ± 0.19</td><td>-3.93 ± 0.11</td><td>$\hat{\Phi}_{\tilde{Q}}$</td><td>-4.75 ± 0.17</td><td>-3.37 ± 0.13</td></tr>
</table>

For each potential function, we report the mean total reward incurred by sampling a task from the domain, running the agent for 500 episodes, and repeating this 100 times.

4.3   Results

Table 1a shows the performance for the scenario without cliff sensor. On this domain, $\hat{\Phi}_d^\mu$ performs very poorly; one reason for this may be that the EA did not find μ^*, but a more likely one is that due to the structure of the domain, even μ^* would incur a very low return for each state and thus lead to a very pessimistic potential function.

As expected, Sarsa outperforms Q-Learning on this domain due to its on-policy nature: because Q-Learning learns the optimal policy directly, it tends to take the path right next to the cliff and is thus more likely to fall in. For both algorithms, $\hat{\Phi}_{\text{avg}}$ does not outperform the baseline agent without shaping. For Q-Learning, $\hat{\Phi}_*$ and $\hat{\Phi}_{\tilde{Q}}(s)$ do better than the baseline, with the latter doing significantly better. Inspection of the learning curves shows that for $\hat{\Phi}_{\tilde{Q}}(s)$, Q-Learning initially incurs high average reward because it is driven away from the cliff by the potential function (remember that this is the average *Sarsa* solution to the domain), but thereafter a regress occurs as it returns to its risky behavior. For Sarsa, there is no significant difference between the baseline and either potential function. This is slightly surprising, since an initial gain would be expected.

The situation changes when the cliff sensor is added (we did not retest the two potential functions that did worse than the baseline), see table 1b. Although the sensor does not help speed learning within a given task, across tasks it provides consistent information. Whenever the grid sensor provides a signal, the agent should not step in that direction. This information is reflected in the average of the value functions and thus in the potential function (technically, what happens is that the state-action space $\mathcal{B}_d$ is enlarged, and fewer state-action pairs are shared between tasks). Under these circumstances, both $\hat{\Phi}_*$ and $\hat{\Phi}_{\tilde{Q}}$ significantly outperform baseline Sarsa, with the latter, again, doing best. The picture for Q-Learning remains largely the same.

5   Shaping Function Representations

We now turn to the case where only a sample of N tasks is available, and our goal is to maximize the expected total return while learning on a number of target tasks, sampled from the same domain. That is, the goal is to find the Φ_L^* that satisfies (1), where $\mathcal{M}$ is the set of possible target tasks.

One option is to compute a $\hat{\Phi}$ that exactly matches the sample data, by letting the set $\mathcal{M}$ of tasks that we average over (see the definitions in section 3) be the set of sample tasks. However, it is not desirable to do so since such an overfit $\hat{\Phi}$ will usually generalize poorly to unseen tasks and state-action pairs. Generalization over state-action pairs requires a function approximator (note that if the state-action space is the same for every task and sufficiently small, this is not necessary). Generalization over tasks requires the identification of state-action pair properties that result in a more consistent (low-variance) estimator $\hat{\Phi}$ across tasks. For factored state spaces, this amounts to doing feature selection; in the general case, it amounts to state abstraction (e.g., [8]).

While the connection with state abstraction methods is interesting, space constraints preclude an in-depth discussion. Therefore, we will focus on factored state spaces, since it is more intuitive and most problems fall in this category. For example, in the cliff domain, which has factored states, the addition of a cliff sensor increases the impact of the potential functions on performance. This is because, *whichever task the agent is in*, the correlation between the cliff sensor feature and expected return is consistent: whenever there is a cliff to the north, large negative return follows when stepping north. The expected return associated with position, on the other hand, varies greatly per task. Thus, the cliff sensor seems like a good feature to use for $\hat{\Phi}$, while position does not.

Both feature selection (FS) and state abstraction methods have been applied to single-task RL, with especially FS seeing a recent surge in interest [11, 12, 4, 9]. Although similar methods have also been applied to transfer learning, most of these seek to reduce the state space of the target task, or find good representations for value functions or policies, by identifying features or abstractions that are useful *within* tasks [5, 19, 15].

None of these approaches applies the insight that features that are not useful for transferring a value function or policy might be useful in some other way, for example for learning a shaping function. Although previous work on shaping in MTRL exists that does apply this insight [6, 13], it circumvents the feature selection problem by manually designing a representation for the value and shaping function. Thus, these papers do not make clear what makes a feature set useful for the shaping function. Snel and Whiteson [14] use an information-theoretic metric to define *task-relevant features* as features that provide information on value within tasks, and *domain-relevant* features to be those that provide information on value across tasks. The latter are the ones that should be used for the shaping function. While experiments validate their definitions, the authors do not demonstrate that automatically finding the right features for shaping can improve generalization performance. Furthermore, their definition might not be the best for a regression setting, since their metric does not take error magnitude into account.

This paper casts the problem as a supervised regression problem to which supervised learning and feature selection algorithms can be applied to identify the relative importance of features. Various learning algorithms are suitable, but for this paper we use a random forest [2] of regression trees. This suits our current purposes nicely since it is a method with a proven track record that is easy to interpret in terms of the original features, and offers ways to estimate relative feature importance in terms of the impact on squared error of the estimate. In the next section, we show how this method can be applied to find task- and domain-relevant features and to find a potential function that generalizes well over tasks.

6   Evaluation of Representations

Consider the episodic domain of which an example task is shown in figure 1b, with a step penalty of -1. In addition to feature x_1, which corresponds to the state numbers shown, the agent perceives two additional features x_2 and x_3. Feature x_2 is the inverse of the square of the shortest distance to the goal, i.e. in the figure, the states would be $(1, 0.25), (2, 1), (3, 0.25)$. Feature x_3 is a binary feature that indicates whether the task is one in which it is undesirable to stay in one place (1) or not (0); if 1, the agent receives a -10 penalty for self-transitions. x_3 is constant within a task, but may change from one task to the next. The goal may be at either of the three states, and the effect of actions 1 and 2 (see figure caption) may be reversed. Action 3 always results in either a self-transition or a goal-transition (when the goal is next to the current state). This amounts to 12 possible tasks in total.

The domain is kept small in order to facilitate computing all potential function types exactly, and to illustrate the main points of the paper. Furthermore, the features are chosen such that they represent three categories: the x_1 feature is task relevant, only providing information within a task (because its relation to return changes in each task); x_3 is useless within a task, but useful to distinguish classes of tasks, and therefore domain relevant. The x_2 feature, finally, is both task and domain relevant. The next subsection will justify and illustrate these claims experimentally.

6.1   Feature Relevance

To compute feature relevance, we choose $\hat{\Phi}_*$, the average over optimal value functions, as the target potential function (experiments showed that for this domain, it does not matter which potential function we choose for this). Thus, we first solve N sample tasks, then use the $N|\mathcal{S}_m||\mathcal{A}_m| = 9N$ state-action-value examples to train a random forest as estimator of $\hat{\Phi}_*$.

Fig. 2a shows how the estimated importance of each feature changes with increasing sample size. For each tree in the forest, feature importance is the weighted (by tree node probability) sum in changes in error for each split on the feature. A change is computed as the difference between the error of the parent node and the total error of the two children. This roughly means that the more

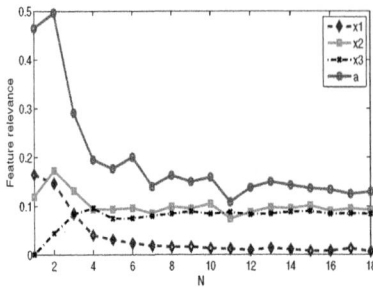 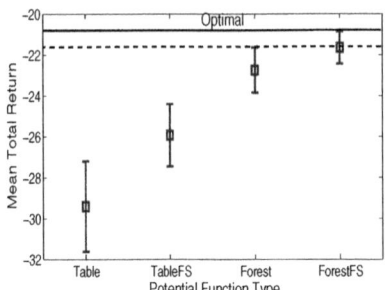

Fig. 2. Left: Feature relevance per sample size N, where a indicates action. Right: Comparison of shaping functions without and with feature selection. Error bars indicate 95% confidence interval. Horizontal solid (dashed) line is mean (95% confidence interval) of a potential function with complete domain knowledge.

a feature contributes to reduction of the squared error in prediction of optimal value, the more important it is. Each feature score then is averaged over all trees in the forest. Final score is computed by repeating each measurement 10 times for each sample size and taking the mean.

The changes in feature relevance with increasing sample size clearly demonstrate the difference between task and domain relevance. For a single task ($N = 1$), x_3 is irrelevant since it is constant within a task. The action is most relevant, followed by x_1 and x_2. As sample size increases, x_2 and x_3 gain in importance, while x_1 loses importance. This pattern makes sense: while x_1 has strong correlation with return in a single task, this correlation decreases for increasing sample size (when all tasks in the domain are sampled perfectly uniformly, it is 0). Thus x_1 is task relevant, but not domain relevant. Feature x_2 is both task *and* domain relevant: it is relevant for $N = 1$, but remains so as sample size increases. This makes sense since decreasing distance to the goal leads to higher expected return both within and across tasks. x_3 is only domain relevant: it can be used to distinguish classes of tasks, but is not useful within a single task. Finally, the importance of the action suggests a shaping function of the form $\Phi(s, a)$; if actions did not provide any information, we might also use a potential over just states, $\Phi(s)$.

6.2   Generalization

To asses the impact on generalization, we next compare four different potential types: a table-based one without FS (*table*) and with FS (*tableFS*), and a random forest without (*forest*) and with (*forestFS*) FS. Before doing so, we computed all potential types of section 3, assuming full knowledge of the domain. This should provide an upper bound on the performance of the potential functions computed based on samples. In this domain, it turns out there is no significant difference between the potential types, so we use $\hat{\Phi}_*$ in all experiments.

Performance of each potential function is assessed by computing the potential function on 4 samples (25% of the domain size), sampling a task from the domain and recording the total reward incurred by a Q-Learning agent run for 10 episodes. Final performance is the average over 100 such trials. Fig. 2b displays the results. The horizontal line indicates performance of the shaping function computed assuming complete domain knowledge, with the dashed line indicating the 95% confidence interval. The table-based function without FS does worst; performance is improved by deselecting the x_1 feature as per fig. 2a, indicating that leaving this task-relevant feature out improves generalization across tasks. The random forests significantly outperform the table-based functions, while the performance difference between them is not significant. The advantage of the forests over the tables is, as noted before, their capacity to also generalize to unseen state-action pairs. The fact that FS does not have a great impact on random forest performance stems from the fact that, by averaging over many independently grown trees, this method has some inherent protection against overfitting.

7   Conclusion

This paper makes two contributions to the multi-task RL (MTRL) literature. First, while shaping functions have been used in MTRL before, this paper is the first to provide an extensive discussion and empirical comparison of different types of shaping functions that could be useful to improve target task performance. While some previous work on shaping in MTRL has used the average over optimal value functions of the source tasks as potential function [14, 6, 17], our results demonstrate that this is not always the best option: the best potential function highly depends on the domain and learning algorithm under consideration.

Second, by casting the problem of finding a shaping function as a supervised regression problem, we can automatically detect which features are relevant for the shaping function, by measuring each feature's influence on the squared error in prediction of cross-task value. We further demonstrate that feature selection improves generalization from the source tasks, which in turn improves the shaping function's ability to speed target task learning. Future work should extend these results to domains with larger feature and state spaces, and assess the impact of shaping functions on more advanced learning algorithms.

References

[1] Asmuth, J., Littman, M.L., Zinkov, R.: Potential-based shaping in model-based reinforcement learning. In: Proceedings of the 23rd AAAI Conference on Artificial Intelligence, pp. 604–609. The AAAI Press (2008)

[2] Breiman, L.: Random forests. Machine Learning 45(1), 5–32 (2001)

[3] Elfving, S., Uchibe, E., Doya, K., Christensen, H.I.: Co-evolution of shaping: Rewards and meta-parameters in reinforcement learning. Adaptive Behavior 16(6), 400–412 (2008)

[4] Hachiya, H., Sugiyama, M.: Feature Selection for Reinforcement Learning: Evaluating Implicit State-Reward Dependency via Conditional Mutual Information. In: Balcázar, J.L., Bonchi, F., Gionis, A., Sebag, M. (eds.) ECML PKDD 2010. LNCS, vol. 6321, pp. 474–489. Springer, Heidelberg (2010)

[5] Jong, N.K., Stone, P.: State abstraction discovery from irrelevant state variables. In: IJCAI 2005 (2005)

[6] Konidaris, G.D., Barto, A.G.: Autonomous shaping: Knowledge transfer in reinforcement learning. In: Proc. 23rd International Conference on Machine Learning, pp. 489–496 (2006)

[7] Lazaric, A., Ghavamzadeh, M.: Bayesian multi-task reinforcement learning. In: ICML, pp. 599–606 (2010)

[8] Li, L., Walsh, T.J., Littman, M.L.: Towards a unified theory of state abstraction for MDPs. In: Aritificial Intelligence and Mathematics (2006)

[9] Mahadevan, S.: Representation discovery in sequential decision making. In: AAAI (2010)

[10] Ng, A.Y., Harada, D., Russell, S.: Policy invariance under reward transformations: Theory and application to reward shaping. In: Proc. 16th International Conference on Machine Learning (1999)

[11] Parr, R., Li, L., Taylor, G., Painter-Wakefield, C., Littman, M.L.: An analysis of linear models, linear value-function approximation, and feature selection for reinforcement learning. In: ICML, pp. 752–759 (2008)

[12] Petrik, M., Taylor, G., Parr, R., Zilberstein, S.: Feature selection using regularization in approximate linear programs for Markov decision processes. In: ICML, pp. 871–878 (2010)

[13] Singh, S., Lewis, R.L., Barto, A.G.: Where do rewards come from? In: Proc. 31st Annual Conference of the Cognitive Science Society, pp. 2601–2606 (2009)

[14] Snel, M., Whiteson, S.: Multi-task evolutionary shaping without pre-specified representations. In: Genetic and Evolutionary Computation Conference, GECCO 2010 (2010)

[15] Sorg, J., Singh, S.: Transfer via soft homomorphisms. In: Proc. 8th Int. Conf. on Autonomous Agents and Multiagent Systems (AAMAS 2009), pp. 741–748 (2009)

[16] Sutton, R.S., Barto, A.G.: Reinforcement Learning: An Introduction. The MIT Press (1998)

[17] Tanaka, F., Yamamura, M.: Multitask reinforcement learning on the distribution of MDPs. In: Proc. 2003 IEEE International Symposium on Computational Intelligence in Robotics and Automation (CIRA 2003), pp. 1108–1113 (2003)

[18] Taylor, M.E., Stone, P.: Transfer learning for reinforcement learning domains: A survey. Journal of Machine Learning Research 10(1), 1633–1685 (2009)

[19] Walsh, T.J., Li, L., Littman, M.L.: Transferring state abstractions between MDPs. In: ICML 2006 Workshop on Structural Knowledge Transfer for Machine Learning (2006)

[20] Wiewiora, E., Cottrell, G., Elkan, C.: Principled methods for advising reinforcement learning agents. In: Proc. 20th International Conference on Machine Learning, pp. 792–799 (2003)

[21] Wilson, A., Fern, A., Ray, S., Tadepalli, P.: Multi-task reinforcement learning: a hierarchical Bayesian approach. In: ICML, pp. 1015–1022 (2007)

Transfer Learning in Multi-Agent Reinforcement Learning Domains

Georgios Boutsioukis, Ioannis Partalas, and Ioannis Vlahavas

Department of Informatics
Aristotle University
Thessaloniki, 54124, Greece
{gampouts,partalas,vlahavas}@csd.auth.gr

Abstract. In the context of reinforcement learning, transfer learning refers to the concept of reusing knowledge acquired in past tasks to speed up the learning procedure in new tasks. Transfer learning methods have been succesfully applied in single-agent reinforcement learning algorithms, but no prior work has focused on applying them in a multi-agent environment. We propose a novel method for transfer learning in multi-agent reinforcement learning domains. We proceed to test the proposed approach in a multi-agent domain under various configurations. The results demonstrate that the method can reduce the learning time and increase the asymptotic performance of the learning algorithm.

1 Introduction

In the Reinforcement Learning (RL) realm, where algorithms often require a considerable amount of training time to solve complex problems, transfer learning can play a crucial role in reducing it. Recently, several methods have been proposed for transfer learning among RL agents with reported success. However, to the best of our knowledge, transfer learning methods have been so far applied only to single-agent RL algorithms.

In a multi-agent system the agent interacts with other agents and must take into account their actions as well[14]. When the agents share a common goal, for example, they must coordinate their actions in order to accomplish it. In this paper we present and attempt to address the specific issues that arise in the application of transfer learning in Multi-agent RL (MARL). We propose a novel method, named **BI**as **Transf ER** (BITER), suitable for transfer learning in agents in MARL. Additionally, we propose an extension of the Q-value reuse algorithm[12] in the multi-agent context.

The core idea behind this method is to use the joint policy that the agents learned in the source task in order to bias the initial policy of the agents in the target task towards it(Section 3). In this work we use the *Joint Action Learning*[1] algorithm as our basic learning mechanism.

The proposed approach can be used regardless of the underlying multi-agent algorithm, but we leave such a scenario for future research. The proposed method

S. Sanner and M. Hutter (Eds.): EWRL 2011, LNCS 7188, pp. 249–260, 2012.
© Springer-Verlag Berlin Heidelberg 2012

is evaluated in the predator-prey multi-agent domain under various configurations (Section 4). The results demonstrate that transfer learning can help to reduce substantially the training time in the target task and improve asymptotic performance. (Section 5 and 6).

2   Transfer Learning in RL

The basic idea behind transfer learning is that knowledge already acquired in a previous task can be used to leverage the learning process in a different (but usually related) task. Several issues must be addressed in a transfer learning method: *a)* how the tasks differ, *b)* if task mappings are required to relate the source and the target tasks and *c)* what knowledge is transferred.

The tasks may have different state spaces, but fixed variables [7,6] or even different state variables [5]. More flexible methods allow the tasks to differ in the state and action spaces (with different variables in both sets) and also in the reward and transition functions [13,12]. These methods use inter-task mappings in order to relate the source and target tasks. More specifically, mappings are usually defined by a pair of functions (χ_S, χ_A) where $\chi_S(s) : S_{target} \to S_{source}$ and $\chi_A(\alpha) : A_{target} \to A_{source}$ [12]. Towards a more autonomous setting, mappings can also be learned [8,11]. A comprehensive survey on transfer learning in single-agent RL can be found in [9].

The level of knowledge that can be transferred across tasks can be low, such as tuples of the form $\langle s, a, r, s' \rangle$ [6,10], value-functions [12] or policies [2]. Higher level knowledge may include rules [7,13], action subsets or shaping rewards [5].

As we already mentioned, so far, no prior work has addressed the issue of transfer learning in the MARL setting. Most similar to our approach, from the methods of single-agent transfer learning, is the one proposed by Madden and Howley [7]. This method uses a symbolic learner to extract rules from the action value functions that were learned in previous tasks. In the target task, the rules are used to bias the action selection. We follow a much simpler approach and avoid using rules. The proposed method provides initial values to the target task learners before the learning process starts.

3   MARL Transfer

Although the difficulty of MARL tasks makes them an attractive target for transfer methods, the presence of multiple agents and the added complexity of MARL algorithms creates new problems specific to the multi-agent setting, which means that the application of transfer learning in these tasks is not a straightforward extension of single agent methods. In the following sections we will present and try to address some of the issues that arise in a MARL context.

In order to focus on the multi-agent aspects that specifically affect transfer learning, we had to set some restrictions. First, we only considered tasks with homogeneous agents, which means that we expect a high degree of similarity between their corresponding action sets. We also assume that the agents behave cooperatively, although this is more of a convention; we do not expect other types of behaviour to be significantly different from the viewpoint of transfer learning, in general.

Agent homogeneity may seem too restrictive; however, tasks with heterogeneous agents can still be viewed as having multiple classes of mutually similar agents; since transfer would generally still take place between these similar agent classes across tasks, the transfer task in this case could be viewed as a series of parallel homogeneous transfers.

3.1   Intertask Mappings across Multi-Agent Tasks

Intertask mappings in single agent tasks map the most similar states and actions between the source and target tasks. A significant difference in the multi-agent case is that the learned knowledge for each task is usually distributed among agents, which means that the mapping functions for the target task have to be defined per-agent. We propose a form for such a function defined for agent i that maps the joint actions of an n-agent task to those of an m-agent task below:

$$\chi_{i,J_n \to J_m}(\vec{\alpha}) : A_1 \times ... \times A_n \to A'_1 \times ... \times A'_m$$

where $J_k = A_1 \times ... \times A_k$. Correspondingly a mapping function that maps states between tasks can be defined per agent. Although states have the same meaning in multi-agent tasks as in a single agent one, they can include parameters that are associated with a specific agent (such as the agent's coordinates). Since it is helpful in a multi-agent setting to make this distinction, we denote these parameters as $\bar{agent}_j$ and as $\bar{s}$ the rest of the state variables in the two tasks. The proposed form of such a mapping function for agent i becomes:

$$\chi_{i,S_n \to S_m}(s) : S_n \to S_m$$

where each state $s \in S_n$ and $s' \in S_m$ of the target and source tasks correspondingly has the form $s : \langle \bar{s}, \bar{agent}_1, ..., \bar{agent}_n \rangle$ and $s' : \langle \bar{s}', \bar{agent}'_1, ..., \bar{agent}'_m \rangle$.

Of course, the source and target tasks can still have different action and state variables and they can be mapped using the same techniques one would use in a single agent task (such as scaling a larger grid to a smaller one).

There are a few different ways to define these mappings, especially when domain specific properties are taken into account. A significant factor is whether the representation of an agent in a source task is considered equivalent to an agent's representation in the target. Intuitively this corresponds to the situation where each agent is thought to retain its "identity" over the two tasks. But it is also possible for a single agent to be mapped to the parameters and actions of different agents. Accordingly, we propose two mapping approaches:

Static agent mapping implements a one-to-one mapping between agents that remain constant. This approach effectively ignores the presence and actions of the extra agents. This dictates that the chosen set of "ignored" agents remains

Fig. 1. Agent perception variations in static mapping when transferring from a two to a three agent task. An arrow from each agent denotes which other agent it is "aware" of.

the same for all states and joint actions[1]. For example, shown below are functions defined for Agent 1 that map a three agent task to a two agent one, effectively ignoring Agent 3:

$$\chi_{1,S_n \to S_m}(\langle \bar{s}_{target}, \overline{agent}_1, \overline{agent}_2, \overline{agent}_3 \rangle) = \langle \bar{s}_{source}, \overline{agent}_1, \overline{agent}_2 \rangle$$

$$\chi_{1,J_n \to J_m}(\langle \alpha_{1,1}, ..., \alpha_{1,i}, \alpha_{2,1}, ..., \alpha_{2,j}, \alpha_{3,1}, ..., \alpha_{3,k} \rangle) = \langle \alpha_{1,1}, ..., \alpha_{1,i}, \alpha_{2,1}, ..., \alpha_{2,j} \rangle$$

where α_{ij} is the j-th action of the i-th agent. It is important to note that these functions are simplified for demonstrative purposes; they make the implicit assumption that $\bar{s}_{target}$ can be mapped directly to $\bar{s}_{source}$ and that each agent has the same associated state variables and actions across tasks. It is also important to keep in mind that these functions are defined per-agent; the set of $n - m$ agents that are ignored in this mapping will be different from the perspective of other agents.

When we transfer from a single agent system to a multi-agent one, there is only one way to pick this "ignored" agent set. But in transfer from multi-agent to multi-agent systems, there is a number of possible variations. Although it might seem that picking between homogeneous agents should make no difference, this is not the case as it will have a different result as to how the agents will perceive each other.

In Figure 1 we present a case where transfer from a task with two agents leads to a three agent one can have two distinct outcomes. Exactly what implications this will have on the behaviour of the agents is not clear and it will depend on the nature of each task [2]; we will not cover this further.

Dynamic or context agent mapping, on the other hand, lifts the restriction that the ignored agents should remain the same for all states and joint actions. Intuitively this means that the agents do not retain an "identity" across the two tasks. There are different ways to implement such a mapping, but typically one would utilise aspects of the domain-specific context. For example, in a gridworld we can map states and actions as to effectively ignore the most distant agents relative to the current agent or the prey. From the viewpoint of agent 1, such mapping functions for a three agent representation mapped to a two agent one using distance as a criterion would be:

---

[1] When the source task is single agent this seems the only sane way to transfer to a multi-agent one, since there is no way to compensate for the lack of perception of other agents.

[2] In our experiments, the two setups produced near-identical results so it proved a non-issue in our case. This may not hold for more complex tasks however.

$$\chi_{1,S_3 \to S_2}(\langle ag\bar{e}nt_1, ag\bar{e}nt_2, ag\bar{e}nt_3 \rangle) = \begin{cases} \langle ag\bar{e}nt_1, ag\bar{e}nt_2 \rangle, & d(x_1, x_2) \le d(x_1, x_3) \\ \langle ag\bar{e}nt_1, ag\bar{e}nt_3 \rangle, & d(x_1, x_2) > d(x_1, x_3) \end{cases}$$

$$\chi_{1,J_3 \to J_2}(s, \langle \alpha_{1i}, ..., \alpha_{2j}, ..., \alpha_{3k} \rangle) = \begin{cases} \langle \alpha_{1i}, ..., \alpha_{2j} \rangle, & d(x_1, x_2) \le d(x_1, x_3) \\ \langle \alpha_{1i}, ..., \alpha_{3k} \rangle, & d(x_1, x_2) > d(x_1, x_3) \end{cases}$$

where $d(x_p, x_q)$ the distance between agents x_p and x_q in the current state. A subtle difference in this case is that the action mapping function is also a function of the current state s being mapped, as in this case it depends on its properties(i.e. the agents' current coordinates). As before, these functions are simplified for demonstration.

3.2 Level of Transferred Knowledge

An important feature of multi-agent systems is that the acquired knowledge is typically distributed across agents instead of residing in a single source. This can be a challenge for transfer methods, since there is no straightforward way to deal with multiple sources in the general case.

We chose to transfer the learned joint policy in order to avoid this issue, since we can use this unified source of knowledge to transfer to each agent. Choosing this relatively higher level of transfer has also the advantage of not having to deal with the internals of each MARL algorithm, since a joint policy contains the effect of all parts of a MARL algorithm – such as the effect of the conflict resolution mechanisms that these algorithms often employ. The trade-off to be made here is that some knowledge that could benefit the target task is discarded, such as the values of suboptimal actions.

3.3 Method of Transfer

Aside from the level of knowledge transferred, we must also decide how to incorporate this knowledge in the target task's learning algorithm. Transfer methods in single agent settings will often modify the learning algorithm in the target task [12]. The usual criterion for convergence in single agent algorithms is to provide a correct estimate of the state or action value function, that can be in turn used to estimate the optimal policy.

We propose a method of transfer that incorporates the transferred knowledge as bias in the initial action value function. Since proofs of convergence do not rely on the specific initial values of this function, we are essentially treating the underlying MARL algorithm as a kind of "black box". We consider the proposed algorithm as a generic transfer method that does not affect the convergence of the underlying RL algorithm. Previous research in biasing the initial Q values[7,3] generally avoids to define the specific intervals that the bias parameter should lie within. This is justified, since an optimal bias parameter value relies on the specific properties of the Q function that is being estimated in the first place. Intuitively, we seek a value high enough such that it will not be overcome by smaller rewards before the goal state is reached a few times, and low enough to not interfere with learning in the later stages. Our experiments have shown that for most problems a relatively small bias (e.g. $b = 1$ when $R_{max} = 1000$) usually has better results and performance will begin to drop as this value is increased. Using a bias value b, Algorithm 1 lists the pseudocode for the generic multi-agent transfer algorithm we propose.

Algorithm 1. BITER for agent i

1: **for all** states s in S_{target} **do**
2:     **for all** joint action vectors $\vec{\alpha_n}$ in $A_1 \times ... \times A_n$ **do**
3:         $Q_{i,target}(s, \vec{\alpha_n}) \leftarrow 0$
4:         **if** $\chi_{i,A,n \to m}(\vec{\alpha_n}) = \pi_{source}(\chi_{i,S,n \to m}(s))$ **then**
5:             $Q_{i,target}(s, \vec{\alpha_n}) \leftarrow b$
6:         **end if**
7:     **end for**
8: **end for**

In this paper we also extended a single agent transfer algorithm, Q-value reuse [12], to the multiagent setting. Q-value reuse adds the Q-values of the source task directly to the Q-values of the target task. In this algorithm, the new Q-values are defined as:

$$Q_{i,target}(s, \vec{\alpha}) \leftarrow Q_{i,target}(s, \vec{\alpha}) + Q_{source}(\chi_{i,S,n \to m}(s), \chi_{i,A,n \to m}(\vec{\alpha_n}))$$

However, unlike the previous method that is only invoked before learning, transfer here takes place during the execution of the target task and becomes a part of the learning algorithm. A significant difference in this case is that one would have to choose which Q_{source} to use. This could be the Q function of an individual agent in the source task, or more elaborate sources such as an average from all agents.

4   Experiments

4.1   Domain

In order to evaluate the proposed methodologies we used the predator-prey domain and more specifically the package developed by Kok and Vlassis [4]. The domain is a discrete grid-world where there are two types of agents: the predators and the preys. The goal of the predators is to capture the prey as fast as possible. The grid is toroidal and fully observable, which means that the predators receive accurate information about the state of the environment.

4.2   Experimental Setup

The learning environment in all cases was a 5×5 grid, where the current state is defined by the locations of the prey and the other predators. The agents can choose their next move from the action set A={NORTH,SOUTH,EAST, WEST,NONE} (where NONE means that they remain in their current location). States in this environment include the x and y coordinates of the prey and the other predators, relative to the current predator, so a state from the viewpoint of predator A in a two agent world with another predator B would be of the form $s = \langle prey_x, prey_y, B_x, B_y \rangle$.

In all cases (for both source and target tasks) the MARL algorithm used is *joint action learning (JAL)*, as described in [1]. The exploration method used is Boltzmann exploration, where in each state the next action is chosen with a probability of

$$Pr(a_i) = \frac{e^{\hat{Q}(s,\alpha_i)/T}}{\sum_{j=1}^{n} e^{\hat{Q}(s,\alpha_j)/T}}$$

where the function $\hat{Q}$ is the estimate of the maximum value of all possible joint actions given an agent's individual action. $T = \frac{\lg(N_s)}{C_t}$ is the *temperature* parameter, where N_s is the number of times the state was visited before and C_t is the difference between the two highest Q-Values for the current state. Boltzmann exploration was fully used in the single and two agent version of the task, but in the three agent version it was more practical to use in 10% of the steps, making it the exploration part of an e-greedy method where $\epsilon = 0.1$ [3] For all experiments we used a constant learning rate $a = 0.1$ and a discount factor $\gamma = 0.9$. When BITER was used, the bias parameter was $b = 1$. The rewards given to each individual agent were $r = 1,000$ for capturing the prey, $r = -100$ when collision with another agent occurs, and $r = -10$ in all other states. For each experiment, 10 independent trials were conducted. The results that we present are averaged over these repetitions.

In all of our experiments the prey follows a random policy, picking an action in each step with uniform probability. Since the prey's policy is fixed and therefore not in our focus, we will use the terms agent and predator interchangeably from now on. The prey is captured when all of the predators move simultaneously to a square adjacent to the prey, ending the current episode. Finally, when two agents collide they are placed in random locations on the grid.

5   Results and Discussion

For each experiment, we also record the initial performance (or *Jumpstart*), averaging capture times over the 1,000 first episodes, the *final average capture time (ACT)* for the last 1,000 episodes, which indicates the final performance of the agents and the *Learning Curve Area Ratio (LCAR)*, defined as $\frac{\sum_{i=1}^{n} c_i}{\sum_{i=1}^{n} d_i}$

---

[3] An exploration parameter of $\epsilon = 0.1$ in a three agent environment means that there is a $(1 - \epsilon)^3 = 0.72$ probability that none of the agents is exploring in the next step.

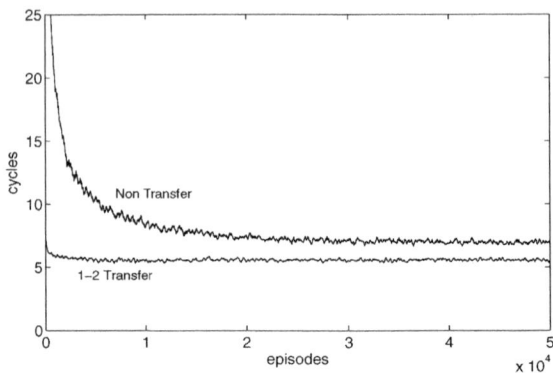

Fig. 2. Average capture times for 1→2 transfer learning

where c_i, d_i the capture time for each compared execution in episode i. Finally, the results do not include the learning time of the source task as it is typically an order of magnitude less than the target task's.

The first batch of transfer experiments involve three tasks of the team capture game, with one, two and three predators respectively. Additionally, we use the static-mapping method for all transfer procedures.

The first transfer case focuses on the two-predator team capture task, where we applied our proposed transfer method using a single-predator capture task as source. In this simple case, the learned policy of the source task is used to bias the initial Q function of the target task. The learning time for the source task is approximately 200 episodes, or about 800 cycles in total. Since the size of the state and action space is relatively small, it can be assumed that the source task's learned policy is optimal. In this case each agent in the target task begins with a policy that is biased towards the learned policy from the single-agent task.

Figure 2 presents the results of BITER compared to the non-transfer case. The x and y axis represent the episodes and capture times (in cycles) respectively. Table 1 presents the recorded metrics for each algorithm.

We first notice that BITER reduces the average capture time. This is evident from the first episodes, where the Jumpstart of the proposed method is substantially better than the case without transfer. Paired t-tests at a confidence level of 95% detected significant differences between the two competing algorithms, for the whole range of learning episodes. Additionally, in Table 1 we notice that BITER achieves better final performance (5.5) than the algorithm without transfer (6.98).

In the second experiment, we use the single-predator and the two-predator versions of the game as source tasks. The target task in both cases is the three-predator game. Learning the two-agent task optimally is a harder task than the single agent one; we used the policy learned after 200,000 episodes which is close to, but may not be the optimal one. A better policy could be achieved by

Table 1. Jumpstart, LCAR and ACT for the two agent team capture task

One to Two-Agent Team Capture Transfer

Method	Jumpstart	LCAR	ACT (50k)
Non Transfer	32.23	1	6.98
BITER	6.17	0.66	5.50

adjusting the learning parameters, but we preferred to use a suboptimal policy as the source, since in practice one would have to settle for one.

Figure 3 illustrates the results of the two instances of BITER along with the non-transfer case. Both instances of BITER reduce learning time to about a third compared to direct learning (Table 2), while they exhibit similar performance. Paired t-tests showed that 1→3 transfer is statistically significantly better than 2→3 after the first 320 episodes, at the 95% level. Both cases are also better than direct learning at the 95% level for all episodes of the simulation. These results verify that transfer improves significantly the performance in the target task. Additionally, BITER improves both the Jumpstart and the ACT of the agents as it is observed in Table 2. Another interesting observation is the fact that transfer from the single-agent case leads to better performance in the target task. The single-agent task is simpler to learn than the two-agent one, which means that a better policy can be found in the source task.

Table 2. Jumpstart, LCAR and ACT for the three agent team capture target task

One and Two to Three-Agent Team Capture Transfer

Method	Jumpstart	LCAR	ACT(200k)
Non Transfer	249.07	1	21.11
2→3 BITER	21.37	0.35	10.72
1→3 BITER	20.36	0.29	8.70

To investigate the performance of the dynamic (or context) agent mapping approach (where the source agent representations are mapped to different agents of the target depending on the context) we set an experiment using the two-agent game as the source task and the three-agent game as the target game. We explore two different dynamic approaches: a) map to the predator closest to the current agent's position and b) map to the predator closest to the prey. The results are shown in Table 3 along with the performance of the static mapping approach. Interestingly, dynamic mapping outperforms the static one in both LCAR and ACT in the prey-distance case, while it performed worse when agent-distance was used. A possible explanation for this behaviour could be this: most collisions in team capture occur near the prey. Since prey-distance mapping improves coordination in this region, it may help to reduce the costly collisions and improve performance.

In the next experiment we evaluate the MARL Q-value reuse algorithm. Boltzmann exploration is unsuitable for this method, so we used a fixed-value

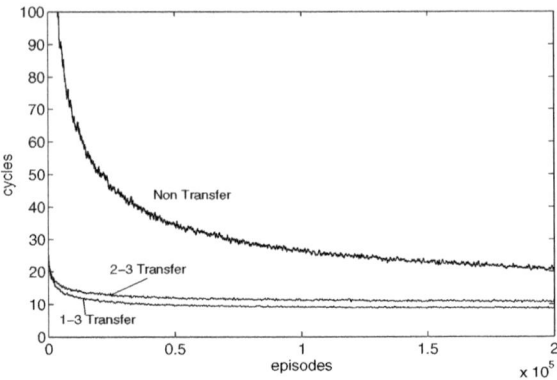

Fig. 3. Average capture times for single and two agent to three agent transfer learning. Results are averaged over 10 runs.

Table 3. Comparison between static and dynamic mapping, using distance from the current agent or distance from the prey

Static and Dynamic mapping performance for 2→3 transfer

Method	Jumpstart	LCAR	ACT (100k)
Static	21.18	1	11.43
Dynamic - agent distance	23.15	1.05	11.96
Dynamic - prey distance	20.58	0.90	10.26

exploration parameter (ϵ−greedy, with $\epsilon = 0.03$). We also run the BITER algorithm with the same exploration parameter value. The results are shown in Table 4. In both experiments, BITER has a clear advantage over MARL Q-value reuse. The paired t-test at 95% confidence level detects significant differences in favour of BITER. While the MARL Q-value reuse helps to reduce the learning time in the target task, it shows that using directly the Q-values from the source task is not the optimal way for transfer in MARL agents.

In our final experiment we test the efficacy of BITER when we alter the action set of the agents in the target task by allowing them to move diagonally; this increases the joint action space by a factor of 4 (per agent). As the source task we use single-agent and two-agent games without diagonal moves.

Table 4. Comparison between Q-value Reuse and BITER for the 1→3 and 2→3 tasks

Method	Jumpstart	LCAR	ACT (100k)
Q-Value Reuse 1→3	20.59	1	19.63
BITER 1→3	19.68	0.57	11.12
Q-Value Reuse 2→3	18.67	1	18.42
BITER 2→3	21.91	0.78	14.35

Table 5. Jumpstart, LCAR and ACT (after 200,000 episodes) for the three-agent team capture task (with diagonal moves)

One and Two to Three Agent Team Capture Transfer (w/diagonals)

Method	Jumpstart(1k)	LC Ratio	Avg. Capture Time(200k)
Non Transfer	206.94	1	16.19
2→3 BITER	21.78	0.35	8.07
1→3 BITER	21.58	0.32	7.30

Mapping the larger joint action set to a smaller one is a problem similar to its single agent version. We simply ignore the new actions, mapping them to a "null" joint-action of zero value. The results are comparable to the non-diagonal version, as transferring from either source task reduces the learning time to about a third. Paired t-tests showed that 1→3 transfer is statistically significantly better than 2→3 after the first 8979 episodes, at the 95% level. Both are also better than direct learning at the 95% level for all episodes of the simulation.

6 Conclusions and Future Work

To the best of our knowledge, transfer learning method have been proposed only for single-agent RL algorithms so far. To address the specific issues that arise in multi-agent environments, we proposed a novel scheme for inter-task mappings between multi-agent tasks and introduced BITER, a generic transfer algorithm that can be adapted to work under the typical restrictions of MARL algorithms. We evaluated BITER in the predator-prey domain under various setups. The results demonstrated that BITER can reduce significantly the learning time and improve the asymptotic performance.

Our exploration of multi-agent intertask mappings revealed a variety of possible mapping schemes, and more research on this subject could explain more about its effect on transfer learning. It could also lead to the development of automated mapping methods, especially in more complex cases, such as cross-domain transfer. Also, while our work focused on discrete and deterministic environments, we believe that transfer learning could be successfully applied in continuous or partially observable environments, although the specific challenges there would be the subject of future work. It would also be interesting to see these methods applied in an adversarial setting, such as simulated soccer.

Our results also indicated the difficulty of applying reinforcement learning methods directly to multi-agent problems, where relatively simple tasks quickly become intractable with the addition of more agents. Transfer learning is a promising method to get around this complexity, and we expect to see a wider adoption of it in the multi-agent setting.

References

1. Claus, C., Boutilier, C.: The dynamics of reinforcement learning in cooperative multiagent systems. In: 15th National Conference on Artificial Intelligence, pp. 746–752 (1998)
2. Fernández, F., Veloso, M.: Probabilistic policy reuse in a reinforcement learning agent. In: 5th International Joint Conference on Autonomous Agents and Multiagent Systems, pp. 720–727 (2006)
3. Hailu, G., Sommer, G.: On amount and quality of bias in reinforcement learning, vol. 2, pp. 728–733 (1999)
4. Kok, J.R., Vlassis, N.: The pursuit domain package. Technical report ias-uva-03-03, University of Amsterdam, The Netherlands (2003)
5. Konidaris, G., Barto, A.: Autonomous shaping: knowledge transfer in reinforcement learning. In: 23rd International Conference on Machine Learning, pp. 489–496 (2007)
6. Lazaric, A.: Knowledge Transfer in Reinforcement Learning. Ph.D. thesis, Politecnico di Milano (2008)
7. Madden, M.G., Howley, T.: Transfer of experience between reinforcement learning environments with progressive difficulty. Artificial Intelligence Review 21(3-4), 375–398 (2004)
8. Soni, V., Singh, S.: Using homomorphisms to transfer options across continuous reinforcement learning domains. In: AAAI Conference on Artificial Intelligence, pp. 494–499 (2006)
9. Taylor, M., Stone, P.: Transfer learning for reinforcement learning domains: A survey. Journal of Machine Learning Research 10, 1633–1685 (2009)
10. Taylor, M.E., Jong, N.K., Stone, P.: Transferring Instances for Model-Based Reinforcement Learning. In: Daelemans, W., Goethals, B., Morik, K. (eds.) ECML PKDD 2008, Part II. LNCS (LNAI), vol. 5212, pp. 488–505. Springer, Heidelberg (2008)
11. Taylor, M.E., Kuhlmann, G., Stone, P.: Autonomous transfer for reinforcement learning. In: 7th International Joint Conference on Autonomous Agents and Multiagent Systems, pp. 283–290 (2008)
12. Taylor, M.E., Stone, P., Liu, Y.: Transfer learning via inter-task mappings for temporal difference learning. Journal of Machine Learning Research 8, 2125–2167 (2007)
13. Torrey, L., Shavlik, J., Walker, T., Maclin, R.: Skill Acquisition Via Transfer Learning and Advice Taking. In: Fürnkranz, J., Scheffer, T., Spiliopoulou, M. (eds.) ECML 2006. LNCS (LNAI), vol. 4212, pp. 425–436. Springer, Heidelberg (2006)
14. Weiss, G.: A Modern Approach to Distributed Artificial Intelligence. MIT Press (1999)

An Extension of a Hierarchical Reinforcement Learning Algorithm for Multiagent Settings

Ioannis Lambrou, Vassilis Vassiliades, and Chris Christodoulou

Department of Computer Science,
University of Cyprus, Nicosia 1678, Cyprus
{cs07li2,v.vassiliades,cchrist}@cs.ucy.ac.cy

Abstract. This paper compares and investigates single-agent reinforcement learning (RL) algorithms on the simple and an extended taxi problem domain, and multiagent RL algorithms on a multiagent extension of the simple taxi problem domain we created. In particular, we extend the Policy Hill Climbing (PHC) and the Win or Learn Fast-PHC (WoLF-PHC) algorithms by combining them with the MAXQ hierarchical decomposition and investigate their efficiency. The results are very promising for the multiagent domain as they indicate that these two newly-created algorithms are the most efficient ones from the algorithms we compared.

Keywords: Hierarchical Reinforcement Learning, Multiagent Reinforcement Learning, Taxi Domain.

1   Introduction

Reinforcement learning (RL) suffers from the curse of the dimensionality, where the addition of an extra state-action variable increases the size of the state-action space exponentially and it also increases the time it takes to reach the optimal policy. In order to deal with this problem, there exist three general classes of methods that exploit domain knowledge by: i) using function approximation to compress the value function or policy and thus generalise from previously experienced states to ones that have never been seen, ii) decomposing a task into a hierarchy of subtasks or learning with higher-level macro-actions, and iii) utilising more compact representations and learn through appropriate algorithms that use such representations.

Hierarchical methods reuse existing policies for simpler subtasks, instead of learning policies from scratch. Therefore they may have a number of benefits such as faster learning, learning from fewer trials and improved exploration [4]. This area has attracted an influx of research work. Singh [17] presents an algorithm and a modular architecture that learns the decomposition of composite Sequential Decision Tasks (SDTs), and achieves transfer of learning by sharing the solutions of elemental SDTs across multiple composite SDTs. Dayan and Hinton [3] show how to create a Q-learning [20] managerial hierarchy in which high level managers learn how to set tasks to their sub-managers who, in turn, learn how

S. Sanner and M. Hutter (Eds.): EWRL 2011, LNCS 7188, pp. 261–272, 2012.

to satisfy them. Kaelbling [10] presents the Hierarchical Distance to Goal (HDG) algorithm, which uses a hierarchical decomposition of the state space to make learning achieve goals more efficiently with a small penalty in path quality. Wiering and Schmidhuber [21] introduced the HQ-learning algorithm, a hierarchical extension of Q-learning, to solve partially observable Markov decision processes. Parr [14] developed an approach to hierarchically structuring Markov Decision Process (MDP) policies called Hierarchies of Abstract Machines (HAMs), which exploit the theory of semi-MDPs [18], but the emphasis is on simplifying complex MDPs by restricting the class of realisable policies, rather than expanding the action choices. Sutton et al. [18] consider how learning, planning, and representing knowledge at multiple levels of temporal abstraction, can be addressed within the mathematical framework of RL and MDPs. They extend the usual notion of action in this framework to include options, i.e., closed-loop policies for taking actions over a period of time. Overall, they show that options enable temporally abstract knowledge and actions to be included in the RL framework in a natural and general way. Dietterich [4] presents the MAXQ decomposition of the value function, which is represented by a graph with Max and Q nodes. Max nodes with no children denote primitive actions and Max nodes with children represent subtasks. Each Q node represents an action that can be performed to achieve its parent's subtask. The distinction between Max nodes and Q nodes is needed to ensure that subtasks can be shared and reused. Andre and Russell [1] extend the HAM method [14] and produced the Programmable HAM (PHAM) method that allows users to constrain the policies considered by the learning process. Hengst [8] presents the HEXQ algorithm which automatically attempts to decompose and solve a model-free factored MDP hierarchically. Ghavamzadeh and Mahadevan [7] address the issue of rational communication behavior among autonomous agents and propose a new multi-agent hierarchical algorithm, called COM-Cooperative Hierarchical RL (HRL) to communication decision. Mehta et al. [11] introduced the multi-agent shared hierarchy (MASH) framework, which generalises the MAXQ framework and allows selectively sharing subtask value functions across agents. They also developed a model-based average-reward RL algorithm for that framework. Shen et al. [16] incorporate options in MAXQ decomposition and Mirzazadeh et al. [13] extend MAXQ-Q [4] producing a new algorithm with less computational complexity than MAXQ-Q. Mehta et al. [12] present the HI-MAT (Hierarchy Induction via Models and Trajectories) algorithm that discovers MAXQ task hierarchies by applying dynamic Bayesian network models to a successful trajectory from a source RL task.

Learning in the presence of multiple learners can be viewed as a problem of a "moving target", where the optimal policy may be changing while the agent learns. Therefore, the normal definition of an optimal policy no longer applies, due to the fact that it is dependent on the policies of the other agents [2]. In many cases, the aim is for a multiagent learning system to reach a balanced state, also known as equilibrium, where the learning algorithms converge to stationary policies. Oftentimes a stationary policy can be deterministic, however, there are situations where we look for stochastic policies. An example from game

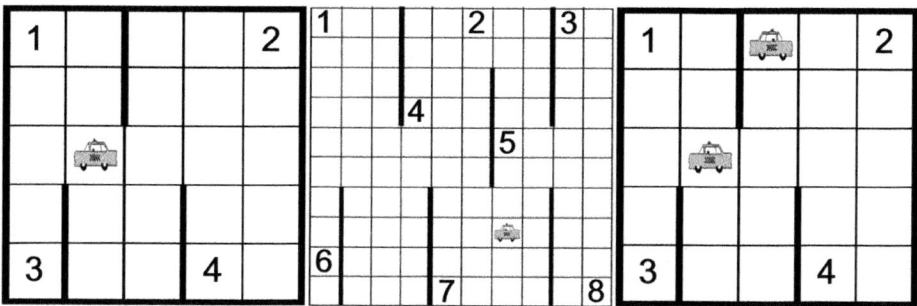

Fig. 1. Single-agent and multiagent taxi domains. (a) Original 5×5 domain (adapted from [4]), (b) Extended 10×10 domain, with different wall distributions and 8 passenger locations and destinations (adapted from [5]), (c) Multiagent domain, where there are 2 or more taxis and passengers.

theory, where pure strategy (i.e., deterministic policy) equilibria do not exist, is the famous Rock-Paper-Scissors game. In this game, the solution is to converge to a mixed strategy (i.e., stochastic policy) equilibrium where the agents play each action with probability $1/3$. Therefore, in multiagent learning settings it is sometimes beneficial for an agent to keep an estimate of its stochastic policy and update it accordingly during the course of learning. This estimate of the stochastic policy is used by the algorithms PHC (Policy Hill Climbing) [2], which is an extension of Q-learning, and WoLF-PHC (Win or Learn Fast-PHC) [2], which is an extension of PHC (see Section 2.3). The WoLF-PHC was empirically shown to converge in self-play to an equilibrium even in games with multiple or mixed policy equilibria [2].

In this paper, we explore a novel combination of the hierarchical method MAXQ with the multiagent RL (MARL) algorithms PHC and WoLF-PHC. We evaluate the performance of Q-learning [20], SARSA (State Action Reward State Action) [15], PHC [2], WoLF-PHC [2], MAXQ-Q [4], as well as the two newly created algorithms MAXQ-PHC and MAXQ-WoLF-PHC, in two single-agent taxi domains [4,5] (shown in Figures 1a and 1b) and a multiagent extension of the taxi domain we created (shown in Figure 1c).

The rest of the paper is organised as follows. Section 2 describes the methodology we followed with a description of the taxi domain, the MAXQ hierarchical decomposition for this domain, and the novel combination of the MAXQ hierarchy with the algorithms PHC and WoLF-PHC. The results are presented and analysed in Section 3, and finally in Section 4, we summarise the conclusions and briefly discuss some issues related to this work.

2   Methodology

For all RL algorithms we investigate in this paper, apart from the PHC and WoLF-PHC variants, ϵ-greedy or Boltzmann exploration is utilised. An ϵ-greedy exploration selects a random action in a given state with a probability ϵ, otherwise it selects the action with the highest Q-value. A Boltzmann exploration selects an action a_i from state s with probability $p(a_i)$ given by equation 1:

$$p(a_i) = \frac{e^{Q(s,a_i)/t}}{\sum\limits_{a \in A} e^{Q(s,a)/t}} \tag{1}$$

where $Q(s, a_i)$ is the value of state s and action a_i, A is the action set, and the temperature t is initialised with a temperature τ_0 at the beginning of an episode, and in subsequent steps is decreased exponentially by a multiplication with a cooling rate.

2.1 Taxi Problems

In the simple taxi problem [4] there is a 5×5 gridworld with four specially-designated locations marked with numbers 1 to 4 (Figure 1a) which represent the possible passenger locations and destinations. The taxi problem is episodic. Initially, in each episode, each taxi starts in a randomly chosen state and each passenger is located at one of the 4 locations (chosen randomly). The episode continues with the execution of the wish of the passenger to be transported to one of the four locations (also chosen randomly). The taxi must go to the passenger's location, pick up the passenger, go to the destination location and put down the passenger there. The episode ends when the passenger is delivered successfully. In the extended version of the simple taxi problem [5], there is a 10×10 gridworld with eight possible passenger locations and destinations marked with numbers 1 to 8 (Figure 1b). The task is the same as the simple taxi problem with the only difference being apart from the different wall distributions, the higher scale due to the larger grid and the increased number of destinations. The multiagent taxi problem (MATP) (Figure 1c) is an extension of the simple taxi problem, that we created to evaluate the MARL algorithms. In this version there exist two or more taxis and two or more passengers. Each taxi could carry at most one passenger at a time. The task is again the same with the only difference being that the episode ends when all the passengers are delivered successfully.

In the single-agent taxi problems there are six primitive actions: (a) four navigation actions that move the taxi one square, North, South, East, or West, (b) a Pickup action, and (c) a Putdown action. In the MATP there is an extra "idle" action (d), which does not modify the position of any agent. Each action is deterministic. There is a negative reward of -1 for each action and an additional reward of +20 for successfully delivering the passenger. There is a negative reward of -10 if the taxi attempts to execute the Putdown or Pickup actions illegally. If a navigation action would cause the taxi to hit a wall, the position of the taxi does not change, and there is only the usual negative reward of -1. In the MATP version, two or more taxis cannot occupy the same cell at the same time. If they collide, they will get an additional negative reward of -20. Then there would be a random selection of which taxi would stay at the current position and the other taxis would go back to their previous positions. If one of the colliding taxis has chosen one of the actions (b), (c) or (d), then surely that taxi would be the one which will stay at the current position and the other taxis would go back to their previous positions. MATP can be seen as a scenario where the state is composed of $n + 2k$ state-variables (where n is the number of taxis and k the number of passengers): the location of each taxi (values 0-24), each passenger location, including the case where the passenger is in the taxi (values 0-4, 0 means in the taxi) and each destination location (values 1-4). Thus there will be $25^n \times 5^k \times 4^k$ states.

Fitch et al. [6] have also explored a multiagent version of the simple taxi domain. In their work, they specifically had two taxis and two passengers and an extra primitive action for picking up the passengers (pickup passenger one and pickup passenger two), whereas our version has only one pickup actions. In [6] the four navigation actions are stochastic: with 92.5% probability they move the taxi in the intended direction and with 7.5% probability they move the taxi randomly in one of the other three directions, compared to our version where the navigation actions are deterministic. Moreover, Fitch et al. [6] consider two versions of the problem, one with and another without taxi collisions. In the version with collisions the taxis cannot occupy the same cell at the same time as is the case in our MATP. In particular, in their MATP version, after a collision, all taxis stay at their previous locations, in contrast with our taxi domain where one taxi always stays at its current position. Finally, a significant difference is that in [6], the agents have full-observability over the joint state-action space, and this is reflected in the Q-value estimates of the agents; in our MATP however, we allow full-observability only over the joint state space, meaning that the agents can observe the state, but cannot observe the other agents' actions. While this reduces the size of the Q-tables, it could make the problem harder.

2.2   MAXQ Hierarchical Decomposition in the Taxi Domain

Dieterich [4] proposed a hierarchical extension of Q-learning, called MAXQ-Q. MAXQ-Q uses the MAXQ hierarchy to decompose a task into subtasks. The MAXQ graph can effectively represent the value function of any hierarchical policy. All taxi problems have the same task as described above and share the same hierarchical structure. This structure contains two sub-tasks, i.e., "get the passenger" and "deliver the passenger", while both of them involve navigation to one of the four locations. The top of the hierarchy starts with the MaxRoot node which indicates the beginning of the task. From the MaxRoot node, the agent has the choice of two "actions" that lead to the MaxGet and MaxPut nodes, corresponding to the two sub-tasks above. From the MaxGet node the agent has the choice of two "actions", the pickup primitive action and the navigate sub-task, while from the MaxPut node the agent has the choice of the navigate sub-task and the putdown primitive action. The navigate sub-task of MaxGet and MaxPut leads to a MaxNavigate node which in turn leads to the four navigation primitive actions of north, east, south and west.

In the graph of the MAXQ hierarchy, each of the Max nodes is a child of a Q node (apart from MaxRoot) that learns the expected cumulative reward of executing its sub-task (or primitive action) which is context dependent. In contrast, the Max nodes learn the context independent expected cumulative reward of their sub-tasks (or primitive actions), which is an important distinction between the functionality of the Max and Q nodes.

2.3   Multiagent Extensions That Use the MAXQ Hierarchy

A multiagent extension of Q-learning to play mixed strategies (i.e., probabilistic policies) in repeated and stochastic games was proposed in [2]. This extension is known as Policy Hill Climbing (PHC) and works by maintaining a policy table in addition to the Q-table. The policy table can be seen as the probability of choosing an action in a given state, therefore for each state, the sum of all policy values is equal to 1. After the Q-values are updated with the Q-learning rule, the policy values are updated using a learning rate parameter δ based on the chosen action, and subsequently normalised in a legal probability distribution. An extension of the PHC algorithm was additionally proposed in [2] that uses the Win or Learn Fast (WoLF) principle, i.e., learn quickly when losing and be cautious when winning. This is achieved by varying the learning rate δ of the policy update. More specifically, two learning rates are used, $\delta_{winning}$ when winning and δ_{losing} when losing, where $\delta_{losing} > \delta_{winning}$. The WoLF-PHC algorithm, uses a separate table to store an estimate of the average policy over time (see [2] for details), and to calculate when the agent is winning or losing. An agent is winning when the expected performance of its current policy is greater than the expected performance of its average policy. It has to be noted that a stochastic policy (i.e., a mapping of states to distributions over the action space) is needed in these algorithms so that the agents are able to reach mixed strategy equilibria, while an estimate of the average stochastic policy over time is needed to check whether the agent is losing or winning.

A hierarchical extension of Q-learning was proposed with the MAXQ decomposition in order to learn the respective value function for each node, thus learning a hierarchical policy [4]. As described above, the PHC and WoLF-PHC algorithms maintain the current policy (and the average policy in the case of WoLF-PHC) along with the value functions. Therefore, a possible combination of the MAXQ hierarchy with the PHC or WoLF-PHC algorithms would most probably enable the learning of a hierarchical multiagent policy. This is the basis of the rationale of the approach we followed in this paper, where we implemented this combination. More specifically, in the case of MAXQ-PHC, each Max node contains additionally the policy of the respective node, while in the case of MAXQ-WoLF-PHC, the average policy was also included. Thus, the selection of "actions" in each node is done based on the stored policy, rather than an ϵ-greedy or a boltzmann exploration schedule. To the best of our knowledge such a combination has not previously been reported in the literature.

3   Results

3.1   Single-Agent Tasks

The Taxi Problem (TP) was run for 50 trials with 3000 episodes per trial and the Extended Taxi Problem (ETP) was run for 30 trials with 5000 episodes per trial. When we use ϵ-greedy exploration, ϵ takes the value 0.1 and when we use Boltzmann exploration, the initial temperature τ_0 takes the value 80 and the cooling rate the value 0.98, except for the MAXQ-Q algorithm; more specifically in the TP, we used a cooling rate of 0.9 at all nodes, while in the ETP, we used cooling rates of 0.9996 at MaxRoot and MaxPut, 0.9939 at MaxGet, and 0.9879 at MaxNavigate. These cooling rate values are the same as the ones used by Dietterich [4] for the TP; we observed, however, that they are more effective in the ETP rather than in the TP in our case. We did some experimentation with these values, but more needs to be done to fine-tune them. For both TP and ETP experiments we used a discount factor γ of 0.95 for all algorithms, and a step size α of 0.3 for Q-learning, SARSA, PHC and WoLF-PHC algorithms and 0.5 for MAXQ-Q, MAXQ-PHC and MAXQ-WoLF-PHC algorithms. The policy learning rate δ was set to 0.2 for the PHC and MAXQ-PHC algorithms. For the WoLF-PHC and MAXQ-WoLF-PHC algorithms, we set $\delta_{winning}$ to 0.05 and δ_{losing} to 0.2 in the TP experiments. In the ETP experiments, these values were set to 0.01 and 0.04 for the WoLF-PHC algorithm, while for the MAXQ-WoLF-PHC were set to 0.1 and 0.4 respectively. These values were found after some preliminary experimentation. Table 1 shows the ranking of the algorithms based on the median of the number of primitive time steps required to complete each successive trial averaged over 50 and 30 trials for TP and ETP respectively, for each algorithm.

As it can be seen, the tested algorithms have different rankings in the two problems. The most efficient algorithm for TP is SARSA using Boltzmann exploration and the most efficient algorithm for ETP is MAXQ-Q using ϵ-greedy exploration with a median of 11.46 and 64 respectively. This is due to the fact that in ETP the state-action space is bigger than the state-action space of TP, so the agents need more time to learn. Using the MAXQ hierarchical method we speed up learning, so the MAXQ-Q is the most efficient algorithm for ETP. We also notice that, both Q-learning and MAXQ-Q have better results when using an ϵ-greedy exploration, in contrast with SARSA which has better results when using Boltzmann exploration. This may be the case because Q-learning and MAXQ-Q are off-policy algorithms, whereas SARSA is an on-policy algorithm. This could also indicate that a more thorough investigation is needed to fine-tune all parameters.

On a closer inspection, when the MAXQ-Q algorithm is paired with a Boltzmann exploration in the TP, the average number of primitive time steps over the last 100 episodes (over all trials) converges around 34.6, whereas when ϵ-greedy exploration is used, the average number of primitive time steps over the last 100 episodes (over all trials) converges around 11. This might suggest that in the case of Boltzmann exploration, regardless of the task, more fine-tuning needs to

Table 1. Ranking of algorithms in the single-agent taxi problem. The efficiency of each algorithm is ranked according to the median of the number of primitive time steps required to complete each successive trial averaged over 50 and 30 trials for TP and ETP respectively. The most efficient algorithm scored 1 and the least efficient scored 10.

Ranking	Taxi Problem		Extended Taxi Problem	
	Algorithm	Median	Algorithm	Median
1	SARSA(boltzmann)	11.46	MAXQ-Q(ϵ-greedy)	64
2	MAXQ-Q(ϵ-greedy)	11.86	SARSA(boltzmann)	94.2
3	Q-learning(ϵ-greedy)	12.1	Q-learning(ϵ-greedy)	95.17
4	SARSA(ϵ-greedy)	12.9	MAXQ-WoLF-PHC	117.18
5	WoLF-PHC	14.35	MAXQ-PHC	125.93
6	PHC	14.74	SARSA(ϵ-greedy)	130.45
7	MAXQ-PHC	16.64	PHC	163.6
8	MAXQ-WoLF-PHC	16.88	MAXQ-Q(boltzmann)	169.28
9	Q-learning(boltzmann)	21.88	Q-learning(boltzmann)	179.98
10	MAXQ-Q(boltzmann)	35.62	WoLF-PHC	220.63

be done to the exploration schedule, while decreasing step sizes could provide an additional advantage. It has to be noted that the exploration schedule of each Max node should be optimised separately.

The PHC and WoLF-PHC algorithms are in the middle of the ranking for TP which means that they have respectable results for TP, even if they were initially created to be used in MARL problems. Moreover, the WoLF-PHC with median of 14.35 is better than the PHC algorithm with median of 14.74. This was expected because WoLF-PHC is an extension of the PHC algorithm which tries to improve it. Something not expected is that PHC seems to be better than WoLF-PHC for ETP. When plotting the graphs of average steps for each episode, we observed that the WoLF-PHC graph is smoother than the PHC graph. Therefore, the median metric of PHC has a lower value than the median value of WoLF-PHC because of oscillations on the PHC curve for average steps that occasionally reach a lower number than the respective WoLF-PHC curve.

Finally, we observe that the hierarchical algorithms MAXQ-PHC and MAXQ-WoLF-PHC are more efficient for ETP rather than for TP. That was expected, as hierarchy speeds up learning, so agents learn faster using these algorithms for ETP where the state-action space is bigger than the state-action space for TP.

3.2   Multiagent Tasks

The multiagent taxi problem (MATP) was run for 10 trials with 1000000 episodes per trial. A preliminary experimentation showed that Boltzmann exploration parameters could not be obtained easily, therefore we used an ϵ-greedy exploration. Exploration parameter ϵ was set to 0.1 in all algorithms except for MAXQ-Q, for which $\epsilon = 0.35$ due to the fact that it could not finish a single trial with

Table 2. Ranking of algorithms in the multiagent taxi problem. The efficiency of each algorithm is ranked according to the median of the number of primitive time steps required to complete each successive trial averaged over 10 trials. The most efficient algorithm scored 1 and the least efficient scored 7.

Ranking	Algorithm	Median
1	MAXQ-WoLF-PHC	102
2	MAXQ-PHC	103
3	WoLF-PHC	133.1
4	PHC	153.4
5	Q-learning (ϵ-greedy)	705.4
6	MAXQ-Q(ϵ-greedy)	901
7	SARSA (ϵ-greedy)	1623.9

lower values. For Q-learning, SARSA, PHC and WoLF-PHC algorithms we used $\alpha = 0.3$ and $\gamma = 0.95$. For the PHC algorithm $\delta = 0.2$, and for WoLF-PHC $\delta_{winning} = 0.05$ and $\delta_{losing} = 0.2$. For MAXQ-PHC and MAXQ-WoLF-PHC algorithms $\alpha = 0.5$ and $\gamma = 0.9$. We also used a $\delta = 0.2$ for MAXQ-PHC, and $\delta_{winning} = 0.05$ and $\delta_{losing} = 0.2$ for MAXQ-WoLF-PHC. In the case of MAXQ-Q, we used $\alpha = 0.05$ and $\gamma = 0.95$. Table 2 shows the ranking of the algorithms with regards to the median of the number of primitive time steps required to complete each successive trial averaged over 10 trials, for each algorithm.

As we can see, the most efficient algorithms are the hierarchical, multiagent algorithms MAXQ-WoLF-PHC and MAXQ-PHC, with medians 102 and 103 respectively. The WoLF-PHC algorithm follows with a median of 133.1, and then the PHC algorithm with a median of 153.4. The least efficient algorithms are the single-agent ones, i.e., Q-learning, MAXQ-Q and SARSA with medians 705.4, 901 and 1623.9 respectively. Figure 2 illustrates the performance of the algorithms in terms of average number of steps over time. From the results in Figure 2 we notice that the initial average number of steps in all hierarchical methods is significantly lower, compared to the corresponding non-hierarchical methods. In addition, we notice that all multiagent algorithms have a significantly lower average number of steps than the single-agent algorithms during all episodes. Interestingly, the behaviour of MAXQ-Q becomes worse over time and this deserves to be investigated further. The results of Figure 2 clearly show that our proposed algorithms are the most efficient for the MATP, due to the combination of the strengths of the MAXQ decomposition and the multiagent learning component that comes from maintaining an estimate of the stochastic policy and updating it accordingly. The single-agent algorithms have the worst performance as expected. The results of SARSA are worse than the results of Q-learning and this might be due to the fact that SARSA does not try to approximate a greedy policy (like Q-learning), but instead to approximate a "target policy" that is the same as its "behaviour policy", which can be disadvantageous in multiagent settings.

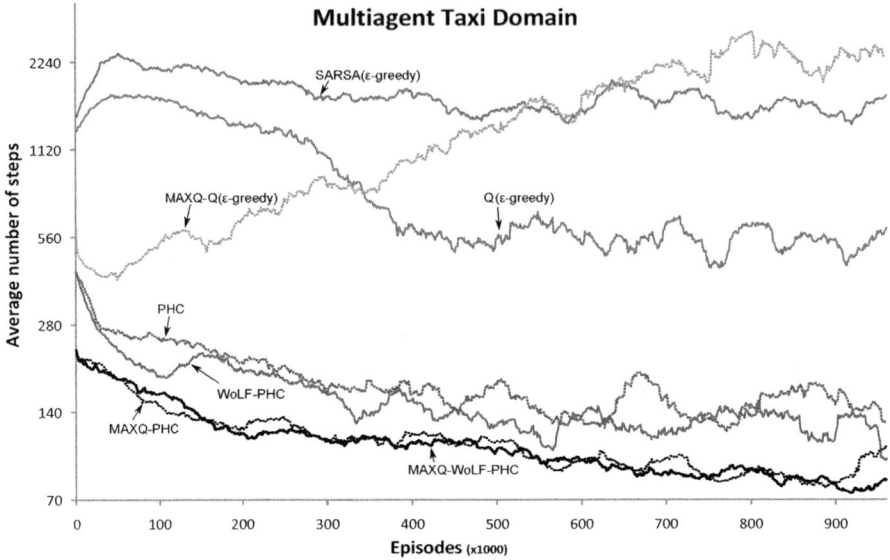

Fig. 2. Multiagent taxi domain results

4   Discussion and Conclusions

In this paper, we investigated and compared some single-agent RL (SARL) algorithms using the taxi domain and the extended taxi domain, and some MARL algorithms using a multiagent extension of the taxi domain we created.

In particular, for the single-agent part, we first compared Q-learning and SARSA using both ϵ-greedy and Boltzmann exploration. We observed that Q-learning is more efficient with ϵ-greedy exploration compared to SARSA, which is more efficient with Boltzmann exploration. Then we compared two multi-agent algorithms, PHC and WoLF-PHC, and we have observed that in the simple TP they have good results, even if they are algorithms for multiagent problems. In addition, WoLF-PHC is more efficient than PHC as we expected, because WoLF-PHC is an extension of PHC, which has been designed to improve it. Subsequently, we investigated and compared hierarchical algorithms and more specifically, algorithms based on the MAXQ decomposition. By trying the MAXQ-Q algorithm for TP and ETP we have observed that MAXQ-Q is more efficient with ϵ-greedy exploration than with Boltzmann exploration. Finally, we extended the PHC and WoLF-PHC algorithms with the MAXQ decomposition and noticed that even if the results are not so good for TP, they improved for ETP where the state-action space increases. For the multiagent part, we first tried the Q-learning, SARSA and MAXQ-Q algorithms using ϵ-greedy exploration and observed that agents do not learn well, something we expected since these are algorithms for SARL. Then we compared the PHC and WoLF-PHC algorithms and observed that WoLF-PHC is more efficient than PHC. Finally, we compared the MAXQ-PHC and MAXQ-WoLF-PHC algorithms and observed that MAXQ-WoLF-PHC is slightly more efficient than MAXQ-PHC. Certainly a direct comparison with other studies using a multiagent taxi domain cannot be made, as their results are reported in a different way (see for example [6]).

In conclusion, our results have demonstrated that as the state-action space grows, the MAXQ decomposition becomes more useful, because MAXQ speeds up learning. Thus, agents learn faster when they use the MAXQ decomposition in both SARL and MARL problems. Additionally, in MARL problems, the introduction of the WoLF-PHC mechanism creates more adaptive agents.

Combining agents which use different algorithms could be further explored (heterogeneous settings). This would demonstrate how the behaviour of each agent would change and if agents can cooperate with other agents that use different algorithms. In the algorithms we used, the agents do not communicate. It would be worthwhile implementing an algorithm which gives the opportunity for agents to communicate, such as WoLF-Partial Sharing Policy (WoLF-PSP) [9] which is an extension of WoLF-PHC. This algorithm could undergo hierarchical extension using the MAXQ decomposition. Moreover, on-policy versions of the PHC and WoLF-PHC algorithms could be implemented and investigated, that combine the on-policy sampling mechanism of the SARSA algorithm, and the policy update mechanisms of the PHC and WoLF-PHC algorithms respectively. Furthermore, they could be extended hierarchically using the MAXQ decomposition. Additionally, the combination of methods that automatically find the hierarchical structure of the task (such as [8], [12]), with the WoLF-PHC mechanism could be explored as well. Finally, directed exploration methods [19] could be utilised and compared with the undirected exploration methods we used in this study (i.e., Boltzmann and ϵ-greedy).

It has to be noted that the curse of dimensionality is still a major problem for the work presented in this paper, since only lookup tables have been used rather than any form of function approximation. To illustrate the explosion of the state space, let us consider using the multiagent task in the extended taxi problem domain. A simple calculation would require 2.703GB of RAM per table (Q-table, policy or average policy), per agent. In addition, as it is well known, the explosion of state-action space does cause generalisation problems. For these reasons, it would be very beneficial if state abstraction or function approximation approaches are employed to our study above in order to reduce the number of parameters needed to achieve generalisation in such large state-action spaces.

References

1. Andre, D., Russell, S.J.: In: Leen, T.K., Dietterich, T.G., Tresp, V. (eds.) Advances in Neural Information Processing Systems 13 (NIPS 2000), pp. 1019–1025. MIT Press, Cambridge (2001)
2. Bowling, M.H., Veloso, M.M.: Artificial Intelligence 136(2), 215–250 (2002)
3. Dayan, P., Hinton, G.E.: In: Hanson, S.J., Cowan, J.D., Giles, C.L. (eds.) Advances in Neural Information Processing Systems 5 (NIPS 1992), pp. 271–278. Morgan Kaufmann, San Francisco (1993)

4. Dietterich, T.G.: Journal of Artificial Intelligence Research 13, 227–303 (2000)
5. Diuk, C., Cohen, A., Littman, M.L.: In: Cohen, W.W., McCallum, A., Roweis, S.T. (eds.) Proceedings of the 25th International Conference on Machine Learning (ICML 2008), pp. 240–247. ACM, New York (2008)
6. Fitch, R., Hengst, B., Šuc, D., Calbert, G., Scholz, J.: Structural Abstraction Experiments in Reinforcement Learning. In: Zhang, S., Jarvis, R.A. (eds.) AI 2005. LNCS (LNAI), vol. 3809, pp. 164–175. Springer, Heidelberg (2005)
7. Ghavamzadeh, M., Mahadevan, S.: In: Proceedings of the 3rd International Joint Conference on Autonomous Agents and Multiagent Systems (AAMAS 2004), vol. 3, pp. 1114–1121. IEEE Computer Society, Washington, DC (2004)
8. Hengst, B.: In: Proceedings of the 19th International Conference on Machine Learning (ICML 2002), pp. 243–250. Morgan Kaufmann, San Francisco (2002)
9. Hwang, K.-S., Lin, C.-J., Wu, C.-J., Lo, C.-Y.: Cooperation Between Multiple Agents Based on Partially Sharing Policy. In: Huang, D.-S., Heutte, L., Loog, M. (eds.) ICIC 2007. LNCS, vol. 4681, pp. 422–432. Springer, Heidelberg (2007)
10. Kaelbling, L.P.: In: Proceedings of the 10th International Conference on Machine Learning (ICML 1993), pp. 167–173. Morgan Kaufmann, San Francisco (1993)
11. Mehta, N., Tadepalli, P., Fern, A.: In: Driessens, K., Fern, A., van Otterlo, M. (eds.) Proceedings of the ICML 2005 Workshop on Rich Representations for Reinforcement Learning, Bonn, Germany, pp. 45–50 (2005)
12. Mehta, N., Ray, S., Tadepalli, P., Dietterich, T.: In: Proceedings of the 25th International Conference on Machine Learning (ICML 2008), pp. 648–655. ACM, New York (2008)
13. Mirzazadeh, F., Behsaz, B., Beigy, H.: In: Proceedings of the International Conference on Information and Communication Technology (ICICT 2007), pp. 105–108 (2007)
14. Parr, R.: Hierarchical control and learning for Markov decision processes. Ph.D. thesis, University of California at Berkeley (1998)
15. Rummery, G.A., Niranjan, M.: On-line Q-learning using connectionist systems. Tech. Rep. CUED/F-INFENG/TR 166, Cambridge University (1994)
16. Shen, J., Liu, H., Gu, G.: In: Yao, Y., Shi, Z., Wang, Y., Kinsner, W. (eds.) Proceedings of the 5th International Conference on Cognitive Informatics (ICCI 2006), pp. 584–588. IEEE (2006)
17. Singh, S.P.: Machine Learning 8, 323–339 (1992)
18. Sutton, R.S., Precup, D., Singh, S.: Artificial Intelligence 112, 181–211 (1999)
19. Thrun, S.B.: Efficient Exploration in Reinforcement Learning. Tech. Rep. CMU-CS-92-102, Carnegie Mellon University, Pittsburgh, PA (1992)
20. Watkins, C.J.C.H.: Learning from delayed rewards. Ph.D. thesis, University of Cambridge (1989)
21. Wiering, M., Schmidhuber, J.: Adaptive Behavior 6, 219–246 (1998)

Bayesian Multitask Inverse Reinforcement Learning

Christos Dimitrakakis[1] and Constantin A. Rothkopf[2]

[1] EPFL, Lausanne, Switzerland
christos.dimitrakakis@epfl.ch
[2] Frankfurt Institute for Advanced Studies, Frankfurt, Germany
rothkopf@fias.uni-frankfurt.de

Abstract. We generalise the problem of inverse reinforcement learning to multiple tasks, from multiple demonstrations. Each one may represent one expert trying to solve a different task, or as different experts trying to solve the same task. Our main contribution is to formalise the problem as statistical preference elicitation, via a number of structured priors, whose form captures our biases about the relatedness of different tasks or expert policies. In doing so, we introduce a prior on policy optimality, which is more natural to specify. We show that our framework allows us not only to learn to efficiently from multiple experts but to also effectively differentiate between the goals of each. Possible applications include analysing the intrinsic motivations of subjects in behavioural experiments and learning from multiple teachers.

Keywords: Bayesian inference, intrinsic motivations, inverse reinforcement learning, multitask learning, preference elicitation.

1 Introduction

This paper deals with the problem of multitask inverse reinforcement learning. Loosely speaking, this involves inferring the motivations and goals of an unknown agent performing a series of tasks in a dynamic environment. It is also equivalent to inferring the motivations of different experts, each attempting to solve the same task, but whose different preferences and biases affect the solution they choose. Solutions to this problem can also provide principled statistical tools for the interpretation of behavioural experiments with humans and animals.

While both inverse reinforcement learning, and multitask learning are well known problems, to our knowledge this is the only principled statistical formulation of this problem. Our first major contribution generalises our previous work [20], a statistical approach for single-task inverse reinforcement learning, to a hierarchical (population) model discussed in Section 3. Our second major contribution is an alternative model, which uses a much more natural prior on the optimality of the demonstrations, in Section 4, for which we also provide computational complexity bounds. An experimental analysis of the procedures is given in Section 5, while the connections to related work are discussed in Section 6. Auxiliary results and proofs are given in the appendix.

S. Sanner and M. Hutter (Eds.): EWRL 2011, LNCS 7188, pp. 273–284, 2012.
© Springer-Verlag Berlin Heidelberg 2012

2   The General Model

We assume that all tasks are performed in an environment with dynamics drawn from the same distribution (which may be singular). We define the environment as a controlled Markov process (CMP) $\nu = (\mathcal{S}, \mathcal{A}, \mathcal{T})$, with state space $\mathcal{S}$, action space $\mathcal{A}$, and transition kernel $\mathcal{T} = \{ \tau(\cdot \mid s, a) : s \in \mathcal{S}, a \in \mathcal{A} \}$, indexed in $\mathcal{S} \times \mathcal{A}$ such that $\tau(\cdot \mid s, a)$ is a probability measure[1] on $\mathcal{S}$. The dynamics of the environment are Markovian: If at time t the environment is in state $s_t \in S$ and the agent performs action $a_t \in A$, then the next state s_{t+1} is drawn with a probability independent of previous states and actions: $\mathbb{P}_\nu(s_{t+1} \in S \mid s^t, a^t) = \tau(S \mid s_t, a_t)$, $S \subset \mathcal{S}$, where we use the convention $s^t \equiv s_1, \ldots, s_t$ and $a^t \equiv a_1, \ldots, a_t$ to represent sequences of variables, with $\mathcal{S}^t, \mathcal{A}^t$ being the corresponding product spaces. If the dynamics of the environment are unknown, we can maintain a belief about what the true CMP is, expressed as a probability measure ω on the space of controlled Markov processes $\mathcal{N}$.

During the m-th demonstration, we observe an agent acting in the environment and obtain a T_m-long sequence of actions and a sequence of states: $\boldsymbol{d}_m \triangleq (a_m^{T_m}, s_m^{T_m})$, $a_m^{T_m} \triangleq a_{m,1}, \ldots, a_{m,T}$, $s_m^{T_m} \triangleq s_{m,1}, \ldots, s_{m,T_m}$. The m-th task is defined via an *unknown utility function*, $U_{m,t}$, according to which the demonstrator selects actions, which we wish to discover. Setting $U_{m,t}$ equal to the total discounted return,[2] we establish a link with inverse reinforcement learning:

Assumption 1. *The agent's utility at time t is defined in terms of future rewards: $U_{m,t} \triangleq \sum_{k=t}^{\infty} \gamma^k r_k$, where $\gamma \in [0,1]$ is a discount factor, and the reward r_t is given by the reward function $\rho_m : \mathcal{S} \times \mathcal{A} \to \mathbb{R}$ so that $r_t \triangleq \rho_m(s_t, a_t)$.*

In the following, for simplicity we drop the subscript m whenever it is clear by context. For any reward function ρ, the controlled Markov process and the resulting utility U define a Markov decision process [17] (MDP), denoted by $\mu = (\nu, \rho, \gamma)$. The agent uses some policy π to select actions $a_t \sim \pi(\cdot \mid s^t, a^{t-1})$, which together with the Markov decision process μ defines a distribution[3] on the sequences of states, such that $\mathbb{P}_{\mu,\pi}(s_{t+1} \in S \mid s^t, a^{t-1}) = \int_{\mathcal{A}} \tau(S \mid a, s_t) \, d\pi(a \mid s^t, a^{t-1})$, where we use a subscript to denote that the probability is taken with respect to the process defined jointly by μ, π. We shall use this notational convention throughout this paper. Similarly, the *expected utility* of a policy π is denoted by $\mathbb{E}_{\mu,\pi} U_t$. We also introduce the family of Q-value functions $\{ Q_\mu^\pi : \mu \in \mathcal{M}, \pi \in \mathcal{P} \}$, where $\mathcal{M}$ is a set of MDPs, with $Q_\mu^\pi : \mathcal{S} \times \mathcal{A} \to \mathbb{R}$ such that: $Q_\mu^\pi(s, a) \triangleq \mathbb{E}_{\mu,\pi}(U_t \mid s_t = s, a_t = a)$. Finally, we use Q^* to denote the optimal Q-value function for an MDP μ, such that: $Q_\mu^*(s, a) = \sup_{\pi \in \mathcal{P}} Q_\mu^\pi(s, a)$, $\forall s \in \mathcal{S}, a \in \mathcal{A}$. With a slight abuse of notation, we shall use Q_ρ when we only

---

[1] We assume the measurability of all sets with respect to some appropriate σ-algebra.

[2] Other forms of the utility are possible. For example, consider an agent who collects gold coins in a maze with traps, and where the agent's utility is the logarithm of the number of coins it has after it has exited the maze.

[3] When the policy is reactive, then $\pi(a_t \mid s^t, a^{t-1}) = \pi(a_t \mid s_t)$, and the process reduces to first order Markov.

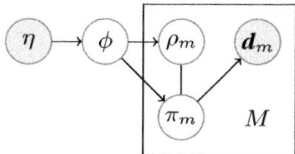

Fig. 1. Graphical model of general multitask reward-policy priors. Lighter colour indicates latent variables. Here η is the hyperprior on the joint reward-policy prior ϕ while ρ_m and π_m are the reward and policy of the m-th task, for which we observe the demonstration $\boldsymbol{d}_m$. The undirected link between π and ρ represents the fact that the rewards and policy are jointly drawn from the reward-policy prior. The implicit dependencies on ν are omitted for clarity.

need to distinguish between different reward functions ρ, as long as the remaining components of μ are clear from the context.

Loosely speaking, our problem is to estimate the sequence of reward functions $\boldsymbol{\rho} \triangleq \rho_1, \ldots, \rho_m, \ldots, \rho_M$, and policies $\boldsymbol{\pi} \triangleq \pi_1, \ldots, \pi_m, \ldots, \pi_M$, which were used in the demonstrations, given the data $\boldsymbol{D} = \boldsymbol{d}_1, \ldots, \boldsymbol{d}_m, \ldots, \boldsymbol{d}_M$ from *all* demonstrations and some prior beliefs. In order to do this, we define a multitask reward-policy prior distribution as a Bayesian hierarchical model.

2.1   Multitask Priors on Reward Functions and Policies

We consider two types of priors on rewards and policies. Their main difference is how the dependency between the reward and the policy is modelled. Due to the multitask setting, we posit that the reward function is drawn from some unknown distribution for each task, for which we assert a hyperprior, which is later conditioned on the demonstrations. The hyperprior η is a probability measure on the set of joint reward-policy priors $\mathscr{J}$. It is easy to see that, given some specific $\phi \in \mathscr{J}$, we can use Bayes' theorem directly to obtain, for any $A \subset \mathcal{P}^M, B \subset \mathcal{R}^M$, where $\mathcal{P}^M, \mathcal{R}^M$ are the policy and reward product spaces:

$$\phi(A, B \mid \boldsymbol{D}) = \frac{\int_{A \times B} \phi(\boldsymbol{D} \mid \boldsymbol{\rho}, \boldsymbol{\pi}) \, \mathrm{d}\phi(\boldsymbol{\rho}, \boldsymbol{\pi})}{\int_{\mathcal{R}^M \times \mathcal{P}^M} \phi(\boldsymbol{D} \mid \boldsymbol{\rho}, \boldsymbol{\pi}) \, \mathrm{d}\phi(\boldsymbol{\rho}, \boldsymbol{\pi})} = \prod_m \phi(\rho_m, \pi_m \mid \boldsymbol{d}_m).$$

When ϕ is not specified, we must somehow estimate some distribution on it. In the *empirical Bayes* case [19] the idea is to simply find a distribution η in a restricted class H, according to some criterion, such as maximum likelihood. In the *hierarchical Bayes* approach, followed herein, we select some prior η and then estimate the *posterior distribution* $\eta(\cdot \mid \boldsymbol{D})$.

We consider two models. In the first, discussed in Section 3 on the following page, we initially specify a product prior on reward functions and on *policy parameters*. Jointly, these determine a unique policy, for which the probability of the observed demonstration is well-defined. The policy-reward dependency is exchanged in the alternative model, which is discussed in Section 4 on page 277. There we specify a product prior on policies and on *policy optimality*. This leads to a distribution on reward functions, conditional on policies.

3   Multitask Reward-Policy Prior (MRP)

Let $\mathcal{R}$ be the space of reward functions ρ and $\mathcal{P}$ the space of policies π. Let $\psi(\cdot \mid \nu) \in \mathcal{R}$ denote a conditional probability measure on the reward functions $\mathcal{R}$ such that for any $B \subset \mathcal{R}$, $\psi(B \mid \nu)$ corresponds to our prior belief that the reward function is in B, when the CMP is known to be ν. For any reward function $\rho \in \mathcal{R}$, we define a conditional probability measure $\xi(\cdot \mid \rho, \nu) \in \mathcal{P}$ on the space of policies $\mathcal{P}$. Let ρ_m, π_m denote the m-th demonstration's reward function and policy respectively. We use a product[4] hyperprior[5] η on the set of reward function distributions and policy distribution $\mathcal{R} \times \mathcal{P}$, such that $\eta(\Psi, \Xi) = \eta(\Psi)\eta(\Xi)$ for all $\Psi \subset \mathcal{R}$, $\Xi \subset \mathcal{P}$. Our model is specified as follows:

$$(\psi, \xi) \sim \eta(\cdot \mid \nu), \qquad \rho_m \mid \psi, \nu \sim \psi(\cdot \mid \nu), \qquad \pi_m \mid \xi, \nu, \rho_m \sim \xi(\cdot \mid \rho_m, \nu), \qquad (3.1)$$

In this case, the joint prior on reward functions and policies can be written as $\phi(P, R \mid \nu) \triangleq \int_R \xi(P \mid \rho, \nu) \, d\psi(\rho \mid \nu)$ with $P \subset \mathcal{P}$, $R \subset \mathcal{R}$, such that $\phi(\cdot \mid \nu)$ is a probability measure on $\mathcal{P} \times \mathcal{R}$ for any CMP ν.[6] In our model, the only observable variables are η, which we select ourselves and the demonstrations $\boldsymbol{D}$.

3.1   The Policy Prior

The model presented in this section involves restricting the policy space to a parametric form. As a simple example, we consider stationary soft-max policies with an inverse temperature parameter c:

$$\pi(a_t \mid s_t, \mu, c) = Softmax(a_t \mid s_t, \mu, c) \triangleq \frac{\exp(cQ^*_\mu(s_t, a_t))}{\sum_a \exp(cQ^*_\mu(s_t, a))}, \qquad (3.2)$$

where we assumed a finite action set for simplicity. Then we can define a prior on policies, given a reward function, by specifying a prior β on c. Inference can be performed using standard Monte Carlo methods. If we can estimate the reward functions well enough, we may be able to obtain policies that surpass the performance of the demonstrators.

3.2   Reward Priors

In our previous work [20], we considered a product-Beta distribution on states (or state-action pairs) for the reward function prior. Herein, however, we develop a more structured prior, by considering reward functions as a measure on the

---

[4] Even if a prior distribution is a product, the posterior may not necessarily remain a product. Consequently, this choice does not imply the assumption that rewards are independent from policies.

[5] In order to simplify the exposition somewhat, while maintaining generality, we usually specify distributions on functions or other distributions directly, rather than on their parameters.

[6] If the CMP itself is unknown, so that we only have a probabilistic belief ω on $\mathcal{N}$, we can instead consider the marginal $\phi(P, R \mid \omega) \triangleq \int_\mathcal{N} \phi(P, R \mid \nu) \, d\omega(\nu)$.

state space $\mathcal{S}$ with $\rho(\mathcal{S}) = 1$. Then for any state subsets $S_1, S_2 \subset \mathcal{S}$ such that $S_1 \cap S_2 = \emptyset$, $\rho(S_1 \cup S_2) = \rho(S_1) + \rho(S_2)$. A well-known distribution on probability measures is a Dirichlet process [11]. Consequently, when $\mathcal{S}$ is finite, we can use a Dirichlet prior for rewards, such that each sampled reward function is equivalent to multinomial parameters. This is more constrained than the Beta-product prior and has the advantage of clearly separating the reward function from the c parameter in the policy model. It also brings the Bayesian approach closer to approaches which bound the L_1 norm of the reward function such as [21].

3.3   Estimation

The simplest possible algorithm consists of sampling directly from the prior. In our model, the prior on the reward function ρ and inverse temperature c is a product, and so we can simply take independent samples from each, obtaining an approximate posterior on rewards an policies, as shown in Alg. 1. While such methods are known to converge asymptotically to the true expectation under mild conditions [12], stronger technical assumptions are required for finite sample bounds, due to importance sampling in step 8.

Algorithm 1. MRP-MC: Multitask Reward-Policy Monte Carlo. Given the data $\boldsymbol{D}$, we obtain $\hat{\eta}$, the approximate posterior on the reward-policy distirbution, and $\hat{\rho}_m$, the $\hat{\eta}$-expected reward function for the m-th task.

1: **for** $k = 1, \ldots, K$ **do**
2: $\quad$ $\phi^{(k)} = (\xi^{(k)}, \psi^{(k)}) \sim \eta$, $\xi^{(k)} = Gamma(g_1^{(k)}, g_2^{(k)})$.
3: $\quad$ **for** $m = 1, \ldots, M$ **do**
4: $\quad\quad$ $\rho_m^{(k)} \sim \xi(\rho \mid \nu)$, $c_m^{(k)} \sim Gamma(g_1^{(k)}, g_2^{(k)})$
5: $\quad\quad$ $\mu_m^{(k)} = (\nu, \gamma, \rho_m^{(k)})$, $\pi_m^{(k)} = Softmax(\cdot \mid \cdot, \mu_m^{(k)}, c_m^{(k)})$, $p_m^{(k)} = \pi_m^{(k)}(a_m^T \mid s_m^T)$
6: $\quad$ **end for**
7: **end for**
8: $q^{(k)} = \prod_m p_m^{(k)} / \sum_{j=1}^{K} \prod_m p_m^{(j)}$
9: $\hat{\eta}(B \mid \boldsymbol{D}) = \sum_{k=1}^{K} \mathbb{I}\left\{\phi^{(k)} \in B\right\} q^{(k)}$, for $B \subset \mathcal{R} \times \mathcal{P}$.
10: $\hat{\rho}_m = \sum_{k=1}^{K} \rho_m^{(k)} q^{(k)}$, $m = 1, \ldots, M$.

An alternative, which may be more efficient in practice if a good proposal distribution can be found, is to employ a Metropolis-Hastings sampler instead, which we shall refer to as MRP-MH. Other samplers, including a hybrid Gibbs sampler, hereafter refered to as MRP-GIBBS, such as the one introduced in [20] are possible.

4   Multitask Policy Optimality Prior (MPO)

Specifying a parametric form for the policy, such as the softmax, is rather awkward and hard to justify. It is more natural to specify a prior on the *optimality* of the policy demonstrated. Given the data, and a prior over a policy class (e.g. stationary policies), we obtain a posterior distribution on policies. Then, via a simple algorithm, we can combine this with the optimality prior and obtain a posterior distribution on reward functions.

As before, let D be the observed data and let ξ be a prior probability measure on the set of policies $\mathcal{P}$, encoding our biases towards specific policy types. In addition, let $\{\psi(\cdot \mid \pi) : \pi \in \mathcal{P}\}$ be a set of probability measures on $\mathcal{R}$, indexed in $\mathcal{P}$, to be made precise later. In principle, we can now calculate the marginal posterior over reward functions ρ given the observations D, as follows:

$$\psi(B \mid D) = \int_{\mathcal{P}} \psi(B \mid \pi) \, d\xi(\pi \mid D), \qquad B \subset \mathcal{R}. \tag{4.1}$$

The main idea is to define a distribution over reward functions, via a prior on the optimality of the policy followed. The first step is to explicitly define the measures on $\mathcal{R}$ in terms of ε-optimality, by defining a prior measure β on $\mathbb{R}_+$, such that $\beta([0, \varepsilon])$ is our prior that the policy is ε-optimal. Assuming that $\beta(\varepsilon) = \beta(\varepsilon \mid \pi)$ for all π, we obtain:

$$\psi(B \mid \pi) = \int_0^\infty \psi(B \mid \varepsilon, \pi) \, d\beta(\varepsilon), \tag{4.2}$$

where $\psi(B \mid \varepsilon, \pi)$ can be understood as the prior probability that $\rho \in B$ given that the policy π is ε-optimal. The marginal (4.1) can now be written as:

$$\psi(B \mid D) = \int_{\mathcal{P}} \left(\int_0^\infty \psi(B \mid \varepsilon, \pi) \, d\beta(\varepsilon) \right) d\xi(\pi \mid D) \tag{4.3}$$

We now construct $\psi(\cdot \mid \varepsilon, \pi)$. Let the set of ε-optimal reward functions with respect to π be: $\mathcal{R}_\varepsilon^\pi \triangleq \{\rho \in \mathcal{R} : \|V_\rho^* - V_\rho^\pi\|_\infty < \varepsilon\}$. Let $\lambda(\cdot)$ be an arbitrary measure on $\mathcal{R}$ (e.g. the counting measure if $\mathcal{R}$ is discrete). We can now set:

$$\psi(B \mid \varepsilon, \pi) \triangleq \frac{\lambda(B \cap \mathcal{R}_\varepsilon^\pi)}{\lambda(\mathcal{R}_\varepsilon^\pi)}, \qquad B \subset \mathcal{R}. \tag{4.4}$$

Then $\lambda(\cdot)$ can be interpreted as an (unnormalised) prior measure on reward functions. If the set of reward functions $\mathcal{R}$ is finite, then a simple algorithm can be used to estimate preferences, described below.

We are given a set of demonstration trajectories D and a prior on policies ξ, from which we calculate a posterior on policies $\xi(\cdot \mid D)$. We sample a set of K policies $\Pi = \{\pi^{(i)} : i = 1, \ldots, K\}$ from this posterior. We are also given a set of reward functions $\mathcal{R}$ with associated measure $\lambda(\cdot)$. For each policy-reward pair $(\pi^{(i)}, \rho_j) \in \Pi \times \mathcal{R}$, we calculate the loss of the policy for the given reward function to obtain a loss matrix:

$$L \triangleq [\ell_{i,j}]_{K \times |\mathcal{R}|}, \qquad \ell_{i,j} \triangleq \sup_s V_{\rho_j}^*(s) - V_{\rho_j}^{\pi^{(i)}}(s), \tag{4.5}$$

where $V^*_{\rho_j}$ and $V^{\pi^{(i)}}_{\rho_j}$ are the value functions, for the reward function ρ_j, of the optimal policy and $\pi^{(i)}$ respectively.[7]

Given samples $\pi^{(i)}$ from $\xi(\pi \mid \boldsymbol{D})$, we can estimate the integral (4.3) accurately via $\hat{\psi}(B \mid \boldsymbol{D}) \triangleq \frac{1}{K} \sum_{i=1}^{K} \int_0^\infty \psi(B \mid \varepsilon, \pi^{(i)}) \, d\beta(\varepsilon)$. In addition, note that the loss matrix L is finite, with a number of distinct elements at most $K \times |\mathcal{R}|$. Consequently, $\psi(B \mid \varepsilon, \pi^{(i)})$ is a piece-wise constant function with respect to ε. Let $(\varepsilon_k)_{k=1}^{K \times |\mathcal{R}|}$ be a monotonically increasing sequence of the elements of L. Then $\psi(B \mid \varepsilon, \pi^{(i)}) = \psi(B \mid \varepsilon', \pi^{(i)})$ for any $\varepsilon, \varepsilon' \in [\varepsilon_k, \varepsilon_{j+1}]$, and:

$$\hat{\psi}(B \mid \boldsymbol{D}) \triangleq \sum_{i=1}^{K} \sum_{k=1}^{K \times |\mathcal{R}|} \psi(B \mid \varepsilon_k, \pi^{(i)}) \beta([\varepsilon_k, \varepsilon_{k+1})). \tag{4.6}$$

Note that for an exponential prior with parameter c, we have $\beta([\varepsilon_k, \varepsilon_{k+1}]) = e^{-c\varepsilon_k} - e^{-c\varepsilon_{k+1}}$. We can now find the optimal policy with respect to the expected utility.

Theorem 1. *Let $\hat{\eta}_k(\cdot \mid \boldsymbol{D})$ be the empirical posterior measure calculated via the above procedure and assume ρ takes values in $[0, 1]$ for all $\rho \in \mathcal{R}$. Then, for any value function V_ρ,*

$$\mathbb{E}_\eta(\|V_\rho - \hat{V}_\rho\|_\infty \mid \boldsymbol{D}) \leq \frac{1}{(1-\gamma)\sqrt{K}} \left(2 + \frac{1}{2}\sqrt{\ln K}\right), \tag{4.7}$$

where the expectation is taken w.r.t the marginal distribution on $\mathcal{R}$.

This theorem, whose proof is in the appendix, bounds the number of samples required to obtain a small loss in the value function estimation, and holds with only minor modifications for both the single and multi-task cases for finite $\mathcal{R}$. For the multi-task case and general $\mathcal{R}$, we can use MPO-MC (Alg. 2 on the next page), to sample N reward functions from a prior. Unfortunately the theorem does not apply directly for infinite $\mathcal{R}$. While one could define an ϵ-net on $\mathcal{R}$, and assume smoothness conditions, in order to obtain in optimality guarantees for that case, this is beyond the scope of this paper.

5 Experiments

Given a distribution on the reward functions ψ, and known transition distributions, one can obtain a stationary policy that is optimal with respect to this distribution via value iteration. This is what single-task algorithms essentially do, but it ignores differences among tasks. In the multi-task setting, we infer the optimal policy $\hat{\pi}^*_m$ for the m-th task. Its L_1-loss with respect to the optimal value function is $\ell_m(\hat{\pi}^*_m) \triangleq \sum_{s \in \mathcal{S}} V^*_{\rho_m}(s) - V^\pi_{\rho_m}(s)$. We are interested in minimising the total loss $\sum_m \ell_m$ across demonstrations. We first examined the efficiency

---

[7] Again, we abuse notation slightly and employ V_{ρ_j} to denote the value function of the MDP (ν, ρ_j), for the case when the underlying CMP ν is known. For the case when we only have a belief ω on the set of CMPs $\mathcal{N}$, V_{ρ_j} refers to the expected utility with respect to ω, or more precisely $V^\pi_{\rho_j}(s) = \mathbb{E}_\omega(U_t \mid s_t = s, \rho_j, \pi) = \int_\mathcal{N} V^\pi_{\nu, \rho_j}(s) \, d\omega(\nu)$.

Algorithm 2. MPO-MC Multitask Policy Optimality Monte Carlo posterior estimate

1: Sample N reward functions $\rho_1, \ldots, \rho_N \sim \psi$.
2: **for** $k = 1, \ldots, K$ **do**
3:    $(\xi^{(k)}, \psi^{(k)}) \sim \eta$, where $\psi^{(k)}$ is multinomial over N outcomes.
4:    **for** $m = 1, \ldots, M$ **do**
5:       $\pi_m^{(k)} \sim \xi^{(k)}(\cdot \mid \boldsymbol{d}_m)$.
6:    **end for**
7: **end for**
8: Calculate $\hat{\phi}_m(\cdot \mid \boldsymbol{d}_m)$ from (4.6) and $\{\pi_m^{(k)} : k = 1, \ldots, K\}$.

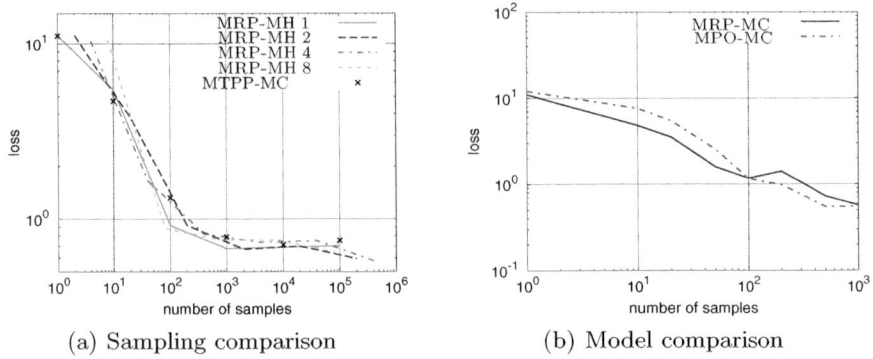

(a) Sampling comparison   (b) Model comparison

Fig. 2. Expected loss for two samplers, averaged over 10^3 runs, as the number of total samples increases. Fig. 2(b) compares the MRP and MPO models using a Monte Carlo estimate. Fig. 2(a) shows the performance of different sampling strategies for the MTPP model: *Metropolis-Hastings* sampling, with different numbers of parallel chains and simple *Monte Carlo* estimation.

of sampling. Initially, we used the *Chain* task [8] with 5 states (c.f. Fig. 3(a)), $\gamma = 0.95$ and a demonstrator using standard model-based reinforcement learning with ϵ-greedy exploration policy using $\epsilon = 10^{-2}$, using the Dirichlet prior on reward functions. As Fig. 2(a) shows, for the MRP model, results slightly favour the single chain MH sampler. Figure 2(b) compares the performance of the MRP and MPO models using an MC sampler. The actual computing time of MPO is larger by a constant factor due to the need to calculate (4.6).

In further experiments, we compared the multi-task perfomance of MRP with that of an imitator, for the generalised chain task where rewards are sampled from a Dirichlet prior. We fixed the number of demonstrations to 10 and varied the nnumber of tasks. The gain of using a multi-task model is shown in Fig. 3(b). Finally, we examined the effect of the demonstration's length, independently of the number of task. Fig. 3(c),3(d) show that when there is more data, then MPO is much more efficient, since we sample directly from $\xi(\pi \mid \boldsymbol{D})$. In that case, the MRP-MC sampler is very inefficient. For reference, we include the performance of MWAL and the imitator.

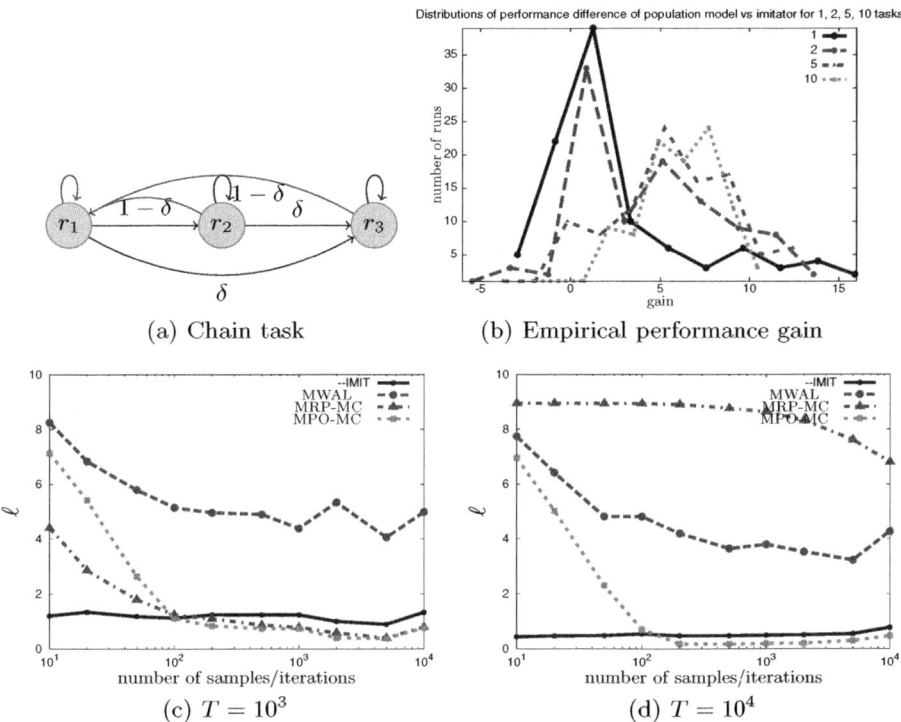

(a) Chain task   (b) Empirical performance gain

(c) $T = 10^3$ (d) $T = 10^4$

Fig. 3. Experiments on the chain task. (a) The 3-state version of the task. (b) Empirical performance difference of MRP-MC and IMIT is shown for $\{1, 2, 5, 10\}$ tasks respectively, with 10 total demonstrations. As the number of tasks increases, so does the performance gain of the multitask prior relative to an imitator. (c,d) Single-task sample efficiency in the 5-state Chain task with $r_1 = 0.2$, $r_2 = 0$, $r_3 = 1$. The data is sufficient for the imitator to perform rather well. However, while the MPO-MC is consistently better than the imitator, MRP-MC converges slowly.

The second experiment samples *variants* of Random MDP tasks [20], from a hierarchical model, where Dirichlet parameters are drawn from a product of *Gamma*(1, 10) and task rewards are sampled from the resulting Dirichlets. Each demonstration is drawn from a softmax policy with respect to the current task, with $c \in [2, 8]$ for a total of 50 steps. We compared the loss of policies derived from MRP-MC, with that of algorithms described in [16, 18, 21], as well as a flat model [20]. Fig. 4(a) on the following page shows the loss for varying c, when the (unknown) number of tasks equals 20. While flat MH can recover reward functions that lead to policies that outperform the demonstrator, the multi-task model MRP-MH shows a clear additional improvement. Figure 4(b) shows that this increases with the number of available demonstrations, indicating that the task distribution is estimated well. In contrast, RP-MH degrades slowly, due to its assumption that all demonstrations share a common reward function.

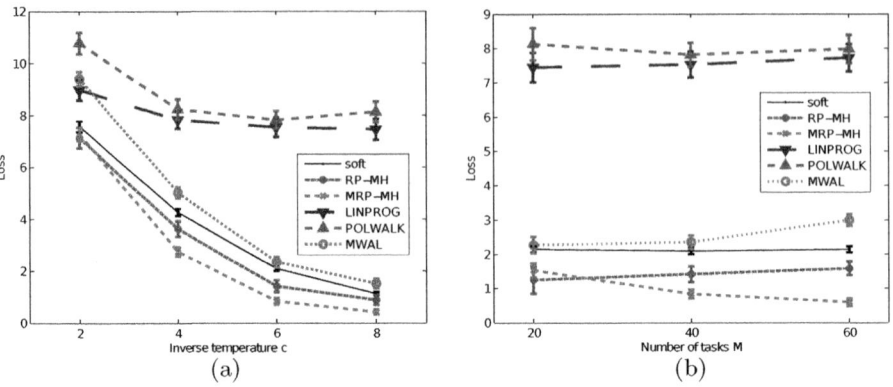

Fig. 4. Experiments on random MDP tasks, comparing MTPP-MH with the original (RP-MH) sampler[20], a demonstrator employing a softmax policy (soft), Policy Walk (POLWALK) [18] and Linear Programming (LINPROG) [16] MWAL [21], averaged over 10^2 runs. Fig. 4(a) shows the loss as the inverse softmax temperature c increases, for a fixed number of $M = 20$ tasks Fig. 4(b) shows the loss relative to the optimal policy as the number of tasks increases, for fixed $c = 8$. There is one 50-step demonstration per task. The error bars indicate standard error.

6   Related Work and Discussion

A number of inverse reinforcement learning [1, 5, 7, 16, 18, 20, 23] and preference elicitation [4, 6] approaches have been proposed, while multitask learning itself is a well-known problem, for which hierarchical Bayesian approaches are quite natural [13]. In fact, two Bayesian approaches have been considered for multitask reinforcement learning. Wilson et al. [22] consider a prior on MDPs, while Lazaric and Ghavamzadeh [14] employ a prior on value functions.

The first work that we are aware of that performs multi-task estimation of utilities is [3], which used a hierarchical Bayesian model to represent relationships between preferences. Independently to us, [2] recently considered the problem of learning for multiple intentions (or reward functions). Given the number of intentions, they employ an expectation maximisation approach for clustering. Finally, a generalisation of IRL to the *multi-agent* setting, was examined by Natarajan et al. [15]. This is the problem of finding a good *joint* policy, for a number of agents acting simultaneously in the environment.

Our approach can be seen as a generalisation of [3] to the dynamic setting of inverse reinforcement learning; of [2] to full Bayesian estimation; and of [20] to multiple tasks. This enables significant potential applications. For example, we have a first theoretically sound formalisation of the problem of learning from multiple teachers who all try to solve the same problem, but which have different preferences for doing so. In addition, the principled Bayesian approach allows us to infer a complete distribution over task reward functions. Technically, the work presented in this paper is a direct generalisation of our previous paper [20], which proposed single task equivalents of the policy parameter priors discussed in Sec. 3, to the multitask setting. In addition to the introduction of multiple tasks, we provide an alternative policy optimality prior, which is a not only a much more natural prior to specify, but for which we can obtain computational complexity bounds.

In future work, we may consider non-parametric priors, such as those considered in [10], for the policy optimality model of Sec. 4. Finally, when the MDP is unknown, calculation of the optimal policy is in general much harder. However, in a recent paper [9] we show how to obtain near-optimal memoryless policies for the unknown MDP case, which would be applicable in this setting.

Acknowledgements. This work was partially supported by the BMBF Project Bernstein Fokus: Neurotechnologie Frankfurt, FKZ 01GQ0840, the EU-Project IM-CLeVeR, FP7-ICT-IP-231722, and the Marie Curie Project ESDEMUU, Grant Number 237816.

A   Auxillary Results and Proofs

Lemma 1 (Hoeffding inequality). *For independent random variables* $X_1, \ldots, X_n$ *such that* $X_i \in [a_i, b_i]$, *with* $\mu_i \triangleq \mathbb{E}\, X_i$ *and* $t > 0$:

$$\mathbb{P}\left(\sum_{i=1}^{n} X_i \geq \sum_{i=1}^{n} \mu_i + nt\right) = \mathbb{P}\left(\sum_{i=1}^{n} X_i \leq \sum_{i=1}^{n} \mu_i - nt\right) \leq \exp\left(-\frac{2n^2 t^2}{\sum_{i=1}^{n} (b_i - a_i)^2}\right).$$

Corollary 1. *Let* $g : X \times Y \to \mathbb{R}$ *be a function with total variation* $\|g\|_{TV} \leq \sqrt{2/c}$, *and let* P *be a probability measure on* Y. *Define* $f : X \to \mathbb{R}$ *to be* $f(x) \triangleq \int_Y g(x, y)\, dP(y)$. *Given a sample* $y^n \sim P^n$, *let* $f^n(x) \triangleq \frac{1}{n} \sum_{i=1}^{n} g(x, y_i)$. *Then, for any* $\delta > 0$, *with probability at least* $1 - \delta$, $\|f - f^n\|_\infty < \sqrt{\frac{\ln 2/\delta}{cn}}$.

Proof. Choose some $x \in X$ and define the function $h_x : Y \to [0, 1]$, $h_x(y) = g(x, y)$. Let h_x^n be the empirical mean of h_x with $y_1, \ldots, y_n \sim P$. Then note that the expectation of h_x with respect to P is $\mathbb{E}\, h_x = \int h_x(y) dP(y) = \int g(x, y) dP(y) = f(x)$. Then $P^n\left(\{y^n : |f(x) - f^n(x))| > t\}\right) < 2e^{-cnt^2}$, for any x, due to Hoeffding's inequality. Substituting gives us the required result.

Proof (Proof of Theorem 1 on page 279). Firstly, note that the value function has total variation bounded $1/(1-\gamma)$. Then corollary 1 applies with $c = 2(1-\gamma)^2$. Consequently, the expected loss can be bounded as follows:

$$\mathbb{E}\,\|V - \hat{V}\|_\infty \leq \frac{1}{1 - \gamma} \left(\sqrt{\frac{\ln 2/\delta}{2K}} + \delta\right).$$

Setting $\delta = 2/\sqrt{K}$ gives us the required result.

References

[1] Abbeel, P., Ng, A.Y.: Apprenticeship learning via inverse reinforcement learning. In: ICML 2004 (2004)

[2] Babes, M., Marivate, V., Littman, M., Subramanian, K.: Apprenticeship learning about multiple intentions. In: ICML 2011 (2011)

[3] Birlutiu, A., Groot, P., Heskes, T.: Multi-task preference learning with gaussian processes. In: ESANN 2009, pp. 123–128 (2009)

[4] Boutilier, C.: A POMDP formulation of preference elicitation problems. In: AAAI 2002, pp. 239–246 (2002)

[5] Choi, J., Kim, K.-E.: Inverse reinforcement learning in partially observable environments. Journal of Machine Learning Research 12, 691–730 (2011)

[6] Chu, W., Ghahramani, Z.: Preference learning with Gaussian processes. In: ICML 2005 (2005)

[7] Coates, A., Abbeel, P., Ng, A.Y.: Learning for control from multiple demonstrations. In: ICML 2008, pp. 144–151. ACM (2008)

[8] Dearden, R., Friedman, N., Russell, S.J.: Bayesian Q-learning. In: AAAI/IAAI, pp. 761–768 (1998)

[9] Dimitrakakis, C.: Robust Bayesian reinforcement learning through tight lower bounds. In: EWRL 2011 (2011)

[10] Doshi-Velez, F., Wingate, D., Roy, N., Tenenbaum, J.: Nonparametric Bayesian policy priors for reinforcement learning. In: NIPS 2010, pp. 532–540 (2010)

[11] Ferguson, T.S.: Prior distributions on spaces of probability measures. The Annals of Statistics 2(4), 615–629 (1974) ISSN 00905364

[12] Geweke, J.: Bayesian inference in econometric models using Monte Carlo integration. Econometrica: Journal of the Econometric Society, 1317–1339 (1989)

[13] Heskes, T.: Solving a huge number of similar tasks: a combination of multi-task learning and a hierarchical Bayesian approach. In: ICML 1998, pp. 233–241. Citeseer (1998)

[14] Lazaric, A., Ghavamzadeh, M.: Bayesian multi-task reinforcement learning. In: ICML 2010 (2010)

[15] Natarajan, S., Kunapuli, G., Judah, K., Tadepalli, P., Kersting, K., Shavlik, J.: Multi-agent inverse reinforcement learning. In: ICMLA 2010, pp. 395–400. IEEE (2010)

[16] Ng, A.Y., Russell, S.: Algorithms for inverse reinforcement learning. In: ICML 2000, pp. 663–670. Morgan Kaufmann (2000)

[17] Puterman, M.L.: Markov Decision Processes: Discrete Stochastic Dynamic Programming. John Wiley & Sons, New Jersey (2005)

[18] Ramachandran, D., Amir, E.: Bayesian inverse reinforcement learning. In: IJCAI 2007, vol. 51, p. 61801 (2007)

[19] Robbins, H.: An empirical Bayes approach to statistics (1955)

[20] Rothkopf, C.A., Dimitrakakis, C.: Preference Elicitation and Inverse Reinforcement Learning. In: Gunopulos, D., Hofmann, T., Malerba, D., Vazirgiannis, M. (eds.) ECML PKDD 2011. LNCS, vol. 6913, pp. 34–48. Springer, Heidelberg (2011)

[21] Syed, U., Schapire, R.E.: A game-theoretic approach to apprenticeship learning. In: NIPS 2008, vol. 10 (2008)

[22] Wilson, A., Fern, A., Ray, S., Tadepalli, P.: Multi-task reinforcement learning: a hierarchical Bayesian approach. In: ICML 2007, pp. 1015–1022. ACM (2007)

[23] Ziebart, B.D., Andrew Bagnell, J., Dey, A.K.: Modelling interaction via the principle of maximum causal entropy. In: ICML 2010, Haifa, Israel (2010)

Batch, Off-Policy and Model-Free Apprenticeship Learning

Edouard Klein[1,3], Matthieu Geist[1], and Olivier Pietquin[1,2]

[1] Supélec, IMS Research Group, France
`firstname.lastname@supelec.fr`
[2] UMI 2958, GeorgiaTech-CNRS, France
[3] Equipe ABC, LORIA-CNRS, France

Abstract. This paper addresses the problem of apprenticeship learning, that is learning control policies from demonstration by an expert. An efficient framework for it is inverse reinforcement learning (IRL). Based on the assumption that the expert maximizes a utility function, IRL aims at learning the underlying reward from example trajectories. Many IRL algorithms assume that the reward function is linearly parameterized and rely on the computation of some associated *feature expectations*, which is done through Monte Carlo simulation. However, this assumes to have full trajectories for the expert policy as well as at least a generative model for intermediate policies. In this paper, we introduce a temporal difference method, namely LSTD-μ, to compute these feature expectations. This allows extending apprenticeship learning to a batch and off-policy setting.

1 Introduction

Optimal control consists in putting a machine in control of a system with the goal of fulfilling a specific task, optimality being defined as how well the task is performed. Various solutions to this problem exist from automation to planification. Notably, the reinforcement learning (RL) paradigm [14] is a general machine learning framework in which an agent learns to control optimally a dynamic system through interactions with it. The task is specified through a reward function, the agent objective being to take sequential decisions so as to maximize the expected cumulative reward.

However, defining optimality (through the reward function) can itself be a challenge. If the system can be empirically controlled by an expert, even though his/her behavior can be difficult to describe formally, apprenticeship learning is a way to have a machine controlling the system by mimicking the expert. Rather than directly mimicking the expert with some supervised learning approach, Inverse Reinforcement Learning (IRL) [8] consists in learning a reward function under which the policy demonstrated by the expert is optimal. Mimicking the expert therefore ends up to learning an optimal policy according to this reward function. A significant advantage of such an approach is that expert's actions can be guessed in states which have not been encountered during demonstration.

S. Sanner and M. Hutter (Eds.): EWRL 2011, LNCS 7188, pp. 285–296, 2012.

Firstly introduced in [13], another advantage claimed by the author would be to find a compact and complete representation of the task in the form of the reward function.

There roughly exists three families of IRL algorithms: feature-expectation-based [1,15,16,17], margin-maximization-based [12,10,11,3] and approaches based on the parameterization of the policy by the reward function [9,7]. The first family assumes a linearly parameterized reward function. This naturally leads to a linearly parameterized value function, the associated feature vector being the so-called feature expectation (see Section 2 for a formal definition). These approaches learn a reward function such that the feature expectation of the optimal policy (according to the learnt reward function) is close to the feature expectation of the expert policy. This is a sufficient condition to have close value functions (for any parameterized reward function, and therefore particularly the optimized one). The second family expresses IRL as a constrained optimization problem in which expert's examples have higher expected cumulative reward than all other policies by a certain margin. Moreover, suboptimality of the expert can be considered through the introduction of slack variables. The last family parameterizes policies with a reward function. Assuming that the expert acts according to a Gibbs policy (respectively to the optimal value function related to the reward function which is optimized), it is possible to estimate the likelihood of a set of state-action pairs provided by the expert. The algorithms differ in the way this likelihood is maximized.

This paper focuses on the first family of algorithms, and more precisely on the seminal work of [1]. All of them rely on the computation of the feature expectation (which depends on policies but not on rewards) of (i) the expert and (ii) some intermediate policies. The expert's feature expectation is computed using a simple Monte Carlo approach (which requires full trajectories of the expert). Other feature expectations are either computed exactly (which requires knowing analytically the dynamics of the system) or with a Monte Carlo approach (which requires simulating the system). The contribution of this paper is LSTD-μ, a new temporal-difference-based algorithm for estimating these feature expectations. It relaxes the preceding assumptions: transitions of the expert are sufficient (rather than full trajectories) and nor the model neither a simulator are necessary to compute intermediate feature expectations. This paper focuses on the algorithm introduced in [1], but the proposed approach can be used in other algorithms based on feature expectation computation.

The rest of this paper is organized as follows. Section 2 provides the necessary background, notably the definition of feature expectation and its use in the seminal IRL algorithm of [1]. Section 3 presents LSTD-μ, our main contribution. Section 4 provides some preliminary experiments and Section 5 opens perspectives.

2   Background

A sequential decision problem is often framed as a Markov Decision Process (MDP) [14], a tuple $\{S, A, P, R, \gamma\}$ with S being the state space, A the action space, $P \in \mathcal{P}(S)^{S \times A}$ the set of Markovian transition probabilities, $R \in \mathbb{R}^S$ the reward function (assumed to be absolutely bounded by 1) and $\gamma \in [0, 1[$ a discounting factor. A policy $\pi \in A^S$ maps states to actions. The quality of a policy is quantified by the associated value function V^π, which associates to each state the expected and discounted cumulative reward: $V^\pi(s) = E[\sum_{t=0}^{\infty} \gamma^t R(s_t) | s_0 = s, \pi]$. Dynamic programming aims at finding the optimal policy π^*, that is one of the policies associated with the optimal value function, $V^* = \mathrm{argmax}_\pi V^\pi$, which maximizes the value for each state. If the model (that is transition probabilities and the reward function) is unknown, learning the optimal control through interactions is addressed by RL.

For IRL, the problem is reversed. It is assumed that an expert acts according to an optimal policy π_E, this policy being optimal according to some unknown reward function R^*. The goal is to learn this reward function from sampled trajectories of the expert. This is a difficult and ill-posed problem [8]. Apprenticeship learning through IRL, which is the problem at hand in this paper, has a somewhat weaker objective: it aims at learning a policy $\tilde{\pi}$ which is (approximately) as good as the expert policy π_E under the unknown reward function R^*, for a known initial state s_0 (this condition can be easily weakened by assuming a distribution over initial states). Now, the approach proposed in [1] is presented.

We assume that the true reward function belongs to some hypothesis space $\mathcal{H}_\phi = \{\theta^T \phi(s), \theta \in \mathbb{R}^p\}$, of which we assume the basis functions to be bounded by 1: $|\phi_i(s)| \leq 1, \forall s \in S, 1 \leq i \leq p$. Therefore, there exists a parameter vector θ^* such that: $R^*(s) = (\theta^*)^T \phi(s)$ In order to ensure that rewards are bounded, we impose that $\|\theta\|_2 \leq 1$. For any reward function belonging to $\mathcal{H}_\phi$ and for any policy π, the related value function $V^\pi(s)$ can be expressed as follows:

$$V^\pi(s) = E[\sum_{t=0}^{\infty} \gamma^t \theta^T \phi(s_t) | s_0 = s, \pi] = \theta^T E[\sum_{t=0}^{\infty} \gamma^t \phi(s_t) | s_0 = s, \pi] \tag{1}$$

Therefore, the value function is also linearly parameterized, with the same weights and with basis functions being grouped into the so-called *feature expectation* μ^π:

$$\mu^\pi(s) = E[\sum_{t=0}^{\infty} \gamma^t \phi(s_t) | s_0 = s, \pi] \tag{2}$$

Recall that the problem is to find a policy whose performance is close to that of the expert's for the starting state s_0, on the unknown reward function R^*. In order to achieve this goal, it is proposed in [1] to find a policy $\tilde{\pi}$ such that $\|\mu^{\pi_E}(s_0) - \mu^{\tilde{\pi}}(s_0)\|_2 \leq \epsilon$ for some (small) $\epsilon > 0$. Actually, this ensures that the value of the expert's policy and the value of the estimated policy (for the starting state s_0) are close for *any* reward function of $\mathcal{H}_\phi$:

$$|V^{\pi_E}(s_0) - V^{\tilde{\pi}}(s_0)| = |\theta^T(\mu^{\pi_E}(s_0) - \mu^{\tilde{\pi}}(s_0))| \leq \|\mu^{\pi_E}(s_0) - \mu^{\tilde{\pi}}(s_0)\|_2 \tag{3}$$

This equation uses the Cauchy-Schwarz inequality and the assumption that $\|\theta\|_2 \leq 1$. Therefore, the approach described here does not ensure to retrieve the true reward function R^*, but to act as well as the expert under this reward function (and actually under any reward function).

Let us now describe the algorithm proposed in [1] to achieve this goal:

1. Starts with some initial policy $\pi^{(0)}$ and compute $\mu^{\pi^{(0)}}(s_0)$. Set $j = 1$;
2. Compute $t^{(j)} = \max_{\theta:\|\theta\|_2 \leq 1} \min_{k \in \{0, j-1\}} \theta^T (\mu^{\pi_E}(s_0) - \mu^{\pi^{(k)}}(s_0))$ and let $\theta^{(j)}$ be the value attaining this maximum. At this step, one searches for the reward function which maximizes the distance between the value of the expert at s_0 and the value of *any* policy computed so far (still at s_0). This optimization problem can be solved using a quadratic programming approach or a projection algorithm [1];
3. if $t^{(j)} \leq \epsilon$, terminate. The algorithm outputs a set of policies $\{\pi^{(0)}, \dots, \pi^{(j-1)}\}$ among which the user chooses manually or automatically the closest to the expert (see [1] for details on how to choose this policy). Notice that the last policy is not necessarily the best (as illustrated in Section 4);
4. solve the MDP with the reward function $R^{(j)}(s) = (\theta^{(j)})^T \phi(s)$ and denote $\pi^{(j)}$ the associated optimal policy. Compute $\mu^{\pi^{(j)}}(s_0)$;
5. set $j \leftarrow j + 1$ and go back to step 2.

There remain three problems: computing the feature expectation of the expert, solving the MDP and computing feature expectations of intermediate policies.

As suggested in [1], solving the MDP can be done approximately by using any appropriate reinforcement learning algorithm. In this paper, we use the Least-Squares Policy Iteration (LSPI) algorithm [4]. There remains to estimate feature expectations. In [1], $\mu^{\pi_E}(s_0)$ is estimated using a Monte Carlo approach over m trajectories: $\hat{\mu}_E(s_0) = \frac{1}{m} \sum_{h=1}^{m} \sum_{t=0}^{\infty} \gamma^t \phi(s_t^{(h)})$. This approach does not hold if only transitions of the expert are available, or if trajectories are too long (in this case, it is still possible to truncate them). For intermediate policies, it is also suggested to estimate associated feature expectations using a Monte Carlo approach (if they cannot be computed exactly). This is more constraining than for the expert, as this assumes that a simulator of the system is available. In order to address these problems, we introduce a temporal-difference-based algorithm to estimate these feature expectations.

3   LSTD-μ

Let us write the definition of the ith component of a feature expectation $\mu^\pi(s)$ for some policy π: $\mu_i^\pi(s) = E[\sum_{t=0}^{\infty} \gamma^t \phi_i(s_t)|s_0 = s, \pi]$. This is exactly the definition of the value function of the policy π for the MDP considered with the ith basis function $\phi_i(s)$ as the reward function. There exist many algorithms to estimate a value function, any of them can be used to estimate μ_i^π. Based on this remark, we propose to use specifically the least-squares temporal difference (LSTD) algorithm [2] to estimate each component of the feature expectation (as each of these components can be understood as a value function related to a specific and known reward function).

Assume that a set of transitions $\{(s_t, r_t, s_{t+1})_{1 \leq t \leq n}\}$ sampled according to the policy π is available. We assume that value functions are searched for in some hypothesis space $\mathcal{H}_\psi = \{\hat{V}_\xi(s) = \sum_{i=1}^{q} \xi_i \psi_i(s) = \xi^T \psi(s), \xi \in \mathbb{R}^q\}$. As reward and value functions are possibly quite different, another hypothesis space is considered for value function estimation. But if $\mathcal{H}_\phi$ is rich enough, one can still consider $\mathcal{H}_\psi = \mathcal{H}_\phi$. Therefore, we are looking for an approximation of the following form: $\hat{\mu}_i^\pi(s) = (\xi_i^*)^T \psi(s)$ The parameter vector ξ_i^* is here the LSTD estimate:

$$\xi_i^* = \left(\sum_{t=1}^{n} \psi(s_t)(\psi(s_t) - \gamma\psi(s_t'))^T \right)^{-1} \sum_{t=1}^{n} \psi(s_t)\phi_i(s_t) \qquad (4)$$

For apprenticeship learning, we are interested more particularly in $\hat{\mu}^\pi(s_0)$. Let $\Psi = (\psi_i(s_t))_{t,i}$ be the $n \times q$ matrix of values predictors, $\Delta\Psi = (\psi_i(s_t) - \gamma\psi_i(s_t'))_{t,i}$ be the related $n \times q$ matrix and $\Phi = (\phi_i(s_t))_{t,i}$ the $n \times p$ matrix of reward predictors. It can be easily checked that $\hat{\mu}^\pi(s_0)$ satisfies:

$$(\hat{\mu}^\pi(s_0))^T = \psi(s_0)^T (\Psi^T \Delta\Psi)^{-1} \Psi^T \Phi \qquad (5)$$

This provides an efficient way to estimate the feature expectation of the expert in s_0.

There remains to compute the feature expectations of intermediate policies, which should be done in an off-policy manner (that is without explicitly sampling trajectories according to the policy of interest). To do so, still interpreting each component of the feature expectation as a value function, we introduce a state-action feature expectation defined as follows (much as the classic Q-function extends the value function): $\mu^\pi(s, a) = E[\sum_{t=0}^{\infty} \gamma^t \phi(s_t) | s_0 = s, a_0 = a, \pi]$ Compared to the classic feature expectation, this definition adds a degree of freedom on the first action to be chosen before following the policy π. With a slightly different definition of the related hypothesis space, each component of this feature expectation can still be estimated using the LSTD algorithm (namely using the LSTD-Q algorithm [4]). The clear advantage of introducing the state-action feature expectation is that this additional degree of freedom allows off-policy learning. Extending LSTD-μ to state-action LSTD-μ is done in the same manner as LSTD is extended to LSTD-Q [4], technical details are not provided here for the clarity of exposition.

Given the (state-action) LSTD-μ algorithm, the apprenticeship learning algorithm presented in Section 2 can be easily extended to a batch and off-policy setting. The solely available data is a set of transitions sampled according to the expert policy (and possibly a set of sub-optimal trajectories). The corresponding feature expectation for the starting state s_0 is estimated with the LSTD-μ algorithm. At step 4 of this algorithm, the MDP is (approximately) solved using

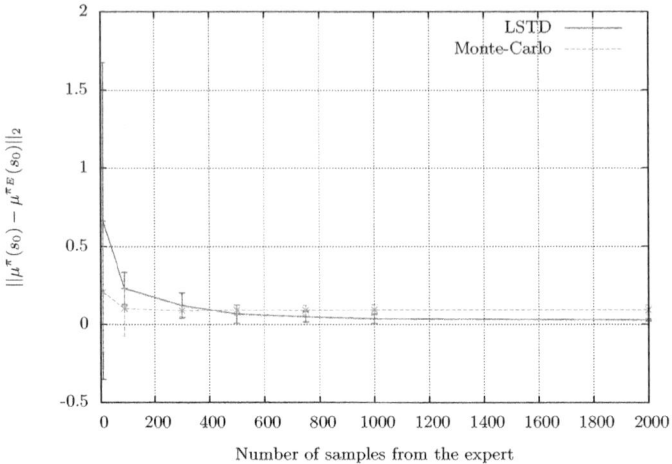

Fig. 1. $||\mu^\pi(s_0) - \mu^{\pi_E}(s_0)||_2$ with respect to the number of samples available from the expert. The error bars represent the standard deviation over 100 runs.

LSPI [4] (an approximate policy iteration algorithm using LSTD-Q as the off-policy Q-function estimator). The corresponding feature expectation at state s_0 is estimated using the proposed state-action LSTD-μ.

Before presenting some experimental results, let us stress that LSTD-μ is simply the LSTD algorithm applied to a specific reward function. Although quite clear, the idea of using a temporal difference algorithm to estimate the feature expectation is new, as far as we know. A clear advantage of the proposed approach is that any theoretical result holding for LSTD also holds for LSTD-μ, such as convergence [6] or finite-sample [5] analysis for example.

4   Experimental Benchmark

This section provides experimental results on two complementary problems. Subsection 4.1 details the protocol and the results while subsection 4.2 inspects the meaning of the different quality criteria.

4.1   Experiment Description and Results

GridWorld. The first experimental benchmark chosen here is one of those proposed in [8], a 5x5 grid world. The agent is in one of the cells of the grid (whose coordinates is the state) and can choose at each step one of the four compass directions (the action). With probability 0.9, the agent moves in the intended direction. With probability 0.1, the direction is randomly drawn. The reward optimized by the expert is 0 everywhere except in the upper-right cell, where it is 1. For every policy, an episode ends when the upper right cell is attained, or after 20 steps. At the start of each episode, the agent is in the lower-left corner of the grid (the opposite of where the reward is). Both the state and action spaces are finite and of small cardinality. Hence, the chosen feature functions ϕ and ψ are the typical features of a tabular representation: 0 everywhere except for the component corresponding to the current state (-action pair).

Both [1]'s algorithm (from now on referred to as the MC variant) and our adaptation (referred to as the LSTD variant) are tested side by side. The MDP solver of the MC variant is LSPI with a sample source that covers the whole grid (each state has a mean visitation count of more than 150) and draws its action randomly. Both $\mu^{\pi^{(j)}}(s_0)$ and $\mu^{\pi_E}(s_0)$ are computed thanks to a Monte-Carlo estimation with enough samples to make the variance negligible. We consider both these computations as perfect theoretical solvers for all intended purpose on this toy problem. We thus are in the case intended by [1]. On the other hand the LSTD variant is used without accessing a simulator. It uses LSPI and LSTD-μ, fed only with the expert's transitions (although we could also use non expert transitions to compute intermediate policies, if available). This corresponds to a real-life setting where data generation is expensive and the system cannot be controlled by an untrained machine.

We want to compare the efficiency of both versions of the algorithm with respect to the number of samples available from the expert, as these samples usually are the bottleneck. Indeed as they are quite costly (in both means and human time) to generate they are often not in abundance hence the critical need for an algorithm to be expert-sample efficient. Having a simulator at one's disposal can also be difficult. For the simple problem we use this is however not an issue. The discussion about the choice of the performance metric has its own dedicated subsection (subsection 4.2). We use here the $||\mu^{\pi^{(j)}}(s_0) - \mu^{\pi_E}(s_0)||_2$ error term. Fig. 1 shows, for some numbers of samples from the expert, the value of $||\mu^{\pi^{(j)}}(s_0) - \mu^{\pi_E}(s_0)||_2$ after the algorithm converged or attained the maximum number of iterations (we used 40). The best policy is found by LSTD variant after one iteration only[1] whereas in the MC variant, convergence happens after at least ten iterations. The best policy is not always the last, and although it has always been observed so far with the LSTD variant, there is absolutely no way to tell whether this is a feature. The score of the best policy (not the last) according to the $||\mu^{\pi^{(j)}}(s_0) - \mu^{\pi_E}(s_0)||_2$ error term is plotted here. We can see that although the LSTD variant is not as good as the MC variant when very few samples are available, both algorithms quickly converge to almost the same value ; our version converges to a slightly lower error value. The fact that our variant can work in a batch, off-policy and model-free setting should make it suitable to a range of tasks inaccessible to the MC variant, the requirement of a simulator being often constraining.

---

[1] Precise reasons why it happens are not clear now, but certainly have something to do with the fact that all the estimations are made along the same distribution of samples.

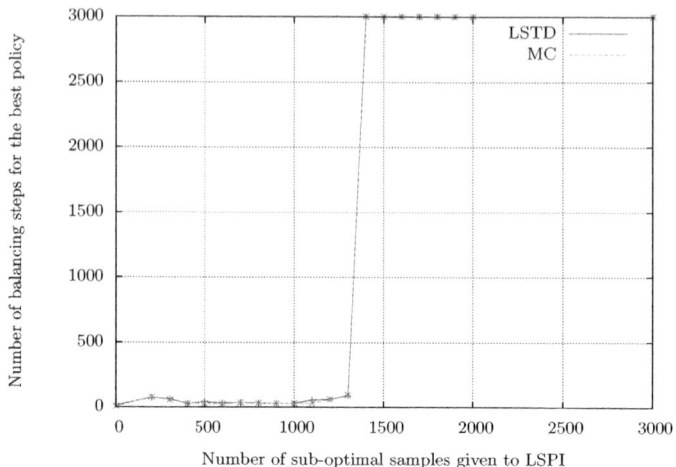

Fig. 2. Number of balancing steps for the policies found by both variants with respect to the number of samples from a sub-optimal policy given to LSPI. Note the absence of middle ground in the quality.

Inverted Pendulum. Another classic toy problem is the *inverted pendulum*, used for example in [4] from where we drew the exact configuration (mass of the pendulum, length of the arm, definition of the feature space, etc.). In this setting, the machine must imitate an expert who maintains the unstable equilibrium of an inverted pendulum allowed to move along only one axis. A balancing step is considered successful (reward 0) when the angle is less than $\frac{\pi}{2}$. It is considered a failure (reward -1) when the pendulum has fallen (i.e. the angle exceeds $\frac{\pi}{2}$). A run is stopped after a failure or after 3000 successful balancing steps.

This problem bears two fundamutal differences with the previous one and thus comes as complementary to it. First, it presents a continuous state space whereas in the GridWorld a tabular representation was possible. Then, the initial state is randomly drawn from a non singleton subspace of the state space. This last point is addressed differently by the LSTD variant and the MC variant. the MC variant will naturally sample the initial state distribution at each new trajectory from the expert. The LSTD variant, on the other hand, still does not need complete trajectories from the expert but mere samples. As it approximates the whole μ^π function and not just $\mu^\pi(s_0)$, it is possible to compute the mean of the approximation $\hat{\mu}^\pi$ over a few states drawn with the same distribution as the initial states.

A striking finding on this toy problem is that the parameter controlling the success or the failure of the experiment is not the number of samples from the expert, but the quality of the samples available to LSPI to solve the MDP. *A contrario* to what happened in the previous toy problem, there are fewer samples from the expert than what LSPI needs to solve the MDP as they do not cover the whole state space. When given only these samples, the LSTD variant fails. The MC variant, however, is not bound by such restrictions and can use as many

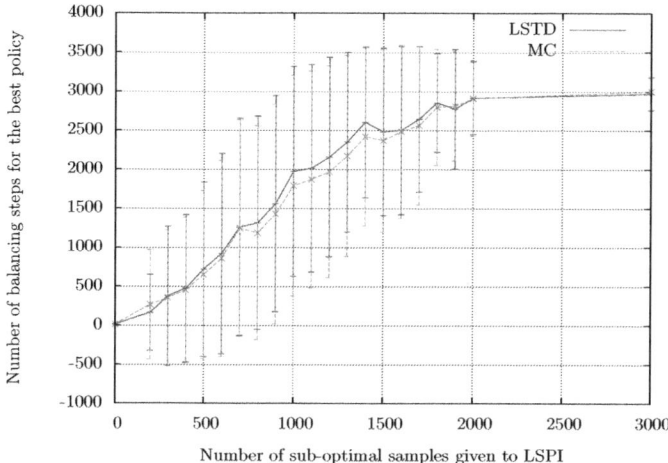

Fig. 3. Same plot as Fig. 2, with mean and standard deviation over 100 runs. The big standard deviation stems from the absence of middle ground in the quality of the policies. The increasing mean w.r.t. the abscissa means that the proportion of good policies increases with the number of samples from a sub-optimal policy given to LSPI.

samples as it needs from a simulator. Thus, with only one trajectory from the expert (100 samples) the MC variant is able to successfully balance the pendulum for more than 3000 steps. The problem here stems from the fact that we don't learn the reward, but a policy that maximizes the reward. If the optimal control problem was solved separately, learning the reward only from the samples of the expert would be possible.

The sensitivity of both algorithms to the number of samples available to LSPI is extreme, as is shown in Fig. 2. It may seem nonsensical to restrict the number of samples available to LSPI for the MC variant as it can access a simulator, but this has been done to show that both variants exhibit the same behavior, hence allowing us to locate the source of this behavior in what both algorithms have in common (the use of LSPI inside the structure of [1]) excluding the point where they differ (LSTD-μ and Monte-Carlo).

Fig. 3 shows that when given samples from a sub-optimal policy, the LSTD variant can sort the problem out, statistically almost always with 3000 sub-optimal samples, and sometimes with as low as 200 sub-optimal samples. Such a setting is still batch, off-policy and model-free. When given enough sub-optimal samples, both variants are successful with only one trajectory from the expert (i.e. 100 samples, this is what is plotted here). Giving more trajectories from the expert does not improve the success rate.

4.2 Discussion about the Quality Criterion

Fig. 4 and 5 illustrate (using the GridWorld problem) the difficulty of choosing the quality criterion ; Fig. 4 shows four different quality criteria during a run of the MC variant. Fig. 5 shows the same criteria for several runs of

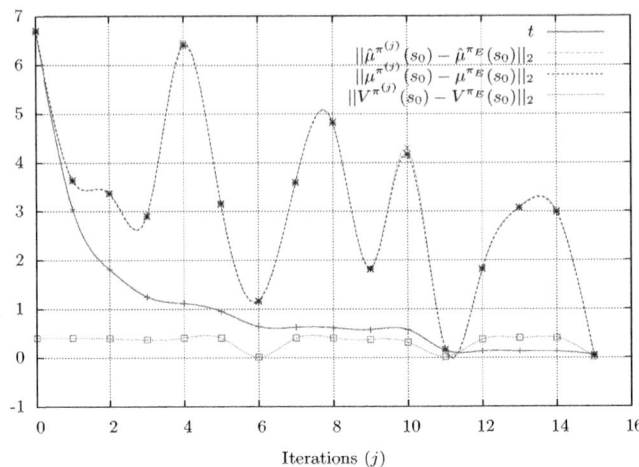

Fig. 4. Different criteria with respect to the number of iterations for a run of the MC variant

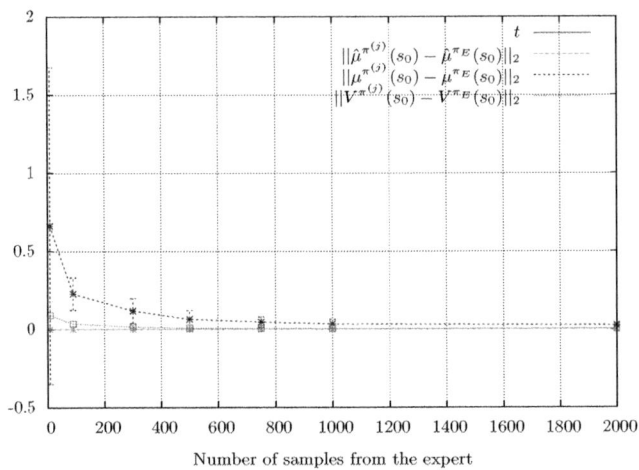

Fig. 5. Different criteria with respect to the number of samples from the expert, for several runs of the LSTD variant. We can see that the algorithm is blind, as all it has access to is always zero. The true error values, however, smoothly converge to something small. Knowing how many expert samples are actually needed in a real world problem is an open question. The error bars represent the standard deviation over 100 runs.

the LSTD variant with a varying number of samples from the expert. The $||\mu^{\pi^{(j)}}(s_0) - \mu^{\pi_E}(s_0)||_2$ term is widely discussed in [1]'s additional material. It bears an interesting relation with the difference between the expert's value function and the current value function in the initial state with respect *to the current reward* (Eq. 3).

The fact that the curves of the true error term $||\mu^{\pi^{(j)}}(s_0) - \mu^{\pi_E}(s_0)||_2$ and its estimate $||\hat{\mu}^{\pi^{(j)}}(s_0) - \hat{\mu}^{\pi_E}(s_0)||_2$ are indistinguishable in Fig. 4 means that, for it has access to a cheap simulator, the MC variant works as if it had access to the exact values. This however cannot be said of the LSTD variant, for which the two curves are indeed different (Fig. 5). Not knowing the true value of $\mu^{\pi_E}(s_0)$ may be a problem for our variant, as it can introduce an error in the stopping criterion of the algorithm.

It shall be noted that although it plays its role, the halt criterion is not a good measure of the quality of the current policy in the MC variant either, as it can be low (and thus halt the algorithm) when the policy is bad. The best policy, however, can be easily chosen among all those created during the execution of the algorithm thanks to the $||\mu^{\pi^{(j)}}(s_0) - \mu^{\pi_E}(s_0)||_2$ term, which the MC variant can compute. When this term is low, the objective performance (that is, $V^{\pi_E}(s_0) - V^{\pi^{(j)}}(s_0)$ with respect to the unknown true reward function) is low too.

5   Conclusion

Given some transitions generated by an expert controlling a system and maximizing in the long run an unknown reward, we ported [1]'s approach to apprenticeship learning via inverse reinforcement learning to a batch, model-free, off-policy setting. Experimentally, there is a need for either a slightly bigger number of samples from the expert or some samples from a sub-optimal policy. We believe this cost is not prohibitive as our approach only requires isolated samples which are often less difficult to get than whole trajectories as needed by the original approach. Furthermore, tranferring the reward and not the policy may overcome this difficulty. We intend to do this in a real life setting. The simple idea of using LSTD to estimate the feature expectation could be applied to other algorithms as well, for example [1,15,16,17].

References

1. Abbeel, P., Ng, A.: Apprenticeship learning via inverse reinforcement learning. In: Proceedings of the Twenty-First International Conference on Machine Learning, p. 1. ACM (2004)
2. Bradtke, S., Barto, A.: Linear least-squares algorithms for temporal difference learning. Machine Learning 22(1), 33–57 (1996)
3. Kolter, J., Abbeel, P., Ng, A.: Hierarchical apprenticeship learning with application to quadruped locomotion. In: Neural Information Processing Systems, vol. 20 (2008)

4. Lagoudakis, M., Parr, R.: Least-squares policy iteration. The Journal of Machine Learning Research 4, 1107–1149 (2003)
5. Lazaric, A., Ghavamzadeh, M., Munos, R.: Finite-sample analysis of lstd. In: Proceedings of the 27th International Conference on Machine Learning (2010)
6. Nedić, A., Bertsekas, D.: Least squares policy evaluation algorithms with linear function approximation. Discrete Event Dynamic Systems 13(1), 79–110 (2003)
7. Neu, G., Szepesvári, C.: Apprenticeship learning using inverse reinforcement learning and gradient methods. In: Proc. UAI, pp. 295–302 (2007)
8. Ng, A., Russell, S.: Algorithms for inverse reinforcement learning. In: Proceedings of the Seventeenth International Conference on Machine Learning, pp. 663–670. Morgan Kaufmann Publishers Inc. (2000)
9. Ramachandran, D., Amir, E.: Bayesian inverse reinforcement learning. In: Proceedings of the International Joint Conference on Artificial Intelligence, pp. 2586–2591 (2007)
10. Ratliff, N., Bagnell, J., Srinivasa, S.: Imitation learning for locomotion and manipulation. In: 2007 7th IEEE-RAS International Conference on Humanoid Robots, pp. 392–397. IEEE (2007)
11. Ratliff, N., Bradley, D., Bagnell, J., Chestnutt, J.: Boosting structured prediction for imitation learning. In: Advances in Neural Information Processing Systems, vol. 19, p. 1153 (2007)
12. Ratliff, N., Bagnell, J., Zinkevich, M.: Maximum margin planning. In: Proceedings of the 23rd International Conference on Machine Learning, p. 736. ACM (2006)
13. Russell, S.: Learning agents for uncertain environments (extended abstract). In: Proceedings of the Eleventh Annual Conference on Computational Learning Theory, p. 103. ACM (1998)
14. Sutton, R., Barto, A.: Reinforcement learning. MIT Press (1998)
15. Syed, U., Bowling, M., Schapire, R.: Apprenticeship learning using linear programming. In: Proceedings of the 25th International Conference on Machine Learning, pp. 1032–1039. ACM (2008)
16. Syed, U., Schapire, R.: A game-theoretic approach to apprenticeship learning. In: Advances in Neural Information Processing Systems, vol. 20, pp. 1449–1456 (2008)
17. Ziebart, B., Maas, A., Bagnell, J., Dey, A.: Maximum entropy inverse reinforcement learning. In: Proc. AAAI, pp. 1433–1438 (2008)

Introduction of Fixed Mode States into Online Profit Sharing and Its Application to Waist Trajectory Generation of Biped Robot

Seiya Kuroda[1], Kazuteru Miyazaki[2], and Hiroaki Kobayashi[3]

[1] Panasonic Factory Solutions Co., Ltd.
[2] National Institution for Academic Degrees and University Evaluation,
1-29-1 Gakuennishimachi, Kodaira, Tokyo 187-8587, Japan
teru@niad.ac.jp
http://svrrd2.niad.ac.jp/faculty/teru/index.html
[3] Meiji University, 1-1-1 Higashimita Tama-ku, Kawasaki, Kanagawa 214-8571, Japan

Abstract. In reinforcement learning of long-term tasks, learning efficiency may deteriorate when an agent's probabilistic actions cause too many mistakes before task learning reaches its goal. The new type of state we propose – *fixed mode* – to which a normal state shifts if it has already received sufficient reward – chooses an action based on a greedy strategy, eliminating randomness of action selection and increasing efficiency. We start by proposing the combining of an algorithm with penalty avoiding rational policy making and online profit sharing with fixed mode states. We then discuss the target system and learning-controller design. In simulation, the learning task involves stabilizing of biped walking by using the learning controller to modify a robot's waist trajectory. We then discuss simulation results and the effectiveness of our proposal.

Keywords: Reinforcement Learning, Exploitation-oriented Learning, Profit Sharing, Improved PARP, biped robot.

1 Introduction

Robots working with human beings in a common living space must adapt autonomously to unknown environments based on sensory input and actions. *Reinforcement learning* (RL)[16], a type of such control, does not assume the existence of a priori knowledge about the environment, focusing instead on goal directed learning from interaction with the environment using rewards and a penalties as teaching signals.

In its focus on applications [14,19,6,15,8,5,7,9,18] to control problems [14,19,6,5,9], RL is used for a wider variety of classes than a control-theory-based approach because it does not assume a priori knowledge about the controlled object. In the *value function* approach [14,19], for example, state values or state-action values are estimated, and in the *direct policy search* approach [17,6], control policy parameters are optimized. Hybrid approaches [5,9] combined with RL and evolutionary computing such as *genetic algorithms* [4] reduce the number of trial-and-error searches.

S. Sanner and M. Hutter (Eds.): EWRL 2011, LNCS 7188, pp. 297–308, 2012.

These "royal road" approaches pursuing optimality, but for this they need much trial-and-error searching to ensure such optimality. This makes them difficult to apply to problems requiring fast decisions in long-term tasks.

We pursue rationality rather than optimality. We likewise seek to reduce the amount of trial-and-error searching by enhancing successful experience in the process termed *exploitation-oriented learning* (XoL) [13]. XoL learning approaches include the profit sharing (PS) [10] rationality theorem, penalty avoiding rational policy making (PARP) algorithm [11], and PS-r$^{\#}$ [13].

We propose an algorithm that learns efficiently in long-term tasks. In applying previous learning methods to long-term tasks, we found that they failed before achieving their purpose due to the influence of randomness in their action selection. To avoid this, we focused on online PS [2] – a type of XoL proposed to obtain rewards efficiently online – and improved PARP) [18] – proposed to avoid penalties more efficiently in real-world applications rather than PARP.

Section 2 defines our problem domain. We treat both discrete and continuous sensory input directly and discretized by a *basis function* [12,18]. Section 3 introduces a *fixed mode state* shifted from a normal state to online PS when it has enough rewards. We also present an overall algorithm to fit our task. Section 4 applies our proposal to a long-term continuous task generating a biped robot's trajectory. Section 5 discusses simulation results. Section 6 frames conclusions.

2   Problem Domain

Consider an agent in an unknown environment.After receiving sensory input from the environment, the agent selects and executes an action. Time is discretized by a single input-action cycle. We assume that sensory input from the environment is the N_s dimensional vector of both continuous and discrete variables. We discretize continuous-discrete sensory input space into discrete state space $\mathcal{S} = \{s_k, \ k = 1, 2, \ldots, N_B\}$ using a *basis function* [12,18].

We also treat N_m dimensional action output in which individual components of output have N_a alternatives to be selected. We define them as follows:

- The set of alternatives of the $i-$th component, $i = 1, 2, \ldots, N_m$ is given as $\mathcal{A}_i = \{a_{i,j} : j = 1, 2, \ldots, N_a\}$.
- Action act_q is constructed as $act_q = (a_{1,j_q}, a_{2,j_q}, \ldots, a_{N_m,j_q})$, in which $a_{i,j_q} \in \mathcal{A}_i, i = 1, 2, \ldots, N_m$. The set of all actions therefore is $\mathcal{A} = \mathcal{A}_1 \times \mathcal{A}_2 \times \cdots \times \mathcal{A}_{N_M}$ and the number of total actions is $|\mathcal{A}| = (N_a)^{N_m}$.

The pair of state s_k and action act_q selected in state (s_k, act_q) termed a *rule*. In PS and most RL, a parameter termed *weight* or *Q-value* is assigned to individual rules to evaluate their importance. A series of rules that begins from a reward/penalty state or an initial state and ends with the next reward/penalty state is termed an *episode*. If an episode contains rules of the same state but paired with different actions, the partial series from one state to the next is termed a *detour*.A rule always existing on a detour is termed an *irrational rule*, but otherwise a *rational rule*.

An agent gets a reward or a penalty based on a result of an action sequence. A reward is given when our purpose is achieved, so the reward reinforces the weights of all rules in the episode using an evaluation function. A penalty is given if our purpose fails. The rule directly responsible for the failure receives the penalty and is termed a *penalty rule*. If all selective rules for a state are penalty or irrational rules, the state is termed a *penalty state*. If a rule brings the current state to a penalty states, the rule is also classified as a penalty rule. A penalty rule is found using a *penalty rule decision* (PRD) procedure [12,18].

We deal with the following penalties:

- There are N_p-1 types of penalties p_ℓ ($\ell = 1, 2, ..., N_p-1$) in which p_1 has the highest priority that should be avoided, and p_{N_p-1} has the lowest priority.
- We also introduce additional penalty p_{N_p} to avoid a penalty state. Its priority is less than that of p_{N_p-1}.

Note that a penalty differs greatly from a negative reward and that our algorithm deals with the penalty and the reward independently of each other.

A function that maps states to actions is termed a *policy*. A policy with a positive amount of reward acquisition expectations is termed a *rational policy*, and a rational policy receiving no penalty is termed a *penalty avoiding rational policy*. Using PRD, we get a penalty avoiding rational policy.

3   Introduction of Fixed Mode States into Online PS

3.1   Profit Sharing

Profit Sharing (PS) reinforces all of the weights of rules on an episode at the same time as an agent gets a reward. It must store the episode $\{(s_1, act_1), ..., (s_i, act_i), ...(s_T, act_T)\}$ in memory, where s_1 an initial state and s_T is a reward-giving state. The number of rules in an episode is termed an *episode length*. If the agent selects action act_T at state s_T and gets reward r, the weight of each rule in the episode is updated as $w(s_i, act_i) \leftarrow w(s_i, act_i) + f(i)$ where $f(i)$ is termed the evaluation function. In general, geometrically decreasing function $f(i) = r\gamma^{T-i}$ is used where γ is termed a *discounted rate* with a range of $\gamma \in (0, 1]$. The rationality theorem about γ has been derived in [10]. The geometrically decreasing function satisfies the theorem. In simulation, we set $\gamma = 0.8$.

3.2   Online PS

PS requires much memories (storage) to save all episodes until an agent receives a reward. Computational complexity required to update weights depends on episode length T because all rules in the episode are updated at once when the agent gets a reward. Online PS proposed in [2] solves these problems using the two ideas detailed below.

Solution for Memory (Storage) Requirement
If we use the geometrically decreasing function, $f(i)$ approaches 0.0 exponentially, implying that it is not necessary to save rules for which $f(i)$ values are close to 0.0. Online PS then needs only to store in memory those rules of a certain length before a reward-giving time and must update only the weights of these rules. We term the length a *queue*. If the episode length exceeds the length of the queue, the rule stored in memory on the first is discarded, so the memory (storage) requirement is bounded by queue length.

Solution for Computational Complexity
For the rule in the queue, $w(s, act) \leftarrow w(s, act) + rc(s, act)$ and $c(s, act) \leftarrow \gamma c(s, act)$ have been executed every step where $c(s, act)$ is termed the *eligibility trace*, and is used in much RL [16,3]. At the start of our learning, all $c(s, act)$s for any state and action are initialized at 0.0 and the queue is emptied. When rule (s, act) is selected and put in the queue, 1.0 is added to $c(s, act)$ of the rule. $c(s, act)$ is updated when the rule is selected and decreases geometrically if the rule is not selected. If the agent selects a rule selected once and already stored in memory in the queue, the previous rule is discarded to avoid updating the same rule twice, so computational complexity is again bounded by queue length.

3.3   Fixed Mode State on Online PS for Long-Term Task

One weak point of PS with the geometrically decreasing evaluation function is learning inefficiency in a state far from a reward-giving state because the contribution of the reward to weight through the function becomes very small.

We therefore used two strategies – learning scheduling and penalty implementation. In learning scheduling, the PS learning task is broken down to a series of subtasks, from shortest term tasks to full term tasks. With penalty implementation, we expect that the number of candidate actions in early stage states will decrease rapidly and learning be speeded up.

Even so, for a very long-term task, the random selection strategy of actions makes PS learning very difficult. We therefore introduce a *fixed mode state*, meaning that we regard the state in which the agent has learned enough as the fixed mode state. In fixed mode, the weight is fixed and the action of the highest weight value is selected deterministically –*greedy selection*.

We consider two strategies regarding a normal state as a fixed mode state:

- **Strategy 1**: If the state gets a reward K times, the state is regarded as fixed mode.
- **Strategy 2**: When one subtask is finished, every state getting a reward in the task is regarded as fixed mode.

If the agent gets a penalty, up to three fixed mode states ahead the state are also returned to non fixed mode states to relearn around these states.

3.4   Rule Decomposition

As stated in Section 2, action space has $(N_a)^{N_m}$ elements, meaning that the agent must choose one action from $(N_a)^{N_m}$ actions for each state and needs three

$(N_B \times (N_a)^{N_m})$-dimensional weight matrices, $\mathbf{Q}$, $\mathbf{P}$, and $\mathbf{E}$, as shown later. If we have N_B discrete states, there are $N_B \times (N_a)^{N_m}$ rules. To reduce the number, we break rule (s_k, act_q) down into a set of sub-rules $\{(s_k, a_1 = a_{1,j_1}), (s_k, a_2 = a_{2,j_2}), \ldots, (s_k, a_{N_m} = a_{N_m,j_{N_m}})\}$, where $a_i \in \mathcal{A}_i$. We choose one value from $\mathcal{A}_i$ for each a_i independently of other components and these sub-rules are simultaneously given a reward or a penalty. In this new framework, the agent needs only three $(N_B \times N_M \times N_a)$-dimensional matrices.

We find a penalty rule efficiently through this new framework, but there remains the possibility of judging a penalty rule that is not a penalty rule as a penalty rule. We introduce the N_0 to avoid the mistake. We execute the judgment of a penalty rule for the rule whose number of times in selection exceeds or equals N_0. For this new framework, penalty threshold λ is introduced and a subrule is classified as a penalty subrule if the failure rate exceeds λ. In the simulation, we set $\lambda = 0.5$. In simulation, we set $N_0 = 7$ based on preliminary experiments.

3.5   Action Selection in Online-PS

The agent selects an action in discrete state s_k as follows:

- If state s_k is neither a penalty state nor a fixed mode state, the agent selects an action from actions not in penalty rules based on the probability distribution proportional to weight $Q_k[i][j]$.
- If state s_k is a fixed mode state, the agent selects the action in the rule that has the largest weight $Q_k[i][j]$.
- If state s_k is a penalty state, the agent avoids a penalty by stochastically accounting for the priority of a penalty. First, we define $\bar{A}_i(\ell)$ as a set of actions whose priority is lower than or equal to p_ℓ. $\bar{A}_i(3)$, for example, is a set of actions not included in penalty rules for p_1, p_2, and p_3. If $\bar{A}_i(N_p) = \phi$, the agent selects an action randomly from $\mathcal{A}_i$. If $\bar{A}_i(\ell) \neq \phi$ but $\bar{A}_i(\ell+1) = \phi$, then the agent selects an action in $\bar{A}_i(\ell)$ based on *ranking selection* with probability $P_r(a_i = a_{i,k_i}) = \frac{2(\bar{N}_{a_i} - k_i + 1)}{\bar{N}_{a_i}(\bar{N}_{a_i}+1)}$ for action $a_{k_i} \in \bar{A}_i(\ell)$, in which $\bar{N}_{a_i}$ is the number of elements of $\bar{A}_i(\ell)$ and it is assumed that actions in $\bar{A}_i(\ell)$ have been reordered sequentially as $k = 1, 2, \ldots, \bar{N}_{a_i}$ from the smallest penalty $P_k[i][j]$ to the largest one.

3.6   Overall Algorithm for Our Proposal

The overall algorithm for our proposal is as follows:

1. We set $Q_k[i][j] = 0$ and $P_k[i][j] = 0$ for all $i = 1, 2, \ldots, N_m$, $j = 1, 2, \ldots, N_a$ and $k = 1, 2, \ldots, N_B$ where N_m, N_a and N_B are the numbers of dimensions and elements for the action, and discrete states. Set $TUC_k = 0$, i.e., the number of times to get a reward, and initialize $life_k$, i.e., the life span for forgetting a useless basis function, for all k. In simulation, we set $life_k = 100$.

2. Repeat the following steps until the end condition for learning is met, e.g., a certain number of episodes.

   (a) Initialize the position of the agent. We make s_0, an arbitrary discrete state, including the initial state.

   (b) Initialize a queue and set $E_k[i][j] = 0$ for all i, j, k where $i = 1, 2, ..., N_m$, $j = 1, 2, ..., N_a$ and $k = 0, 1, 2, ..., N_B$.

   (c) Repeat the following steps until the end condition of an episode is met.

      i. The agent selects action act_q in discrete state s_k at $epoc = t$ based on the method in Section 3.5.

      ii. For each j at dimension i, $E_k[i][j] \leftarrow E_k[i][j] + 1$.

      iii. The rule is added to the queue as stated in Section 3.2.

      iv. After the agent selects action act'_q, it transits to next state s'_k generated by the method in [12,18].

      v. If the agent gets reward rwd, the following updates are used for all states and actions in the queue:
      $Q_k[i][j] \leftarrow Q_k[i][j] + rwd \cdot E_k[i][j]$ and $TUC_k \leftarrow TUC_k + 1$
      If we use Strategy 1, the state that is $TUC_k \geq K$ changes to the fixed mode state.

      vi. If the agent gets penalty p_ℓ ($\ell = 1, 2, ..., N_p$) whose value is $pnlty$ for dimension i, for three states, that have times $epoc = t, t - 1, t - 2$, from the state receiving a penalty, the following is done:

      $$P_k[l][i][j] \leftarrow P_k[l][i][j] + pnlty \cdot E_k[i][j] \qquad (1)$$

      If there is a fixed mode state in three states, the state changes to a non fixed mode state.
      We execute PRD [12,18] for the state getting a penalty directly to find a new penalty state.
      If a new penalty state is found, the penalty matrix of the rule transiting to the new penalty state is updated by equation (1).
      If $\ell = N_p$, go to step 2(c)i. Otherwise, go to step 2a.

      vii. We execute discount by $E_k[i][j] \leftarrow \gamma E_k[i][j]$, and the following, execute $life_k \leftarrow life_k - 1, k = 0, 1, 2, ..., N_B$. If $life_k < 0$, state s_k is eliminated from memory (storage).

      viii. We set $s_k \leftarrow s'_k$ and $epoc = t + 1$. If the end condition of learning is satisfied, the next trial is started. If we use Strategy 2, the state that is $TUC_k > 0$ excluding a penalty state changes to a fixed mode state.

      ix. If we satisfy the end condition about an episode, go to step 2a. Otherwise, go to step 2(c)i.

4   Learning of Biped Walking Robot Waist Trajectory

We discuss a successful application of XoL to the inverted pendulum problem[13], applying it to the waist trajectory generator of a biped robot with a learning controller as shown in Fig.1. The trajectory generator learns a waist trajectory

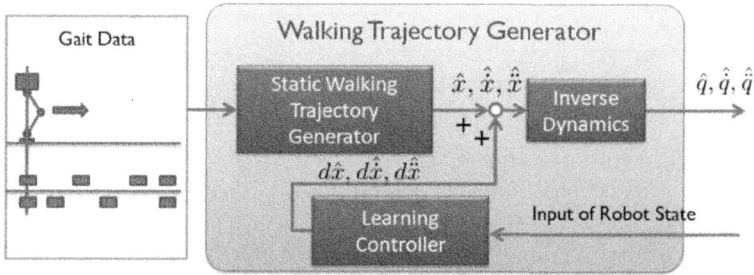

Fig. 1. Trajectory generation of biped robot

for stable dynamic walking by modifying the waist trajectory for static walking. The waist trajectory for static walking is relatively easily designed. The design for dynamic stable walking is usually formulated as an optimization problem with two-point constraints for a nonlinear dynamic system, and the solution is very difficult.

This learning task has features the following: (1) Sensory input consists of both piecewise continuous variables and discrete variables. Joint angular velocities, for example, are continuous but jump when the robot's swing leg touches the ground. One example of a discrete variable is the index for the single-leg support phase and the two-leg support phase. (2) Generator output consists of continuous waist coordinates. (3) There are multiple types of failure in learning runs. The zero moment point (ZMP), for example, went out from a support polyhedron, inverse kinematics could not be solved, or joint torque exceeded the limit. (4) The task is long-term, i.e., from starting a walk to stopping it after several steps.

This task is therefore a good test bed for determining the effect of the fixed mode state in Section 3.

The simulated robot consists of two legs and a torso. Each leg has 6 degree of freedom (dof) and total dof is 12. Parameters resemble an upper grade elementary school student in size and mass – 1.1 [m] tall and weighing 28.87 [kg]. The control sampling time is 2[ms] and the learning sampling time 20[ms].

4.1 States for Learning

We prepare the following sensory input as states for learning. (1) 10 for the right-leg support gate, and -10 for the left-leg support gate. (2) 10 for the two-leg support phase, and -10 for the single-leg support phase. (3) Coordinate (x, y) of the waist for the support leg. The height of the waist is kept constant. (4) Coordinate (x, y) of ZMP error. (5) Acceleration $(\ddot{x}, \ddot{y})$ of the waist.

Dimensions of the state is $N_s = 8$. Variables from (3) to (5) are scaled with appropriate factors.

4.2 Definition of Actions and Modifying Waist Trajectory

The action specifies the modification values (dx, dy) from the static walking waist trajectory, so the dimension of the action is $N_m = 2$ and selected from

Table 1. Definition of penalties

p_ℓ	pnlty	Conditions
0	10	When $\|$ZMP error $\| \geq 6.5 cm$
1	10	When $\|$ the joint velocity $\| \geq 20\pi$
2	10	When the state transits to a penalty state

Table 2. Definition of rewards

Name	rwd	Condition
RWDforGD	10	When the agent reaches to the target time t_f but $Terr \geq \widehat{Terr}$
RWDforSuc	10000	When the agent reaches to the goal time and $Terr < \widehat{Terr}$

-0.024[m] to 0.024[m], which is discretized to 15 alternatives, i.e., $N_a = 15$. The number of actions is $15^2 = 225$. In our new framework, however, the number of subtasks is reduced to 30.

Learning takes place every 20 ms. We use 3-order B-spline functions to generate a smooth waist trajectory from modified values.

4.3   Rewards and Penalties

We use three types of penalties in Table 1, i.e., $N_p = 3$.

If the agent gets penalty p_ℓ, each of the subrules taken at the last epoc is given penalty $pnlty$ and is decided using λ for whether it should be classified as a penalty rule. This helps accelerate learning. In general, however, we must be careful when using this method because it has the possibility of generating too many penalties. Threshold λ controls this problem.

To define rewards, we use the following cost function:

$$Terr = \frac{1}{t_f} \int_0^{t_f} \left\{ \left(\frac{\|\boldsymbol{Zerr}(t)\|}{zmpLimit} \right)^2 t \right\} dt + 50 \left(|Zerr_x(t_f)| + |Zerr_y(t_f)| \right)$$

$$+ \frac{1}{t_f} \sqrt{1000 \int_0^{t_f} \|\hat{\boldsymbol{q}}(t) - \boldsymbol{q}(t)\|^2 dt + \frac{1}{1000} \int_0^{t_f} \|\boldsymbol{u}(t)\|^2 dt} + 10 \frac{\sum_{i=1}^{dof} |\hat{\boldsymbol{q}}_i(t_f) - \boldsymbol{q}_i(t_f)|}{dof}$$

$$+ 10 \frac{\sum_{i=1}^{dof} |\dot{\hat{\boldsymbol{q}}}_i(t_f) - \dot{\boldsymbol{q}}_i(t_f)|}{dof} \tag{2}$$

where t_f is target walking time, $Zerr_x$ and $Zerr_y$ are x and y components of ZMP error, and $\boldsymbol{u}(t) \in \mathcal{R}^{dof}$ is a joint torque vector. $zmpLimit$ is the allowable ZMP error limit, i.e., $zmpLimit = 0.065$ in simulation, and $dof = 12$.

When the agent reaches to target time t_f, we calculate $\widehat{Terr}$ that is the estimation of final cost $Terr$ as follows:

$$Initialization: \ \widehat{Terr} = 1.05 Terr \tag{3}$$

$$After\ initialization: \ \widehat{Terr} = \begin{cases} 0.5\widehat{Terr} + 0.5 Terr & (Terr < \widehat{Terr}) \\ 0.9\widehat{Terr} + 0.1 Terr & (Terr \geq \widehat{Terr}) \end{cases} \tag{4}$$

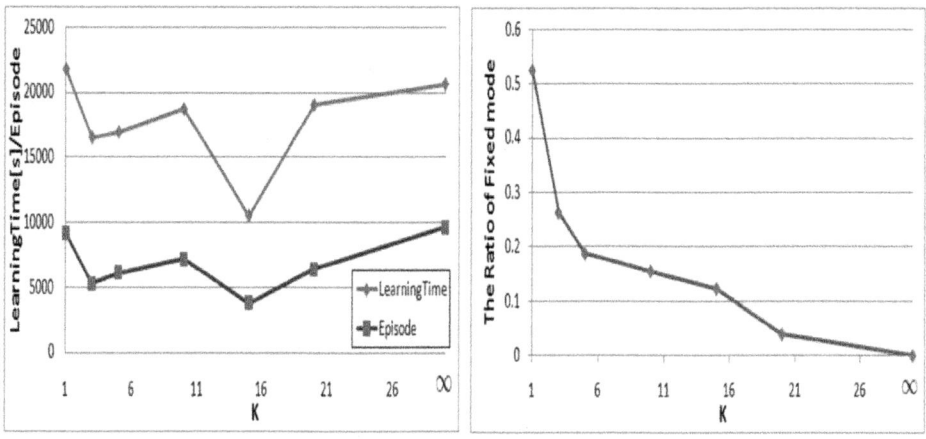

Fig. 2. Strategy1: Total learning time and **Fig. 3.** Strategy1: Ratio of number of fixed number of episodes   mode states to N_B

Rewards are defined in Table 2. RWDforSuc is given to the agent when it reaches the target walking time the first time or, after that, when $Terr < \widehat{Terr}$. In this case, the counter of success is updated as $TUC + 1$. If the agent reaches it but $Terr \geq \widehat{Terr}$, then the agent gets RWforGD.

5   Simulation Results

5.1   Learning Schedule

Since our learning task is over rather long-term, we schedule it as follows:

(1) in the initial stage, target walking time t_f is set to $t_f = Tstep$, where $Tstep$ is one step time. (2) t_f is extended by $Tstep/2$ after the agent receives RWDforSuc LT times until t_f reaches full term t_w.

The robot starts from a two-leg standing pose, walks 5 steps, and stops with a two-leg standing pose, again, as shown in Fig.1. It takes $t_w = 11.25[s]$, including preparation time.

The agent could not walk using the initial walking trajectory designed for stable static walking. The following results are averages for 15 trials with different random seeds.

5.2   Simulation (1) : Effect of Strategy 1 for Fixed Mode State

We first conducted simulation to confirm the effect of Strategy 1 for the fixed mode state, where we use $LT = 15$ and set $K = 1, 3, 5, 10, 15, 20$ and ∞. Fig.2 shows normalized total learning time and the number of episodes for each K. We confirm an improved number of episodes and learning time compared to $K = \infty$, meaning that we do not use the fixed mode state. In $K = 15(= LT)$, we see a

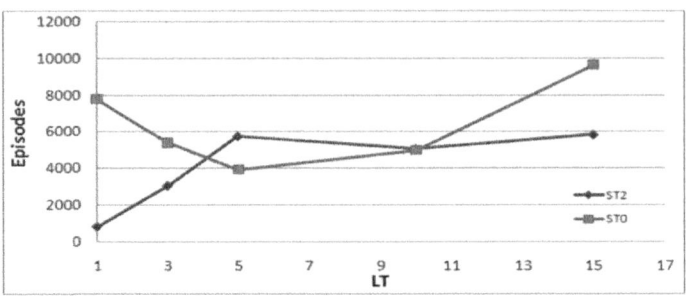

Fig. 4. Strategy2: Number of episodes for LT

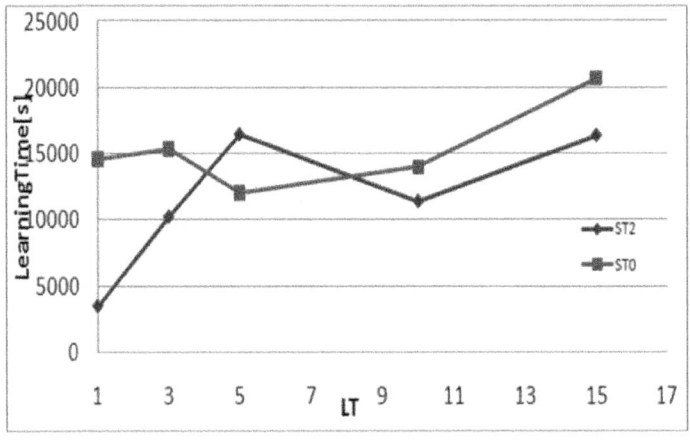

Fig. 5. Strategy2: Total learning time for LT

two times higher performance. For $K = 1$, the number of episodes and learning time are near the case of $K = \infty$.

Fig.3 shows the ratio of the number of fixed mode mode states to N_B for each K. We confirmed a large change in the ratio of the value of K and the change in learning speed. For $K = 1$, states change to fixed mode very easily, so the agent has difficulty in finding undiscovered states, degrading performance.

5.3   Simulation (2) : Effect of Strategy 2 for Fixed Mode State

We conducted simulations to confirm the effect of Strategy 2 for fixed mode states for $LT = 1, 3, 5, 10$ and 15. Fig.4 shows the number of episodes for each LT, and Fig.5 is the total learning time. Red (ST2) corresponds to Strategy 2 and blue (ST0) to a case in which no fixed mode state is used. We confirmed the effectiveness of Strategy 2 for $LT = 1$ and 3 in both the number of episodes and total learning time.

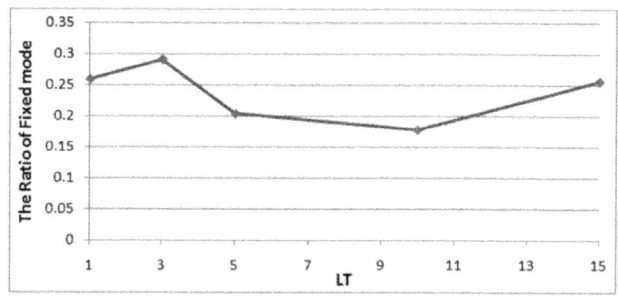

Fig. 6. Strategy2: Ratio of number of fixed mode states to N_B

Fig.6 shows the ratio of the number of fixed mode state to N_B and that the ratio is almost constant for Strategy 2 over a large range of LT, so Strategy 2 is easy to use in practice compared to Strategy 1.

6   Conclusions

We have proposed new exploitation-oriented learning (XoL) having *fixed mode state* shifted from normal when the state has already received sufficient reward. We used online PS to get rewards efficiently in long-term tasks.

We have adapted our method to generating a biped robot waist trajectory. We have discussed simulation results and confirmed our method's effectiveness. Now we are studying the relationship between K and LT to determine how to choose appropriate K.

We plan to apply our method to other control problems, e.g., the bicycle balancing problem[14] nonholonomic systems[6,9]. We also plan to extend our method to use with a real-world biped robot and explore real-world applications.

Acknowledgements. This work was supported by Grant-in-Aid for Scientific Research(C) 21560275.

References

1. Chrisman, L.: Reinforcement Learning with perceptual aliasing: The Perceptual Distinctions Approach. In: Proc. of the 10th National Conf. on Artificial Intelligence, pp. 183–188 (1992)
2. Hasemi, K., Suyari, H.: A Proposal of algorithm that reduces computational complexity for Online Profit Sharing. Report of the Institute of Electronics, Information and Communication Engineers NC-105(657), 103–108 (2006)(in Japanese)
3. Kimura, H., Kobayashi, S.: An analysis of actor/critic algorithm using eligibility traces: reinforcement learning with imperfect value function. In: Proc. of the 15th Int. Conf. on Machine Learning, pp. 278–286 (1998)
4. Goldberg, D.E.: Genetic Algorithms in Search, Optimization, and Machine Learning. Addison-Wesley Professional (1989)

5. Heidrich-Meisner, V., Igel, C.: Evolution Strategies for Direct Policy Search. In: Rudolph, G., Jansen, T., Lucas, S., Poloni, C., Beume, N. (eds.) PPSN 2008. LNCS, vol. 5199, pp. 428–437. Springer, Heidelberg (2008)
6. Ikeda, K.: Exemplar-Based Direct Policy Search with Evolutionary Optimization. In: Proc. of 2005 Congress on Evolutionary Computation, CEC 2005, pp. 2357–2364 (2005)
7. Matsui, T., Goto, T., Izumi, K.: Acquiring a Government Bond Trading Strategy Using Reinforcement Learning. J. of Advanced Computational Intelligence and Intelligent Informatics 13(6), 691–696 (2009)
8. Merrick, K., Maher, M.L.: Motivated Reinforcement Learning for Adaptive Characters in Open-Ended Simulation Games. In: Proc. of the Int. Conf. on Advanced in Computer Entertainment Technology, pp. 127–134 (2007)
9. Miyamae, A., Sakuma, J., Ono, I., Kobayashi, S.: Instance-based Policy Learning by Real-coded Genetic Algorithms and Its Application to Control of Nonholonomic Systems. J. of the Japanese Society for Artificial Intelligence 24(1), 104–115 (2009) (in Japanese)
10. Miyazaki, K., Yamamura, M., Kobayashi, S.: On the Rationality of Profit Sharing in Reinforcement Learning. In: Proc. of the 3rd Int. Conf. on Fuzzy Logic, Neural Nets and Soft Computing, pp. 285–288 (1994)
11. Miyazaki, K., Kobayashi, S.: Reinforcement Learning for Penalty Avoiding Policy Making. In: Proc. of the 2000 IEEE Int. Conf. on Systems, Man and Cybernetics, pp. 206–211 (2000)
12. Miyazaki, K., Kobayashi, S.: A Reinforcement Learning System for Penalty Avoiding in Continuous State Spaces. J. of Advanced Computational Intelligence and Intelligent Informatics 11(6), 668–676 (2007)
13. Miyazaki, K., Kobayashi, S.: Exploitation-Oriented Learning PS-r$^{\#}$. J. of Advanced Computational Intelligence and Intelligent Informatics 13(6), 624–630 (2009)
14. Randløv, J., Alstrøm, P.: Learning to Drive a Bicycle Using Reinforcement Learning and Shaping. In: Proc. of the 15th Int. Conf. on Machine Learning, pp. 463–471 (1998)
15. Stone, P., Sutton, R.S., Kuhlamann, G.: Reinforcement Learning toward RoboCup Soccer Keepaway. Adaptive Behavior 13(3), 0165–0188 (2005)
16. Sutton, R.S., Barto, A.G.: Reinforcement Learning: An Introduction. A Bradford Book. MIT Press (1998)
17. Sutton, R.S., McAllester, D., Singh, S.P., Mansour, Y.: Policy Gradient Methods for Reinforcement Learning with Function Approximation. In: Advances in Neural Information Processing Systems, vol. 12, pp. 1057–1063 (2000)
18. Watanabe, T., Miyazaki, K., Kobayashi, H.: A New Improved Penalty Avoiding Rational Policy Making Algorithm for Keepaway with Continuous State Spaces. J. of Advanced Computational Intelligence and Intelligent Informatics 13(6), 675–682 (2009)
19. Yoshimoto, J., Nishimura, M., Tokita, Y., Ishii, S.: Acrobot control by learning the switching of multiple controllers. J. of Artificial Life and Robotics 9(2), 67–71 (2005)

MapReduce for Parallel Reinforcement Learning

Yuxi Li[1] and Dale Schuurmans[2]

[1] College of Computer Science and Engineering
Univ. of Electronic Science and Technology of China
Chengdu, China
[2] Department of Computing Science
University of Alberta
Edmonton, Alberta, Canada

Abstract. We investigate the parallelization of reinforcement learning algorithms using MapReduce, a popular parallel computing framework. We present parallel versions of several dynamic programming algorithms, including policy evaluation, policy iteration, and off-policy updates. Furthermore, we design parallel reinforcement learning algorithms to deal with large scale problems using linear function approximation, including model-based projection, least squares policy iteration, temporal difference learning and recent gradient temporal difference learning algorithms. We give time and space complexity analysis of the proposed algorithms. This study demonstrates how parallelization opens new avenues for solving large scale reinforcement learning problems.

1 Introduction

Reinforcement learning (RL) can solve a wide range of problems in science, engineering and economics that are modeled as Markov or Semi-Markov Decision Processes (MDPs) [1, 14, 16]. For large problems, however, one encounters the "curse of dimensionality". In such cases, function approximation, and in particular, linear function approximation, have proved to be a critical strategy for generalizing and scaling-up. Significant research effort is still devoted to developing efficient algorithms for large-scale problems.

For large problems, parallelism appears to be another promising approach to scaling up, particularly since multi-core, cluster and cloud computing are becoming increasingly the norm. Among parallel computing frameworks, MapReduce [5] has recently been attracting much attention in both industry and academia. There are numerous successes of MapReduce [4, 8, 10], including applications for machine learning and data mining problems. Among these, the most prominent is the complete rewrite of the production indexing system for the Google web search service. However, there has yet to be significant research effort on designing MapReduce parallel algorithms for reinforcement learning.

In this paper, we propose MapReduce algorithms to parallelize several important reinforcement learning algorithms. We identify maximum, vector, and matrix operations, such as matrix-vector multiplications, as the core enabling

S. Sanner and M. Hutter (Eds.): EWRL 2011, LNCS 7188, pp. 309–320, 2012.

techniques for parallelizing reinforcement learning algorithms in MapReduce. Using these primitives, we show how MapReduce parallel algorithms can be designed for classical dynamic programming (DP) algorithms, including policy evaluation, policy iteration and off-policy updates, as well as tabular reinforcement learning algorithms. To cope with large scale problems via linear function approximation, we show how MapReduce parallel algorithms can be developed for approximate reinforcement learning algorithms, including the model-based projection method, least squares policy iteration, temporal difference (TD) learning, and the recent gradient TD algorithms. We present time and space complexity analysis for the proposed algorithms.

With such MapReduce reinforcement learning algorithms, one can handle complex sequential decision problems more efficiently by exploiting parallelism. MapReduce reinforcement learning algorithms are able to provide solutions to problems that would be infeasible otherwise. For example, it is mentioned in [18] that for the game of Go [15] over one million features were used when building a value function, making least-squares methods infeasible. However, parallelism can address such issues. Moreover, with MapReduce the programmer does not have to explicitly manage the complexity of underlying parallel/distributed issues, such as data distribution, load balancing and fault tolerance. In fact, parallel computing broadens the range of problems that can be feasibly tackled with tabular reinforcement learning methods, athough it enhances function approximation methods as well, as we will demonstrate.

The remainder of the paper is organized as follows. Section 2 first provides a brief introduction of MapReduce and discuss a design of matrix-vector multiplication. Then in Section 3, we discuss MapReduce algorithms for DP algorithms and RL algorithms in the tabular setting. Finally, we discuss MapReduce algorithms for RL with linear function approximation in Section 4, and conclude.

2   MapReduce

MapReduce is a framework introduced by Google for handling large data sets via distributed processing on clusters of commodity computers. It allows one to express parallel computations without worrying about the messy details of parallelism, including data distribution, load balancing and fault tolerance.

In this model, computation is expressed as taking a set of input (key, value) pairs and producing a set of output (key, value) pairs via two functions: Map and Reduce. The Map function takes an input pair and produces a set of intermediate (key, value) pairs. The MapReduce library groups together all intermediate values associated with the same intermediate key and passes them to the Reduce function. The Reduce function accepts an intermediate key and a set of values for that key, and merges these values to form a possibly smaller set of values. Both the Map and Reduce functions are specified by the user. It is possible to further improve performance via partition and combine functions. The partition function divides the intermediate key space and assigns them to reducers. The combine function allows for local aggregation before the shuffle and sort phase.

Input Matrix $M = \{(i, (j, mval))\}$,
    Vector $V = \{(i, vval)\}$
Output Vector $V = \{(i, mval * vval)\}$

MV-Map1(Key i, Value v)
if (i,v) is of type Vector **then**
  Output(i, v)
end if
if (i,v) is of type Matrix **then**
  $(j, mval) \leftarrow v$
  Output$(j, (i, mval))$
end if

MV-Reduce1(Key j, Value $v'[1..m]$)
$mv \leftarrow []$; $vv \leftarrow []$
for v in $v'[1..m]$ **do**
  if (j,v) is of type Vector **then**
    $vv \leftarrow v$
  else
    if (j,v) is of type Matrix **then**
      Add v to mv;
    end if
  end if
end for
for $(i', mval')$ in mv **do**
  Output$(i', mval' * vv)$
end for

Fig. 1. Matrix-Vector multiplication, step 1

Input Partial vector $V' = \{(i, vval')\}$
Output Result vector $V = \{(i, vval)\}$

MV-Map2(Key i, Value v)
Output(i, v)

MV-Reduce2(Key i, Value $v[1..m]$)
$sum \leftarrow 0$
for v' in $v[1..m]$ **do**
  $sum \leftarrow sum + v'$
end for
Output(i, v)

Fig. 2. Matrix-Vector multiplication, step 2

Iterative MapReduce. Most RL algorithms are iterative by nature. There are several research efforts to design and implement iterative MapReduce, e.g., Haloop [3] and Twister [6]. Haloop extends Hadoop [20] for iterative programs by providing new programming model and architecture, loop-aware scheduling, caching for loop-invariant data and caching to support fixpoint evaluation.

MapReduce for matrix-vector multiplication. We now present a MapReduce implementation for multiplying an $M \times N$ matrix $\{a_{i,j}\}$ with an $N \times 1$ vector $\{b_i\}$. The basic idea follows. We need two MapReduce jobs for the multiplication. In MapReduce Step 1, we multiply the j-th vector element b_j with each element $a_{i,j}$ of the j-th column of the matrix to obtain a set of $(key : i, value : b_j \times a_{i,j})$ pairs, for $i \in \{1, 2, \ldots, M\}$; In Step 2, we sum up values with key i to get the i-th vector element. Figures 1 and 2 present the MapReduce pseudo code for matrix-vector multiplication. The sorting phase dominates the algorithms complexity. Assuming the load is uniformly distributed on P processing units, the running time complexity is $O(\frac{MN}{P} log(\frac{MN}{P}))$, and the space complexity is $O(\frac{MN}{P})$.

Such implementation assumes that a processing unit can not multiply a row of the matrix and the vector. Otherwise, a single map function can handle the matrix-vector multiplication. In the sequel, we decompose matrix operations into matrix-vector multiplications; and we use vector and maximum operations directly, since they are straightforward to parallelize in MapReduce. See more discussions about designing matrix operations in [8, 11].

3   MapReduce for Tabular DP and RL

First, we introduce the notation we will use. An MDP is defined by a finite set of states S, a finite set of actions A, an $|S||A| \times |S|$ transition matrix P and an $|S||A| \times 1$ reward vector $\mathbf{r}$. The entry $P_{(sa,s')}$ specifies the conditional probability of transiting to state s' starting from state s and taking action a. The entry $\mathbf{r}_{(sa)}$ specifies the reward obtained when taking action a in state s. A standard objective is to maximize the infinite horizon discounted reward $\sum_{t=1}^{\infty} \gamma^{t-1} r_t$. In this case, one can always have an optimal stationary, deterministic policy [1], denoted as $\boldsymbol{\pi}$. The entry $\pi_{(sa)}$ specifies the probability of taking action a in state s. A policy is stationary if the selection probability does not change over time. In a deterministic policy, there is an optimal action for each state, i.e., the probabilities are either 0 or 1. For convenience, let Π denote an $|S| \times |S||A|$ matrix with $\Pi(s, (s, a)) = \pi(s, a)$.

3.1   Policy Evaluation

We have $\mathbf{v} = \Pi(\mathbf{r} + \gamma P \mathbf{v})$ for state value function $\mathbf{v}$. A state-based policy evaluation algorithm can be defined by repeatedly applying an operator $\mathcal{O}$ defined by $\mathcal{O}\mathbf{v} = \Pi(\mathbf{r} + \gamma P \mathbf{v}) = \Pi\mathbf{r} + \gamma\Pi P\mathbf{v}$. For a fixed policy $\boldsymbol{\pi}$, a MapReduce policy evaluation algorithm can first calculate $\tilde{\mathbf{v}} = \Pi\mathbf{r}$ and $H = \Pi P$ using parallel matrix-vector and matrix-matrix multiplications respectively. Then $\mathbf{v} \leftarrow \tilde{\mathbf{v}} + \gamma H\mathbf{v}$ can be iteratively calculated until convergence, which includes iterative calls of parallel operations for matrix-vector multiplication and vector-vector addition. State-action policy evaluation can be dealt with similarly.

For calculating $H = \Pi P$, the time complexity is $O(|S| \frac{|S|^2 |A|}{P} log(\frac{|S|^2 |A|}{P}))$, and the space complexity is $O(\frac{|S|^2 |A|}{P})$. For the iteration, $\mathbf{v} \leftarrow \tilde{\mathbf{v}} + \gamma H\mathbf{v}$, the time complexity is $O(T \frac{|S|^3}{P} log(\frac{|S|^3}{P}))$, where T is the number of iterations; and the space complexity is $O(\frac{|S|^3}{P})$. Thus the complexity of the whole policy evaluation depends on which part dominates.

One can also consider an iterative MapReduce implementation of policy evaluation using Haloop [3], as shown in Figure 3. In Haloop, one has the additional functionality: AddMap and AddReduce to specify Map and Reduce function(s) in each iteration; SetInput to specify the input to each iteration; AddInvariantTable to specify loop-invariant data; SetDistanceMeasure to specify a distance for results for fixed-point check; and SetFixedPointThreshold and/or SetMaxNumOfIterations to specify the loop termination condition. Thus, for policy evaluation one can first calculate $H = \Pi P$ using MapReduce matrix-matrix multiplication; this remains invariant across iterations. Then we introduce a map function (without reduce) to update $\mathbf{v}$ after the matrix-vector multiplication $\mathbf{v}' = H\mathbf{v}_{i-1}$ in the main function.[1] IterationInput specifies the most recent $\mathbf{v}_{i-1}$

---

[1]   Note, to make the implementation more efficient, one should integrate this update step with the step for matrix-vector multiplication. We present it in this separated manner for clarity.

IterationInput
Input: int i
return the *i*-th column of matrix P
Main
Job job = new Job()
add job for matrix-vector multiplication

job.AddInvariantTable(Π)
job.SetInput(IterationInput)
job.setMaxNumOfIteration($|S|$)
job.Submit()

(Above: Calculating $H = \Pi P$)

IterationInput
Input: int i
return the $(i-1)$-th step of **v**

UpdateV-Map
Input: Key k, vector r, constant γ, **v**'
Output(0, $\mathbf{v} = \tilde{\mathbf{v}} + \gamma \mathbf{v}'$)
ResultDistance
Input: $\mathbf{v}_{i-1}, \mathbf{v}_i$
return $\|\mathbf{v}_{i-1} - \mathbf{v}_i\|$
Main
Job job = new Job()
add job for matrix-vector multiplication

job.AddMap(UpdateQ-Map)
job.AddInvariantTable(M)
job.SetInput(IterationInput)
job.SetDistanceMeasure(ResultDistance)

job.setFixedPointThreshold(ϵ)
job.setMaxNumOfIteration(T)
job.Submit()

Fig. 3. Haloop-based policy evaluation

as the input to each iteration; see Figure 3. In ResultDistance, we use L_2-norm to calculate the difference of consecutive **v**'s to test convergence. We pre-define the number of iterations and the fixed point threshold respectively as T and ϵ.

3.2 Policy Iteration

Policy iteration is a standard MDP planning algorithm that iteratively conducts policy evaluation and policy improvement. In policy improvement, given a current policy π, whose state value function **v** or state-action value function **q** have already been determined, one can obtain an improved policy π' by setting: $a^*(s) = \arg\max_a \mathbf{q}_{(sa)} = \arg\max_a \mathbf{r}_{(sa)} + \gamma P_{(sa,:)}\mathbf{v}$, $\pi'_{(sa)} = 1$, if $a = a^*(s)$; or, 0, if $a \neq a^*(s)$. The policy improvement theorem verifies that the above update leads to an improved policy; that is, $\Pi \mathbf{q} \leq \Pi' \mathbf{q}$ implies $\mathbf{v} \leq \mathbf{v}'$.

It is straightforward to design parallel algorithms for policy improvement in MapReduce. For example, given **q**, for each state s, one finds the maximum state-action values **q** for all actions. Similarly, given **v**, for each state s and action a, one calculates $\mathbf{r}_{(sa)} + \gamma P_{(sa,:)}\mathbf{v}$, which involves a vector-vector multiplications and an addition. Then, for state s, we find the maximum over the resulting values for all actions.

Vector-vector multiplication, vector-vector addition and maximum operators are linear in both time and space. Thus, policy evaluation dominates time and space complexity in policy iteration.

3.3 Off-policy Updates

Off-policy updates form the basis for many RL algorithms, in particular, value iteration and Q-learning. For a state value function **v**, the off-policy operator $\mathcal{M}$ is

defined as $\mathcal{M}\mathbf{v} = \Pi^*(\mathbf{r} + \gamma P\mathbf{v})$, where, $\Pi^*(\mathbf{r} + \gamma P\mathbf{v})_{(s)} = \max_a(\mathbf{r}_{(sa)} + \gamma P_{(sa,:)}\mathbf{v})$
For a state-action value function $\mathbf{q}$, the off-policy operator $\mathcal{M}$ is defined as $\mathcal{M}\mathbf{q} = \mathbf{r} + \gamma P\Pi^*\mathbf{q}$, where, $\Pi^*\mathbf{q}_{(s)} = \max_a \mathbf{q}_{(sa)}$. The above off-policy updates aim at making the corresponding value functions closer to their respective Bellman equations; namely, $\mathbf{v} = \Pi^*(\mathbf{r} + \gamma P\mathbf{v})$ and $\mathbf{q} = \mathbf{r} + \gamma P\Pi^*\mathbf{q}$. It is straightforward to design parallelized algorithms for off-policy updates using MapReduce's vector-vector multiplication and max operation, which have linear complexity.

3.4   Tabular Online Algorithms

There are several tabular online algorithms, such as, TD(λ), Sarsa(λ), and Q(λ)-learning, for $\lambda \in [0, 1]$, where each online sample updates one entry of the value function, and the current update is affected by previous updates. For simplicity, we give updates for TD(0), Sarsa(0) and Q(0)-learning respectively.

$$V(s) \leftarrow V(s) + \alpha[r + \gamma V(s') - V(s)]$$
$$Q(s, a) \leftarrow Q(s, a) + \alpha[r + \gamma Q(s', a') - Q(s, a)]$$
$$Q(s, a) \leftarrow Q(s, a) + \alpha[r + \gamma \max_{a'} Q(s', a') - Q(s, a)]$$

For large-scale problems, we can take advantage of parallel computing: in Q-learning for picking the action that gives the largest state-action value given the state; and in Sarsa and Q-learning for selecting an action given the current state and state-action value function, for example, with ϵ-greedy policy. Both involve a maximum operation. These operations have linear complexity.

4   MapReduce for RL: Linear Function Approximation

In the linear architecture, the approximate value function is represented by: $\hat{Q}(s, a; \mathbf{w}) = \sum_{i=1}^{k} \phi_i(s, a) w_i$, where $\phi_i(\cdot, \cdot)$ is a basis function, w_i is its weight, and k is the number of basis functions. Define

$$\phi^\mathsf{T}(s, a) = \{\phi_1(s, a), \phi_2(s, a), \ldots, \phi_k(s, a)\},$$
$$\Phi^\mathsf{T} = \{\phi(s_1, a_1), \ldots, \phi(s, a), \ldots, \phi(s_{|S|}, a_{|A|})\},$$
$$\mathbf{w}^\mathsf{T} = \{w_1, w_2, \ldots, w_k\}.$$

where T denotes matrix transpose. Then we have $\hat{Q} = \Phi\mathbf{w}$.

4.1   Model-Based Projection

Define an operator $\mathcal{P}$ that projects a state-action value function $\mathbf{q}$ to the column span of Φ, $\mathcal{P}\mathbf{q} = \text{argmin}_{\hat{q} \in span(\Phi)} \|\mathbf{q} - \hat{\mathbf{q}}\|_{\mathbf{z}} = \Phi(\Phi^\mathsf{T} Z\Phi)^{-1}\Phi^\mathsf{T} Z\mathbf{q}$. Here, $Z = \text{diag}(\mathbf{z})$, where $\mathbf{z}$ is the stationary state-action visit distribution for $P\Pi$.

Approximate dynamic programming composes the on-policy operator $\mathcal{O}$ and the subspace projection operator $\mathcal{P}$ to compute the best approximation of one-step update of the value function representable with the basis functions. Such combined operator is guaranteed to converge to a fixed point.

We discuss how to parallelize such a projection. Since $\Phi(\Phi^{\mathsf{T}}Z\Phi)^{-1}\Phi^{\mathsf{T}}Z$ remains fixed, we can calculate it using a matrix inversion and several matrix-matrix multiplications. A matrix inversion can be calculated with singular value decomposition (SVD)[2], whose MapReduce implementation is available in the open source Apache Hama project. In the following, we discuss matrix inversion with MapReduce based on the algorithm in [2].

We first describe briefly the fast iterative matrix inversion algorithm in [2, Section 2.9]. Suppose A is an invertible square matrix. The algorithm is based on the classical Newton method, which is motivated by the desire of low computational complexity and good numerical robustness. The algorithm starts with some B_0 such that $\|I - B_0 A\| < 1$. Then it iteratively improve B_k by letting $B_{k+1} = 2B_k - B_k A B_k$. This iteration can be interpreted as solving $X^{-1} - A = 0$ with Newton's method. A choice for B_0 is $B_0 = A^{\mathsf{T}}/\mathrm{tr}(A^{\mathsf{T}}A)$, where $\mathrm{tr}(A^{\mathsf{T}}A)$ is the trace of $A^{\mathsf{T}}A$, the sum of the diagonal entries of $A^{\mathsf{T}}A$. It can be shown that such a choice for B_0 is suitable and the above iterative method is efficient.

Each iteration involves matrix multiplications and subtraction, so it is straightforward to parallelize in MapReduce. Since $\Phi^{\mathsf{T}}Z\Phi$ is a square matrix, we can use this iterative method to invert it. For calculating $\Phi^{\mathsf{T}}Z\Phi$, the time complexity is $O((k+|S||A|)\frac{k|S||A|}{P}log(\frac{k|S||A|}{P}))$, and the space complexity is $O(\frac{k|S||A|}{P})$, assuming we use matrix-vector multiplication. For each iteration, the time complexity is $O(\frac{k^5}{P}log(\frac{k^3}{P}))$, and the space complexity is $O(\frac{k^3}{P})$. The complexity can be improved by customizing a MapReduce algorithm for matrix multiplication.

4.2   Least-Squares Policy Iteration

LSPI [9] combines the data efficiency of the least squares temporal difference (LSTD) method and the policy search efficiency of policy iteration. As shown in [9], for a given $\boldsymbol{\pi}$, the weighted least squares fixed point solution is: $\mathbf{w} = \left(\Phi^{\mathsf{T}}\Delta_\mu(\Phi - \gamma P\Pi\Phi)\right)^{-1}\Phi^{\mathsf{T}}\Delta_\mu R$, where Δ_μ is the diagonal matrix with the entries of $\mu(s,a)$, which is a probability distribution over state-action pairs $(S \times A)$. This can be written as $A\mathbf{w} = b$, where $A = \Phi^{\mathsf{T}}\Delta_\mu(\Phi - \gamma P\Pi\Phi)$ and $\mathbf{b} = \Phi^{\mathsf{T}}\Delta_\mu\mathbf{r}$.

Without a model of the MDP—that is, without the full knowledge of P, Π and $\mathbf{r}$—one needs a learning method to discover an optimal policy. It is shown in [9] that $\mathbf{A}$ and $\mathbf{b}$ can be learned incrementally as, for a sample (s,a,r,s'):

$$A \leftarrow A + \phi(s,a)(\phi(s,a) - \gamma\phi(s',\pi(s')))^{\mathsf{T}}$$
$$\mathbf{b} \leftarrow \mathbf{b} + \phi(s,a)r$$

With the new weight vector $\mathbf{w}'$ a new policy is obtained. Thus, one can iteratively improve the policy until convergence.

To parallelize LSPI in MapReduce, consider the following. To calculate A and $\mathbf{b}$, one can parallelize the algorithm over samples (s_t, a_t, r_t, s'_t). Moreover,

---

[2] For a real number matrix A, if its singular value decomposition is $A = U\Sigma V^{\mathsf{T}}$, the pseudo-inverse is $A^+ = V\Sigma^+U^{\mathsf{T}}$. To get the pseudoinverse of the diagonal matrix Σ, we first transpose Σ, and then take the reciprocal of each non-zero element on the diagonal, and leave the zeros intact.

Input: $samples = \{(s, a, r, s')\}$
Output: A and b
LSPI-Ab-Map(Key k, Value v)
$A \leftarrow []$; $\mathbf{b} \leftarrow []$
for (s, a, r, s') in samples **do**
    $\phi_0 \leftarrow \phi(s, a)$
    $\phi_d \leftarrow \phi_0 - \gamma\phi(s', \pi(s'))$
    for $i \leftarrow 1$ **to** k **do**
        $A(i, :) \leftarrow A(i, :) + \phi_0(i)\phi_d^\mathsf{T}$
    end for
    $\mathbf{b} = \mathbf{b} + \phi_0 r$
end for
Output(1, $\{A, \mathbf{b}\}$)

LSPI-Ab-Reduce(Key k, Value Ab')
$A \leftarrow []$; $\mathbf{b} \leftarrow []$
for $(A', \mathbf{b}')$ in Ab' **do**
    $A \leftarrow A + A'$; $\mathbf{b} \leftarrow \mathbf{b} + \mathbf{b}'$
end for
Output $(0, \{A, \mathbf{b}\})$

Input: Samples *Samples*
Output: Weight vector **w**
Main
while w not convergent **do**
    Calculate A, b
    Calculate $D^{-1}b, D^{-1}B$
    Calculate **w** (Figure 5)
end while

Fig. 4. Parallel LSPI with MapReduce

there is a special structure in $\phi(s_t, a_t)(\phi(s_t, a_t) - \gamma\phi(s'_t, \pi(s'_t)))^\mathsf{T}$; that is, it is a vector-vector multiplication, resulting in a matrix. Thus, we can design a parallel algorithm which guarantees that each element in one vector multiplies each element in another vector. In this way, we obtain matrix A. The time complexity is $O(\frac{k^2}{P}log(\frac{k^2}{P}))$, and the space complexity is $O(\frac{k^2}{P})$. It is straightforward to obtain vector **b**, since it involves only vector-scalar multiplication.

In Figure 4, we provide a MapReduce algorithm to calculate A and **b**, then present an implementation for LSPI. We divide the (s, a, r, s') samples into groups. Each MapReduce job handles one group of samples. We use $\phi(s, a)$ to denote the value vector of basis functions for state-action pair (s, a), which involves an underlying evaluation of basis functions. We output the key as 0 and value as a pair of $\{A, \mathbf{b}\}$ so that all outputs (that is, the partial results of A and **b**) will go to a single reduce function. The reduce function collects partial results of A and **b** and sums them, respectively For LSPI, we choose not to use Haloop since the current version cannot support iterations with "loop-within-loop". This remains an interesting potential future extension of Hadoop; namely, a framework to support "loop-within-loop" iterations.

To solve $A\mathbf{w} = \mathbf{b}$, we note that the obvious solution of computing $\mathbf{w} = A^{-1}\mathbf{b}$ may be inefficient. Instead, we deploy Jacobi iteration to solve this linear system. This can be done efficiently in the proposed framework as follows. Set $A = B + D$, where D is a diagonal matrix, making it easy to obtain $\mathbf{w} = D^{-1}(\mathbf{b} - B\mathbf{w})$. We then obtain the following iteration:

$$\mathbf{w}_{t+1} = D^{-1}(\mathbf{b} - B\mathbf{w}_t)$$

Note that $D^{-1}b$ and $D^{-1}B$ each only need to be calculated once, and each is easy since D is diagonal. The time and space complexity will be linear in k. Thus, the matrix inversion problem $\mathbf{w} = A^{-1}b$ is converted into iterations that require

Update-w-Map
Input: Key k, vectors $D^{-1}b$, $\mathbf{w}$
Output(1,$\mathbf{w} = D^{-1}b - \mathbf{w}$)
IterationInput
Input: int i
return the $(i-1)$-th step of $\mathbf{w}$
ResultDistance
Input: $\mathbf{w}_{i-1}$, $\mathbf{w}_i$
return $\|\mathbf{w}_{i-1} - \mathbf{w}_i\|$

Main
Job job = new Job()
add job for matrix-vector multiplication

job.AddMap(Update-w-Map)
job.SetInput(IterationInput)
job.SetDistanceMeasure(ResultDistance)

job.setFixedPointThreshold(ϵ)
job.setMaxNumOfIteration(T)
job.Submit()

Fig. 5. Solve $Aw = b$

only matrix-vector multiplications. Thus, for each iteration, the time complexity is $O(\frac{k^2}{P}log(\frac{k^2}{P}))$, and the space complexity is $O(\frac{k^2}{P})$.

In Figure 5, we give an iterative MapReduce implementation in the Haloop style for calculating $\mathbf{w}$ after calculating A and b. As above, $D^{-1}b$ and $D^{-1}B$ are given so we can add a map function (without reduce) to update $\mathbf{w}$ after the matrix-vector multiplication $\mathbf{w}' = D^{-1}B\mathbf{w}_{i-1}$. IterationInput specifies that the most recent $\mathbf{w}_{i-1}$ as the input to each iteration. We use the L_2-norm to calculate the difference of consecutive $\mathbf{w}$'s to test the convergence, as specified in ResultDistance. We set the number of iterations as T, and the fixed point threshold as ϵ, two predefined numbers.

An alternative method for solving $Ax = b$ is to use a conjugate gradient method [7]. If A is symmetric and positive semi-definite, then we can apply the conjugate gradient method directly. Since A is usually not symmetric in our problem, we need to solve the normal equations, $A^\mathsf{T}Ax = A^\mathsf{T}b$. Unfortunately, the condition number $\kappa(A^\mathsf{T}A) = \kappa^2(A)$ might be significantly increased, which results in slow convergence. To address such a problem, choosing a good preconditioner would be important.

Such techniques form the basis for parallelizing similar least squares RL methods in MapReduce, e.g., the backward approach and the fitted-Q iteration in [19].

4.3 Temporal Difference Learning

In TD learning with linear function approximation, with k basis functions, we define $\phi^\mathsf{T}(s) = \{\phi_1(s), \cdots, \phi_k(s)\}$. The approximate value function is then given by $\hat{v}(s) = \phi(s)^\mathsf{T}\mathbf{w}$, where $\mathbf{w}$ is the weight vector.

The update procedure for approximate TD(0) is then

$$\mathbf{w}_{t+1} = \mathbf{w}_t + \alpha_t\phi(s_t)[r_t + \gamma\phi^\mathsf{T}(s_{t+1})\mathbf{w}_t - \phi^\mathsf{T}(s_t)\mathbf{w}_t]$$

With linear function approximation, the update rule for TD(λ), $\lambda \in [0,1]$ is

$$d_t = r + \gamma\phi^\mathsf{T}(s_{t+1})\mathbf{w}_t - \phi^\mathsf{T}(s_t)\mathbf{w}_t$$
$$z_t = \gamma\lambda z_{t-1} + \phi(s_t)$$
$$\mathbf{w}_{t+1} = \mathbf{w}_t + \alpha_t d_t z_t$$

Input: Samples *Samples*
Output: Weight vector **w**
for sample (s, a, r, s') in *Samples* **do**
 Calculate $\phi(s)$, $\phi(s')$, $\phi(s') - \phi(s)$, $[\phi(s') - \phi(s)]\mathbf{w}_t$
 Calculate the scalar $d_t = r + [\phi(s') - \phi(s)]\mathbf{w}_t$
 Calculate $z_t = \gamma \lambda z_{t-1} + \phi(s)$
 Calculate $\mathbf{w}_{t+1} = \mathbf{w}_t + \alpha_t d_t z_t$
end for

Fig. 6. MapReduce TD(λ)

where d_t is the temporal difference, z_t is the eligibility trace [16].

TD updates involve vector-vector additions, subtractions and multiplications (resulting in scalars), which are straightforward to parallelize in MapReduce. It is also straightforward to calculate basis functions for a state with MapReduce, which needs only a map function. It is particular interesting to parallelize TD learning algorithms when the feature dimension is huge, that is, when k is very large, for example, at the scale of millions in Go [15].

To design parallel TD(λ) MapReduce, we give the pseudo-code of a driver in Algorithm 6. In each iteration, it calculates the basis functions, the temporal difference d_t, the eligibility trace z_t and updates the weight vector **w** with samples (s, a, r, s'). We see that it is not complicated to design parallel TD(λ) in MapReduce. We choose to design a driver to implement the iterations in TD learning, since we realize that the current Haloop or Twister does not support the kind of iteration in TD. That is, in TD, before updating the weight vector **w**, some operations are needed to calculate basis functions, d_t and z_t, which themselves need MapReduce operations, and the inputs to each of them are different. A future work is to extend Hadoop to support more general iterations, for example, that for TD learning.

Significantly, there are a series of recent papers about gradient TD algorithms, e.g. [17, 13], addressing the instability and divergence problem of function approximation with off-policy TD learning/control. In the following, we give TDC [17](linear TD with gradient correction) algorithm directly.

$$d_t = r + \gamma \phi^\mathsf{T}(s_{t+1})\mathbf{w}_t - \phi^\mathsf{T}(s_t)\mathbf{w}_t$$
$$\mathbf{w}_{t+1} = \mathbf{w}_t + \alpha_t d_t \phi(s_t) - \alpha_t \gamma \phi(s_{t+1})(\phi^\mathsf{T}(s_t)\eta_t)$$
$$\eta_{t+1} = \eta_t + \beta_t(d_t - \phi^\mathsf{T} s_t)\eta_t)\phi(s_t)$$

We can design parallel TDC algorithm with MapReduce similar to that for TD algorithm, since the updates involve vector-vector additions, subtractions and multiplications (resulting in scalars), so we do not discuss it further. As well, other gradient TD algorithms can be dealt with similarly.

The above parellelization of TD and gradient TD algorithms have linear time and space complexity in the number of basis functions, and thus are suitable for problems with a large dimension or with a large number of features, e.g., for the game of Go [15], over a million features are used when building a value function.

Recently, Zinkevich et al. propose a parallelized stochastic gradient descent algorithm and gave theoretical analysis [21]. Both TD and gradient TD algorithms can exploit such an algorithm for parallelization, so that the parallel learning algorithm collects many sets of samples, then assigns each processing unit to conducting learning with TD or gradient TD update rules, after that, takes the average as the final result. Such an approach of parallelization assumes that a TD or gradient TD can be handled by a processor, and averages results from many processors to achieve high efficiency and accuracy.

5   Conclusions

We have investigated techniques for parallelizing reinforcement learning algorithms with MapReduce. In particular, we have provided parallel dynamic programming algorithms with MapReduce, including policy evaluation, policy iteration and off-policy updates, as well as tabular reinforcement learning algorithms. Furthermore, we proposed parallel algorithms with MapReduce for reinforcement learning, to cope with large scale problems with linear function approximation; namely the model-based projection method, least squares policy iteration, temporal difference learning, and the very recent gradient TD algorithms. We emphasize that iterative MapReduce is critical for parallel reinforcement learning algorithms, and provided algorithms in the Haloop style for tabular policy evaluation and for solving a linear system $Aw = b$. We give time and space complexity analysis of the proposed algorithms. These algorithms show that reinforcement learning algorithms can be significantly parallelized with MapReduce and open new avenues for solving large scale reinforcement learning problems.

We also observe that the current Hadoop and its iterative proposals, including Haloop and Twister, are not general enough to support certain iteration styles in a natrual way, for example, for LSPI and TD(λ). It is desirable to extend Hadoop for more general iterative structures, for example, the "loop-in-loop" iterations. Our preliminary experiments show that a cluster of machines running Hadoop can solve large scale linear systems $Aw = b$ efficiently; while Matlab on a single computer can easily encounter "Out of memory". It is desirable to further study the performance of proposed parallel DP and RL algorithms with MapReduce empiricaly. It would also be interesting to study alternative parallel frameworks, e.g., GraphLab [12], which address potential inefficiency with MapReduce, e.g., for problems with data-dpendancy due to MapReduce's share-nothing feature.[3]

References

[1] Bertsekas, D.P., Tsitsiklis, J.N.: Neuro-Dynamic Programming. Athena Scientific, Massachusetts (1996)
[2] Bertsekas, D.P., Tsitsiklis, J.N.: Parallel and Distributed Computation: Numerical Methods. Athena Scientific, Massachusetts (1997)

---

[3] This research is partially supported by "the Fundamental Research Funds for the Central Universities" in China.

[3] Bu, Y., Howe, B., Balazinska, M., Ernst, M.D.: Haloop: Efficient iterative data processing on large clusters. In: The 36th International Conference on Very Large Data Bases (VLDB 2010), Singapore (September 2010)

[4] Chu, C.-T., Kim, S.K., Lin, Y.-A., Yu, Y., Bradski, G., Ng, A., Olukotun, K.: Map-reduce for machine learning on multicore. In: Advances in Neural Information Processing Systems 19 (NIPS 2006), pp. 281–288 (December 2006)

[5] Dean, J., Ghemawat, S.: Mapreduce: Simplied data processing on large clusters. In: OSDI 2004, San Francisco, USA, pp. 137–150 (December 2004)

[6] Ekanayake, J., Li, H., Zhang, B., Gunarathne, T., Bae, S.-H., Qiu, J., Fox, G.: Twister: A runtime for iterative mapreduce. In: The First International Workshop on MapReduce and its Applications, Chicago, USA (June 2010)

[7] Golub, G.H., Van Loan, C.F.: Matrix Computations. The Johns Hopkins University Press, Baltimore (1996)

[8] Kung, U., Tsourakakis, C.E., Faloutsos, C.: Pegasus: A peta-scale graph mining system - implementation and observations. In: ICDM 2009, Miami, pp. 229–238 (December 2009)

[9] Lagoudakis, M.G., Parr, R.: Least-squares policy iteration. The Journal of Machine Learning Research 4, 1107–1149 (2003)

[10] Lin, J., Dyer, C.: Data-Intensive Text Processing with MapReduce. Morgan & Claypool (2009)

[11] Liu, C., Yang, H.-C., Fan, J., He, L.-W., Wang, Y.-M.: Distributed nonnegative matrix factorization for web-scale dyadic data analysis on mapreduce. In: Proceedings of the 19th International World Wide Web Conference (WWW 2010), Raleigh, North Carolina, USA, April 26-30, pp. 681–690 (2010)

[12] Low, Y., Gonzalez, J., Kyrola, A., Bickson, D., Guestrin, C., Hellerstein, J.M.: Graphlab: A new parallel framework for machine learning. In: Uncertainty in Artificial Intelligence (UAI), Catalina Island, USA (July 2010)

[13] Maei, H.R., Sutton, R.S.: GQ(λ): A general gradient algorithm for temporal-difference prediction learning with eligibility traces. In: Proceedings of the Third Conference on Artificial General Intelligence, Lugano, Switzerland (2010)

[14] Puterman, M.L.: Markov decision processes: discrete stochastic dynamic programming. John Wiley & Sons, New York (1994)

[15] Silver, D., Sutton, R.S., Muller, M.: Reinforcement learning of local shape in the game of Go. In: Proceedings of the 20th International Joint Conference on Artificial Intelligence (IJCAI), Hyderabad, India, pp. 1053–1058 (January 2007)

[16] Sutton, R.S., Barto, A.G.: Reinforcement Learning: An Introduction. MIT Press (1998)

[17] Sutton, R.S., Maei, H.R., Precup, D., Bhatnagar, S., Silver, D., Szepesvari, C., Wiewiora, E.: Fast gradient-descent methods for temporal-difference learning with linear function approximation. In: Proceedings of ICML 2009, Montreal, Canada, pp. 993–1000 (June 2009)

[18] Szepesvari, C.: Algorithms for Reinforcement Learning. Morgan Kaufmann & Claypool (2010)

[19] Tsitsiklis, J.N., Van Roy, B.: Regression methods for pricing complex American-style options. IEEE Transactions on Neural Networks (Special Issue on Computational Finance) 12(4), 694–703 (2001)

[20] White, T.: Hadoop: The Definitive Guide. O'Reilly (2009)

[21] Zinkevich, M., Weimer, M., Smola, A., Li, L.: Parallelized stochastic gradient descent. In: Proceedings of Advances in Neural Information Processing Systems 24 (NIPS 2010), Vancouver, Canada, pp. 2217–2225 (December 2010)

Compound Reinforcement Learning: Theory and an Application to Finance

Tohgoroh Matsui[1], Takashi Goto[2], Kiyoshi Izumi[3,4], and Yu Chen[3]

[1] Chubu University, Kasugai, Japan
TohgorohMatsui@tohgoroh.jp
[2] Bank of Tokyo-Mitsubishi UFJ, Ltd., Tokyo, Japan
takashi_6_gotou@mufg.jp
[3] The University of Tokyo, Tokyo, Japan
{izumi@sys.t,chen@k}.u-tokyo.ac.jp
[4] JST PRESTO, Tokyo, Japan

Abstract. This paper describes compound reinforcement learning (RL) that is an extended RL based on the compound return. Compound RL maximizes the logarithm of expected double-exponentially discounted compound return in return-based Markov decision processes (MDPs). The contributions of this paper are (1) Theoretical description of compound RL that is an extended RL framework for maximizing the compound return in a return-based MDP and (2) Experimental results in an illustrative example and an application to finance.

Keywords: Reinforcement learning, compound return, value functions, finance.

1 Introduction

Reinforcement learning (RL) has been defined as a framework for maximizing the sum of expected discounted rewards through trial and error [14]. The key ideas in RL are, first, defining the value function as the sum of expected discounted rewards and, second, transforming the optimal value functions into the Bellman equations. Because of these techniques, some good RL methods, such as temporal difference learning, that can find the optimal policy in Markov decision processes (MDPs) have been developed. Their optimality, however, is based on the expected discounted rewards. In this paper, we focus on the compound return[1]. The aim of this research is to maximize the compound return by extending the RL framework.

In finance, the compound return is one of the most important performance measures for ranking financial products, such as mutual funds that reinvest their gains or losses. It is related to the geometric average return, which takes into account the cumulative effect of a series of returns. In this paper, we consider tasks that we would face a hopeless situation if we fail once. For example, if we were to reinvest the interest or dividends in a financial investment, the effects of compounding interest would be great, and a large negative return would have serious consequences. It is therefore important to consider the compound returns in such tasks.

---

[1] Notice that the "return" is used in financial terminology in this paper, whereas the return is defined as the sum of the rewards by Sutton and Barto [14] in RL.

S. Sanner and M. Hutter (Eds.): EWRL 2011, LNCS 7188, pp. 321–332, 2012.

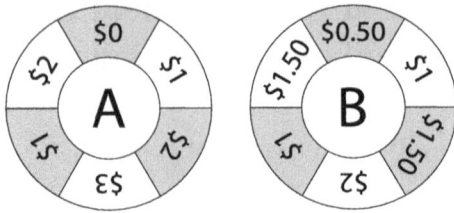

Fig. 1. Two-armed bandit with two wheels, A and B. The stated values are the amount of payback on a $1 bet. The average payback of wheel A is $1.50, and that of wheel B is $1.25.

The gains or losses, that is, the rewards, would be increased period-by-period, if we reinvested those gains or losses. In this paper, we consider return-based MDPs instead of traditional reward-based MDPs. In return-based MDPs, the agent receives the simple net returns instead of the rewards, and we assume that the return is a random variable that has Markov properties. If we used an ordinary RL method for return-based MDPs, it would maximize the sum of expected discounted returns. However, the compound return could not be maximized.

Some variants of the RL framework have been proposed and investigated. Average-reward RL [6,12,13,15] maximizes the arithmetic average rewards in reward-based MDPs. Risk-sensitive RL [1,2,5,7,9,11] not only maximizes the sum of expected discounted rewards, it also minimizes the risk defined by each study. While they can learn risk-averse behavior, they do not take into account maximizing the compound return.

In this paper, we describe an extended RL framework, called "compound RL", that maximizes the compound return in return-based MDPs. In addition to return-based MDPs, the key components of compound RL are double exponential discounting, logarithmic transformation, and bet fraction. In compound RL, the value function is based on the logarithm of expected double-exponentially discounted compound return and the Bellman equation of the optimal value function. In order to avoid the values diverging to negative infinity, a bet fraction parameter is used.

The key contributions of this paper are: (1) Theoretical description of compound RL that is an extended RL framework for maximizing the compound return in a return-based MDP and (2) Experimental results in an illustrative example and an application to finance. Firstly, we illustrate the difference between the compound return and the rewards in the next section. We then describe the framework of compound RL and a compound Q-learning algorithm that is an extension of Q-learning [17]. Section 5 shows the experimental results, and finally, we discuss our methods and conclusions.

2   Compound Return

Consider a two-armed bandit problem. This bandit machine has two big wheels, each with six different paybacks, as shown in Figure 1. The stated values are the amount of payback on $1 bet. The average payback for $1 from wheel A is $1.50, and that from wheel B is $1.25. If we had $100 at the beginning and played 100 times with $1 for each bet, wheel A would be better than wheel B. The reason is simply that the average profit

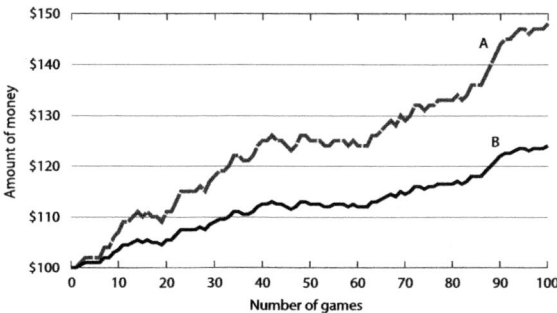

Fig. 2. Two example performance curves when we have $100 at the beginning, and we bet $1, 100 times, on either wheel A or B

of wheel A is greater than that of wheel B. Figure 2 shows two example performance curves when we bet $1 on either wheel A or B, 100 times. The final amounts of assets are near the total expected payback.

However, if we bet all money for each bet, then betting on wheel A would not be the optimal policy. The reason is that wheel A has a zero payback and the amount of money will become zero in the long term. In this case, we have to consider the compound return that is correlated to the geometric average rate of return:

$$G = \left(\prod_{i=1}^{n} (1 + R_i) \right)^{1/n} - 1, \tag{1}$$

where R_i is the i-th rate of return, and n represents the number of periods. Let P_t be the price of an asset that an agent has in time t. Holding the asset from one period, from time step $t - 1$ to t, the "simple net return" or "rate of return" is calculated by

$$R_t = \frac{P_t - P_{t-1}}{P_{t-1}} = \frac{P_t}{P_{t-1}} - 1 \tag{2}$$

and $1 + R_t$ is called the "simple gross return." The compound return is defined as follows [3] :

$$(1 + R_{t-n+1})(1 + R_{t-n+2}) \ldots (1 + R_t) = \prod_{i=1}^{n} (1 + R_{t-n+i}). \tag{3}$$

Whereas the geometric average rate of return of wheel A is -1, that of wheel B is approximately 0.14. Figure 3 shows two example performance curves when we have $100 at the beginning, and bet all money, 100 times, on either wheel A or B. The reason the performance curve of wheel A stops is that the bettor lost all his or her money when the payback was zero. Note that it has a logarithmic scale vertically. If we choose wheel B, then the expected amount of money at the end would be as much as $74 million.

As we see here, considering the compound return, maximizing the sum of expected discounted rewards is not useful. It is a general idea that the compound return is important in choosing mutual funds for financial investment [10]. Therefore, the RL agent

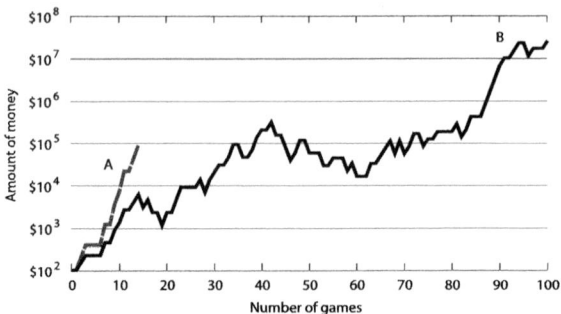

Fig. 3. Two example performance curves when we have \$100 at the beginning, and bet all money, 100 times, on either wheel A or B

should maximize the compound return instead of the sum of the expected discounted rewards in such cases.

3 Compound RL

Compound RL is an extension of the RL framework to maximize the compound return in return-based MDPs. We firstly describe return-based MDP, and then the framework of compound RL.

Consider the next return at time step t:

$$R_{t+1} = \frac{P_{t+1} - P_t}{P_t} = \frac{P_{t+1}}{P_t} - 1. \tag{4}$$

In other words, $R_{t+1} = r_{t+1}/P_t$, where r_{t+1} is the reward. The future compound return is written as

$$\rho_t = (1 + R_{t+1})(1 + R_{t+2}) \ldots (1 + R_T), \tag{5}$$

where T is a final time step. For continuing tasks in RL, we consider that T is infinite; that is,

$$\rho_t = (1 + R_{t+1})(1 + R_{t+2})(1 + R_{t+3}) \ldots$$
$$= \prod_{k=0}^{\infty} (1 + R_{t+k+1}). \tag{6}$$

In return-based MDPs, R_{t+k+1} is a random variable, $R_{t+k+1} \geq -1$, that has Markov properties.

In compound RL, double-exponential discounting and bet fraction are introduced in order to prevent the logarithm of the compound return from diverging. The double-exponentially discounted compound return with bet fraction is defined as follows:

$$\rho_t = (1 + R_{t+1}f)(1 + R_{t+2}f)^{\gamma}(1 + R_{t+3}f)^{\gamma^2} \ldots$$
$$= \prod_{k=0}^{\infty} (1 + R_{t+k+1}f)^{\gamma^k}, \tag{7}$$

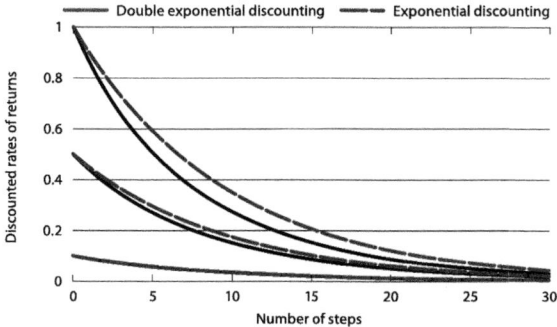

Fig. 4. Double exponential discounting and exponential discounting

where f is the bet fraction parameter, $0 < f \leq 1$. The logarithm of ρ_t can be written as

$$
\begin{aligned}
\log \rho_t &= \log \prod_{k=0}^{\infty} \left(1 + R_{t+k+1}f\right)^{\gamma^k} \\
&= \sum_{k=0}^{\infty} \log \left(1 + R_{t+k+1}f\right)^{\gamma^k} \\
&= \sum_{k=0}^{\infty} \gamma^k \log \left(1 + R_{t+k+1}f\right).
\end{aligned}
\tag{8}
$$

The right-hand side of Equation (8) is same as that of simple RL, in which the reward, r_{t+k+1}, is replaced with the logarithm of simple gross return; $\log(1 + R_{t+k+1}f)$. If $\gamma < 1$, then the infinite sum of the logarithm of simple gross return has a finite value as long as the return sequence $\{R_k\}$ is bounded. In compound RL, the agent tries to select actions in order to maximize the logarithm of double-exponentially discounted compound return it gains in future. It is equal to maximizing the double-exponentially discounted compound return.

Discounting is also a financial mechanism and it is called time preference in economics. Discounting in simple RL is called exponential discounting in economics. The double-exponentially discounted return can be considered as a kind of risk-adjusted returns in finance and it also can be considered as a kind of temporal discounting in economics. Figure 4 shows the difference between double exponential discounting and ordinary exponential discounting when $R_{t+k+1} = 1, 0.5$, and 0.1. The compound RL's double exponential discounting curve is very similar to the simple RL's exponential discounting curve when $|R_{t+k+1}|$ is small.

The bet fraction is the fraction of our asset that we place on a bet or in an investment. The Kelly criterion [8], which is well known in finance, is a formula used to determine the bet fraction that maximizes the expected logarithm of wealth when the accurate win probability and return are known. Since we cannot know the accurate win probability and return a priori, we use a parameter for the bet fraction.

In compound RL, the value of state s under a policy π is defined as the expected logarithm of double-exponentially discounted compound return under π:

$$
\begin{aligned}
V^\pi(s) &= \mathrm{E}_\pi \left[\log \rho_t \middle| s_t = s \right] \\
&= \mathrm{E}_\pi \left[\log \prod_{k=0}^{\infty} (1 + R_{t+k+1}f)^{\gamma^k} \middle| s_t = s \right] \\
&= \mathrm{E}_\pi \left[\sum_{k=0}^{\infty} \gamma^k \log(1 + R_{t+k+1}f) \middle| s_t = s \right],
\end{aligned}
$$

this can be written in a similar fashion as simple RL:

$$
\begin{aligned}
&= \mathrm{E}_\pi \left[\log(1 + R_{t+1}f) + \gamma \sum_{k=0}^{\infty} \gamma^k \log(1 + R_{t+k+2}f) \middle| s_t = s \right] \\
&= \sum_{a \in \mathcal{A}(s)} \pi(s,a) \sum_{s' \in \mathcal{S}} \mathcal{P}^a_{ss'} \left(\mathrm{R}^a_{ss'} + \gamma \mathrm{E}_\pi \left[\sum_{k=0}^{\infty} \gamma^k \log(1 + R_{t+k+2}f) \middle| s_{t+1} = s' \right] \right) \\
&= \sum_a \pi(s,a) \sum_{s'} \mathcal{P}^a_{ss'} \left(\mathrm{R}^a_{ss'} + \gamma V^\pi(s') \right),
\end{aligned} \tag{9}
$$

where $\pi(s,a)$ is selection probability and $\pi(s,a) = \Pr\left[a_t = a \middle| s_t = s \right]$, $\mathcal{P}^a_{ss'}$ is transition probability, $\mathcal{P}^a_{ss'} = \Pr\left[s_{t+1} = s' \middle| s_t = s, a_t = a \right]$, and $\mathrm{R}^a_{ss'}$ is the expected logarithm of simple gross return, $\mathrm{R}^a_{ss'} = \mathrm{E}\left[\log(1 + R_{t+1}f) \middle| s_t = s, a_t = a, s_{t+1} = s' \right]$. Equation (9) is the Bellman equation for V^π in compound RL. Similarly, the value of action a in state s can be defined as follows:

$$
\begin{aligned}
Q^\pi(s,a) &= \mathrm{E}_\pi \left[\log \rho_t \middle| s_t = s, a_t = a \right] \\
&= \sum_{s' \in \mathcal{S}} \mathcal{P}^a_{ss'} \left(\mathrm{R}^a_{ss'} + \gamma V^\pi(s') \right).
\end{aligned} \tag{10}
$$

4   Compound Q-Learning

As we have seen above, in compound RL, the Bellman optimality equations are the same in form as simple RL. The difference in the Bellman optimality equation between compound RL and simple RL is that the expected simple gross return, $\log(1 + \mathrm{R}^a_{ss'})$, is used in compound RL instead of the expected rewards, $\mathcal{R}^a_{ss'}$. Therefore, most of the algorithms and techniques for simple RL are applicable to compound RL, by replacing the reward, r_{t+1}, with the logarithm of simple gross return, $\log(1 + R_{t+1}f)$. In this paper, we show the Q-learning algorithm for compound RL and the convergence in return-based MDPs.

Q-learning [17] is one of the most well-known basic RL algorithms, which is defined by

$$
Q(s_t, a_t) \leftarrow Q(s_t, a_t) + \alpha \left(r_{t+1} + \gamma \max_a Q(s_{t+1}, a) - Q(s_t, a_t) \right), \tag{11}
$$

Algorithm 1. Compound Q-Learning

---

input: discount rate γ, step size α, bet fraction f
Initialize $Q(s, a)$ arbitrarily, for all s, a
loop {for each episode}
    Initialize s
    repeat {for each step of episode}
        Choose a from s using policy derived from Q (e.g., ϵ-greedy)
        Take action a, observe return R, next state s'
        $Q(s, a) \leftarrow Q(s, a) + \alpha \left[\log(1 + Rf) + \gamma \max_{a'} Q(s', a') - Q(s, a) \right]$
        $s \leftarrow s'$
    until s is terminal
end loop

---

where α is a parameter, called step-size, $0 \leq \alpha \leq 1$. We extend the Q-learning algorithm for traditional RL to one for compound RL, which we have called "compound Q-learning." In this paper, traditional Q-learning is called "simple Q-learning," to distinguish it from compound Q-learning.

Compound Q-learning is defined by

$$Q(s_t, a_t) \leftarrow Q(s_t, a_t) + \alpha \left(\log(1 + R_{t+1}f) + \gamma \max_a Q(s_{t+1}, a) - Q(s_t, a_t) \right). \tag{12}$$

Equation (12) is same as Equation (11), replacing r_t with $\log(1 + R_t f)$. The procedural form of the compound Q-learning algorithm is shown in Algorithm 1.

In this paper, we focus on return-based MDPs; we assume that the rate of return R_{t+1} has Markov properties, that is, R_{t+1} depends only on s_t and a_t. In return-based MDPs, we can show the convergence of compound Q-learning. Compound Q-learning replaces the rewards, r_{t+1}, in simple Q-learning with the logarithm of simple gross return, $\log(1 + R_{t+1}f)$. On the other hand, the Bellman equation for the optimal action value function Q^* in compound RL also replaces the expected rewards, $\mathcal{R}^a_{ss'}$, in simple RL with the expected logarithm of simple gross return, $\mathsf{R}^a_{ss'}$. Therefore, considering the logarithm of simple gross return in compound RL as rewards in simple RL, the action values Q approach to the optimal action values Q^* in compound RL.

More strictly, rewards are limited to be bounded in the Watkins and Dayan's convergence of simple Q-learning. We, therefore, have to limit the logarithm of the simple gross return to be bounded; that is, $1 + R_{t+1}f$ is greater than 0, and has an upper bound, in a return-based MDP. Thus, we will prove the following theorem.

Theorem 1. *Given bounded return* $-1 \leq R_t \leq \mathsf{R}$, *bet fraction* $0 < f \leq 1$, *step size* $0 \leq \alpha_t < 1$, *and* $0 < 1 + R_t f$, $\sum_{i=1}^{\infty} \alpha_{t^i} = \infty$, $\sum_{i=1}^{\infty} [\alpha_{t^i}]^2 < \infty$, *then* $\forall s, a[Q_t(s, a) \rightarrow Q^*(s, a)]$ *with probability 1, in compound Q-learning, where* R *is the upper bound of* R_t.

Proof. Let $r_{t+1} = \log(1 + R_{t+1}f)$, then the update equation of compound Q-learning shown in Equation (12) is equal to that of simple Q-learning. Since $\log(1 + R_t f)$ is

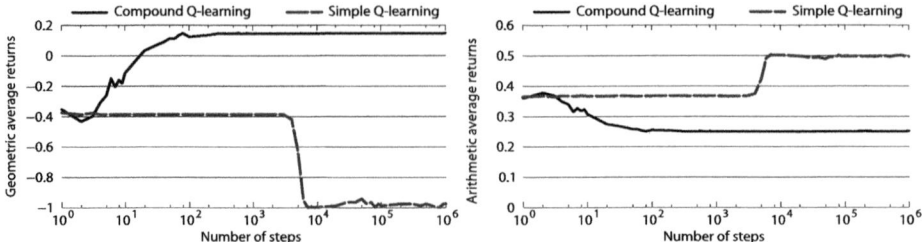

Fig. 5. Results for the two-armed bandit experiment

bounded, we can prove Theorem 1 by replacing r_t with $\log(1 + R_t f)$ in the Watkins and Dayan's proof [17] .

5   Experimental Results

5.1   Two-Armed Bandit

Firstly, we compared compound Q-learning and simple Q-learning, using the two-armed bandit problem described in Section 2. Each agent has $100 at the beginning and plays 100 times. The reward for simple Q-learning is the profit for betting $1, that is, the payback minus $1. The rate of return for compound Q-learning is the profit divided by the bet value of $1, that is, the same value as the rewards for simple Q-learning in this task. We set the discount rate of $\gamma = 0.9$ for both. The agents used ϵ-greedy selection, with $\epsilon = 0.1$, while learning, and chose actions greedily while evaluating them. The step-size parameter was $\alpha = 0.01$, and the bet fraction was $f = 0.99$. These parameters were selected empirically. For each evaluation, 251 trials were independently ran in order to calculate the average performance. We carried out 101 runs with different random seeds and got the average.

The results are shown in Figure 5. The left graph compares the geometric average returns, which means the compound return per period. The right graph compares the arithmetic average rewards. Compound Q-learning converged to a policy that chose wheel B, and simple Q-learning converged to one that chose wheel A. Whereas the arithmetic average return of the simple Q-learning agent was higher than that of compound Q-learning, the geometric average return was better from compound Q-learning than from simple Q-learning.

5.2   Global Bond Selection

Secondly, we investigated the applicability of compound Q-learning to a financial task: global government bonds selection. Although government bonds are usually considered as risk-free bonds, they still have default risks, that is, governments may fail to pay back its debt in full when economic or financial crisis strikes the country. Therefore, we have to choose bonds considering the yields and the default risks.

In this task, an agent learns to choose one from three 5-year government bonds: USA, Germany, and UK. The yields and default probabilities are shown in Table 1. We

Table 1. The yields and default probabilities in the global bond selection

Country	Yields	Default Prob.
USA	0.01929	0.036
Germany	0.02222	0.052
UK	0.02413	0.064

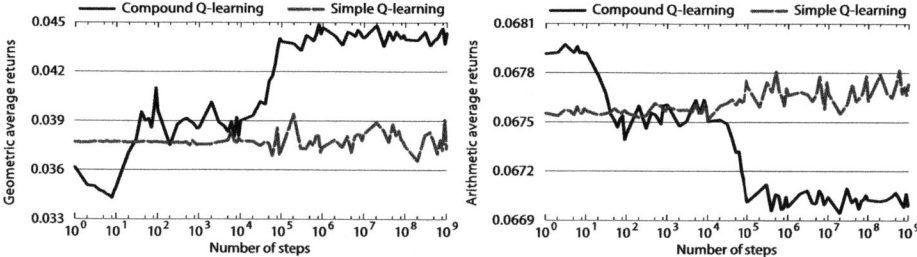

Fig. 6. Results in the global bond selection

obtained the yields of 5-year government bonds on 31st December 2010 from the web site of the Wall Street Journal, WSJ.com. The 5-year default probabilities were obtained from the CMA global sovereign credit risk report [4], which were calculated based on the closing values on 31st December 2010 by CMA.

Because the interest of government bonds are paid every half year, we calculated the half-year default probabilities based on the 5-year default probabilities, assuming that it occurs uniformly. In this task, when a default occurs, the principal is reduced by 75% and the rest of the interest is not paid. For example, when you choose German government bond and its default occurs in the second period, the return would be $0.01111 - 0.75 = -0.73889$, where 0.01111 is the interest for the first half-year. For simplicity, time-varying of the yields, the default probabilities, and the foreign exchange rates are not considered. We thus formulated a global government bonds selection task as a three-armed bandit task. The parameters for compound RL and simple RL were $\gamma = 0.9$, $f = 1.0$, $\alpha = 0.001$, and $\epsilon = 0.2$.

Figure 6 shows the learning curves of the geometric average returns (left) and the arithmetic average returns (right). Although the geometric average return of simple Q-learning did not increase, that of compound Q-learning increased. The proportion of learned policies are shown in Figure 7. Simple Q-learning could not converge a definite policy because of the very nearly equal action values based on the arithmetic average returns. On the other hand, it shows compound Q-learning acquired policies that choose U.S. government bond in most cases. It was the optimal policy based on the compound return.

6   Discussion and Related Work

In compound RL, the simple rate of return, R, is transformed into the reinforcement defined by the logarithm of simple gross return, $\log(1 + Rf)$, where f is a bet fraction

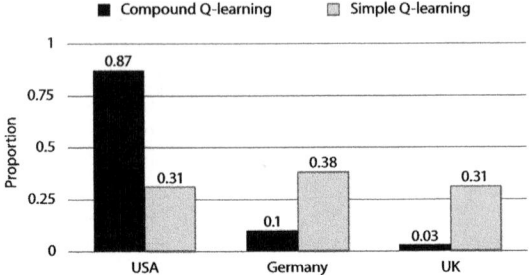

Fig. 7. Proportion of acquired policies in the global government bonds selection

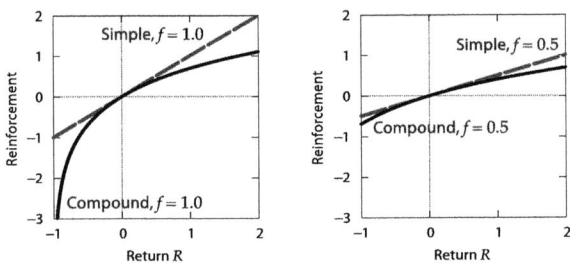

Fig. 8. The difference between the reinforcement for compound RL and simple RL

parameter, $0 < f \leq 1$. Compared with the reinforcement for simple RL, the logarithmic transformation suppresses positive reinforcement and increases negative reinforcement. Figure 8 shows the difference between the reinforcement of compound RL and that of simple RL. The effect of logarithmic transformation becomes larger when the bet fraction f increases.

On the other hand, the bet fraction is a well-known concept in avoiding over-investing in finance. The bet fraction that maximizes the investment effect can be calculated if the probability distribution is known [8]. It is called the Kelly criterion. Vince proposed a method, called "optimal f," which estimates the Kelly criterion and chooses a bet fraction based on the estimation [16]. It is, therefore, a natural idea to introduce a bet fraction parameter to compound RL.

There are some related work. Risk-sensitive RL [1,2,5,7,9,11] not only maximizes the sum of discounted rewards, but it also minimizes the risk. Because the investment risk is generally defined as the variance of the returns in finance, the expected-value-minus-variance criterion [7,11] seems to be suitable for financial applications. Schwartz's R-learning [12] maximizes the average rewards instead of the sum of discounted rewards and Singh modified the Bellman equation [13]. Tsitsiklis and Van Roy analytically compared the discounted and average reward temporal-difference learning with linearly parameterized approximations [15]. Gosavi proposed a synchronous RL algorithm for long-run average reward [6]. However, risk-sensitive RL and average-reward RL are not effective for maximizing the compound return.

7   Conclusion

In this paper, we described compound RL that maximizes the compound return in return-based MDPs. We introduced double exponential discounting and logarithmic transformation of the double-exponentially discounted compound return, and defined the value function based on these techniques. We formulated the Bellman equation for the optimal value function using the logarithmic transformation with a bet fraction parameter. The logarithmic reinforcement results in inhibiting positive returns and enhancing negative returns. We also extended Q-learning into compound Q-learning and showed its convergence. Compound RL maintains the advantages of traditional RL, because it is a natural extension of traditional RL. The experimental results in this paper indicate that compound RL could be more useful in financial applications. Although compound RL theoretically works in general return-based MDPs, the both environments in this paper were single-state return-based MDPs. We have to investigate the performance of compound RL in multi-state return-based MDPs and compare with risk-sensitive RL in the next.

We aware that many RL methods and techniques, for example policy gradient, eligibility traces, and function approximation, can be introduced to compound RL. We plan to explore these methods in the future.

Acknowledgements. This work was supported by KAKENHI (23700182).

References

1. Basu, A., Bhattacharyya, T., Borkar, V.S.: A learning algorithm for risk-sensitive cost. Mathematics of Operations Research 33(4), 880–898 (2008)
2. Borkar, V.S.: Q-learning for risk-sensitive control. Mathematics of Operations Research 27(2), 294–311 (2002)
3. Campbell, J.Y., Lo, A.W., Graig MacKinlay, A.: The Econometrics of Financial Markets. Princeton University Press (1997)
4. CMA. Global sovereign credit risk report, 4th quarter 2010. Credit Market Analysis, Ltd. (CMA) (2011)
5. Geibel, P., Wysotzki, F.: Risk-sensitive reinforcement learning applied to control under constraints. Journal of Artificial Intelligence Research 24, 81–108 (2005)
6. Gosavi, A.: A reinforcement learning algorithm based on policy iteration for average reward: Empirical results with yield management and convergence analysis. Machine Learning 55(1), 5–29 (2004)
7. Heger, M.: Consideration of risk in reinforcement learning. In: Proc. of the Eleventh International Conference on Machine Learning, ICML 1994, pp. 105–111 (1994)
8. Kelly Jr., J.L.: A new interpretation of information rate. Bell System Technical Journal 35, 917–926 (1956)
9. Mihatsch, O., Neuneier, R.: Risk-sensitive reinforcement learning. Machine Learning 49(2-3), 267–290 (2002)
10. Poundstone, W.: Fortune's Formula: The untold story of the scientific betting system that beat the casinos and wall street. Hill and Wang (2005)
11. Sato, M., Kobayashi, S.: Average-reward reinforcement learning for variance penalized Markov decision problems. In: Proc. of the Eighteenth International Conference on Machine Learning, ICML 2001, pp. 473–480 (2001)

12. Schwartz, A.: A reinforcement learning method for maximizing undiscounted rewards. In: Proc. of the Tenth International Conference on Machine Learning (ICML 1993), pp. 298–305 (1993)
13. Singh, S.P.: Reinforcement learning algorithms for average-payoff Markovian decision processes. In: Proc. of the Twelfth National Conference on Artificial Intelligence (AAAI 1994), vol. 1, pp. 700–705 (1994)
14. Sutton, R.S., Barto, A.G.: Reinforcement Learning: An Introduction. The MIT Press (1998)
15. Tsitsiklis, J.N., Van Roy, B.: On average versus discounted reward temporal-difference learning. Machine Learning 49, 179–191 (2002)
16. Vince, R.: Portfolio management formulas: mathematical trading methods for the futures, options, and stock markets. Wiley (1990)
17. Watkins, C.J.C.H., Dayan, P.: Technical note: Q-learning. Machine Learning 8(3/4), 279–292 (1992)

Proposal and Evaluation of the Active Course Classification Support System with Exploitation-Oriented Learning

Kazuteru Miyazaki and Masaaki Ida

National Institution for Academic Degrees and University Evaluation,
1-29-1 Gakuennishimachi, Kodaira, Tokyo 187-8587, Japan
teru@niad.ac.jp

Abstract. The National Institution for Academic Degrees and University Evaluation (NIAD-UE) is an exclusive institution in Japan which can award academic degrees based on the accumulation of academic credits for non-university students. An applicant who wishes to be awarded a degree from NIAD-UE must classify his obtained credits according to pre-determined *criteria* for each disciplinary field. The criteria are constructed by several items. A sub-committee of experts in the relevant field judges whether the course credit classification by the applicant is appropriate or not paying attention on the first item in the criteria. Recently, the Active Course Classification Support (ACCS) system has been proposed to support the sub-committee for the validation of applicant's course classification. Entering a classification item number into ACCS, ACCS suggests an appropriate course that belongs to the set of classification item number. However, some difficulties of deciding appropriate item numbers still remain. This study aims to improve the method of determining item numbers, which should be judged by the sub-committee, by using machine learning. We use *Exploitation-oriented Learning* as the learning method for improving ACCS, and present a numerical example to show the effectiveness of our proposed method.

Keywords: reinforcement learning, exploitation-oriented learning, course classification, recommender system.

1 Introduction

Recently, there has been a wealth of impressive empirical results in *Reinforcement Learning* (RL)[13], as well as significant theoretical advances. Both types of advances in RL will lead to fruitful progress in real-world applications, such as decision making in social systems.

As globalization progresses, quality assurance for higher education is becoming an increasingly important issue in recent reforms of higher education worldwide. There is increasing demand for measuring the outcomes of student learning to improve the quality of higher education.

NIAD-UE(http://www.niad.ac.jp/) in Japan is an incorporated administrative agency of the Ministry of Education, Culture, Sports, Science and Technology (MEXT). The institution's mission is to contribute to the development

S. Sanner and M. Hutter (Eds.): EWRL 2011, LNCS 7188, pp. 333–344, 2012.

Fig. 1. The application guidebook "Alternative Routes to a Bachelors Degree"

Fig. 2. The Committee for Validation and Examination of Degrees

Table 1. Criteria for the information engineering sub-field

Item 1	Information Engineering Basic Theory (4+ credits)
Item 2	Computer System (4+ credits)
Item 3	Information Processing (4+ credits)
Item 4	Electrical and Electronics, Communications, or Systems
Item 5	Exercises and Experiments in Information Engineering (6+ credits)
Item 6	Related Courses (4+ credits)
Item 7	Others

of higher education, in order to create a society in which the achievements of students at various higher education institutions, including non-university institutions, are duly appreciated. Assessing the results of various learning provided at the higher education level, NIAD-UE awards Bachelor's degrees to learners who have acquired sufficient academic credits and have made sufficient academic achievements in the particular disciplinary field.

In this degree-awarding process, an applicant who wishes to be awarded a degree from NIAD-UE must classify his earned credits according to pre-determined criteria for various disciplinary fields (Fig. 1). Table 1 is an example of the criteria. The validity of the applicant's course classification is judged by the sub-committee of the *Committee for Validation and Examination of Degrees* (Fig. 2) whose members are experts in the field. The members of the sub-committee take much time to read the syllabus of each course to judge whether the classification of the applicant is appropriate or not. NIAD-UE awards degrees to about 2,500 applicants each year, and to save labor, there is a need for using information technology to assist the course classification work. As a most promising method, application of RL in this classification process is strongly expected.

The *Course Classification Support* (CCS) system [5], which is a prototype *recommender systems* [10,2], has been proposed to support the course classification. However, CCS has the drawback of excluding the occurrence frequency of technical terms, which plays a key role in the course classification. To overcome this problem, the *Active Course Classification Support* (ACCS) system has recently been proposed [6]. By using ACCS, we can control the number of candidates of courses that should be judged by the sub-committee, which would reduce the course classification workload for the sub-committee along the criteria for the chosen disciplinary field.

Each criterion for each disciplinary field is constructed by several items that characterize the field. In this paper, we discuss the criteria for the *information engineering* sub-field listed in Table 1. Items 1, 2, 3 and 4 are called *specialized courses* which are important for the information engineering sub-field, and are considered in this paper. Each item in specialized courses has several course examples to support the course classification, and usually has minimum number of credits that must be acquired by the applicant.

In the existing course classification process, the sub-committee usually starts judging the applicant's course classification based on the courses that are classified in the first item, for example, item 1 in Table 1. After checking the criteria of item 1, the sub-committee then checks the criteria of item 2, and then those of item 3. Though the sub-committee has classified all specialized courses, we do not need to check the criteria of item 4 since it has no condition for the number of credits to be acquired by the applicant.

Generally, the difficulty of appropriate course classification depends on the disciplinary field. Therefore, in this paper, we aim to obtain the characteristics of course classification by machine learning, so that we can apply the results of the learning for the determining the item number and thus assist the judgments by the sub-committee. We use *Exploitation-oriented Learning* (XoL)[7], which is suitable in this situation as a machine learning method for ACCS, and present a numerical example to show the effectiveness of our method.

2   Outline of the Degree-Awarding by NIAD-UE

NIAD-UE's academic degree-awarding has two different schemes. One is based on a combination of the successful completion of a junior college program (or an equivalent level of education) and the accumulation of sufficient credits upon the completion of such a program or education level (Scheme I). The other scheme is based on the successful completion of an NIAD-UE-approved program provided by an educational institution that has been established legally, yet operates under the jurisdiction of ministries other than MEXT (Scheme II). A Bachelors degree can be obtained under either of these two schemes, whereas Masters and Doctoral degrees are awarded only under Scheme II. In this paper, we focus on Scheme I.

Although there exist similar systems as Scheme II in other countries, the Scheme I of NIAD-UE is a unique system in the world [8,9,11]. NIAD-UE annually publishes an application guidebook for Scheme I, *Alternative Routes to a Bachelors Degree* (ARBD) (Fig. 1). In this paper, we refer to the 2010 version which is available on NIAD-UE's Website at:
http://www.niad.ac.jp/ICSFiles/afieldfile/2010/03/19/no7_5_gakushiH22.pdf

Under Scheme I, Bachelors degrees were awarded in 26 disciplinary fields in 2010, i.e. Art, Engineering, Literature, Nursing, Theology, and so on. Some disciplinary fields are divided into sub-fields. For example, the engineering field has the following eight sub-fields: Mechanical Engineering, Electrical and Electronics Engineering, Information Engineering, Applied Chemistry, Biological Technology, Materials Engineering, Civil Engineering, and Architecture. In this paper, we focus on the information engineering sub-field in the engineering field.

An applicant has to certify that he has accumulated the appropriate number of credits according to the criteria pre-determined for each field. For example, the criteria for the information engineering sub-field are presented in Table 1. The course classification by an applicant is not always valid, and so the validity of course classification is judged by the sub-committee of the Committee for Validation and Examination of Degrees (Fig. 2), the members of which take much time to read the syllabus of each course. In recent years the number of applicants has been increasing, and so there is a need to assist the judging work of the sub-committee by using information technology. In the next section, we explain how CCS and ACCS are used to support course classification during degree-awarding by NIAD-UE.

3 Course Classification Support System

3.1 Construction of myDB

The Course Classification Support (CCS) system receives syllabus data as an input and outputs a candidate course classification according to ARBD. It is important to construct a key database called the *term-classification database* to ensure an efficient course classification. In the following, we describe how to make **myDB**, which is one of the term-classification databases.

In 2002, the authors collected 962 syllabuses of information engineering courses offered by 13 universities in Japan [5]. They read these syllabuses and classified them into the items listed in Table 1. One syllabus may be classified into two or more of the items listed in the table.

Based on the classification, they constructed a *set of special technical terms* for each item, consisting of technical terms contained only in that item The technical terms are extracted by the method of reference [12]. This aims to find the characteristic terms that are contained only in each item. myDB is constructed by the set $\{myDB(1),...,myDB(M)\}$ where M is the number of items and $myDB(i)$ $(i=1,2,...,M)$ is the set of special technical terms for item i.

3.2 CCS and Its Features

CCS is a system that matches myDB and the syllabus data of each course. After matching between each term in the syllabus data and myDB, the matched terms and their corresponding item numbers are outputted.

The effectiveness of CCS in some examples has been confirmed [5]. However, it has the drawback of excluding the *frequency of occurrence* of technical terms, which plays a key role in the classification of courses. The Active Course Classification Support (ACCS) system [6] has been proposed to overcome this problem.

3.3   The Active Course Classification Support System

myDB is a data set of terms by item that are present only in each item. However, the frequency of occurrence of technical terms is not considered. The specificity of a given technical term in a chosen item should change in a manner that is dependent on the frequency of the term's occurrence in syllabuses relevant to the item. Therefore, in reference [6], the authors reconfigured myDB by considering the frequency of occurrence of technical terms and incorporating this aspect into the CCS.

First, the authors extracted technical terms from all the syllabuses of the above-mentioned 13 universities by the method described in reference [12]. Subsequently, they carried out *interval estimation* of the probability that an extracted technical term is contained in item i ($i = 1, 2, 3, 4$). In order to increase specificity, the upper and lower probability can be calculated as follows;

$$\frac{\frac{x}{n} + \frac{Z_\gamma^2}{2n} \pm \frac{Z_\gamma}{\sqrt{n}} \sqrt{\frac{x}{n}\left(1 - \frac{x}{n}\right) + \frac{Z_\gamma^2}{4n}}}{1 + \frac{Z_\gamma^2}{n}}, \tag{1}$$

where x is "the number of syllabuses that include the term and in which the term is contained in item i", n is "the number of syllabuses that include the term", and Z_γ is 1.96 and 2.58 at the confidence levels of 95% and 99%, respectively. We set $Z_\gamma = 1.96$ in Section 5.

Taking this into account, they created a database that accumulates data sets of each technical term and the confidence interval estimated for the terms to be contained in item i ($i = 1, 2, 3, 4$). Hereafter, this database is referred to as **myDBc**. myDBc is constructed by the set $\{myDBc(1), ..., myDBc(N)\}$ where $myDBc(i)$ (i=1,2,...,N) is a technical term and N is the number of technical terms. Furthermore, we can get the lower and the upper limits of the confidence interval about the term x on the item number i ($i = 1, 2, ..., M$) by the functions $LL_i(myDBc(x))$ and $UL_i(myDBc(x))$, respectively.

In ACCS, myDB is replaced by myDBc. The paper [6] shows that we input syllabus data of the course submitted by the applicant into ACCS. ACCS matches terms [1] between the syllabus data and myDBc. Then we can get "a matched technical term and the lower limit of the confidence interval of the probability that the term is contained in the item i (i =1,2,...,M)." The sub-committee classifies the course in which the lower limit of the confidence interval about the item i is greater than *a threshold value*. If we gradually lower the threshold value from 1.0, we can identify those courses that are likely to be classified to the item number. We have called this method *threshold learning*.

3.4   Features on ACCS

The ACCS method can control the number of candidates of courses that should be judged by the sub-committee, thus reducing the workload for sub-committee members.

---

[1] In Section 5, we attempt to input syllabus data directly into ACCS by scanning and OCR processing of the syllabuses.

On the other hand, we do not understand the detail of how to decide the item number to be judged by the sub-committee. The sub-committee usually starts the judgment as to whether the course classification by an applicant is good based on the courses that are classified in the first item by the applicant. Ideally, it is desirable to learn the characteristics of the items that seem to be different for each field, and to determine the item number to be judged by the sub-committee. RL and XoL are machine learning methods that are suitable for such learning, so we examined combining them with ACCS.

4   Proposal of ACCS with Exploitation-Oriented Learning

4.1   Incompleteness of Thereshold Learning

If the threshold learning about ACCS is perfect, that is, for example, the course that has the highest threshold value for an item number is actually classified to that item, then we can start from any item number. However, the threshold learning is not perfect, that is, the course that has the highest threshold value for an item number is not always classified to that item. For example, in the information engineering sub-field, though there are many courses belonging to *Computer System* (item 2), the number of courses belonging to *Information Processing* (item 3) tends to be smaller than item 2. Therefore, some courses that were regarded to be classified to item 3 by ACCS may be classified to item 2 by the sub-committee. This means that it is difficult to find courses belonging to *Information Processing* (item 3).

The incompleteness of threshold learning depends on the design of the criteria for each field. Therefore, in this paper, we aim to learn the characteristics of the criteria and to use it for our course classification. For example, in the information engineering sub-field, we aim to be able to find courses that belong to item 3 in preference to item 1 or 2.

4.2   Learning by Exploitation-Oriented Learning

We aim to learn the characteristics of the criteria for each field by RL or XoL, and to use them for determining the item number to be judged by the sub-committee. This means that we give ACCS a learning system that can determine the item number.

In XoL and RL, the learning agent senses a set of discrete attribute-value pairs and performs an action in some discrete varieties. The environment provides a reward signal to the agent as a result of some sequence of actions. A sensory input and an action constitute a pair that is termed a *rule*. The function that maps sensory inputs to actions is termed a *policy*. When there are n sensory inputs and m actions, we can make n^m types of policies.

In many RL and some XoL methods, a scalar weight that indicates the importance of the rule is assigned to each rule. The learning progresses by updating the weight in the method.

Sensory Inputs and Actions

We should change the item number to be judged depending on whether the requirements for each item are satisfied. We regard the satisfaction of requirements, that is, the number of credits required to satisfy the criteria, as sensory inputs. On the other hand, we regard the item number that we want to satisfy the criteria as an action output.

Design of Rewards

The design of a reward influences whether the learning is successful. The most basic reward design is to reward the learning agent when it has satisfied the criteria. However, if we use this design, we cannot expect to reduce the number of courses to be judged by the sub-committee, because the reward does not have information about the number of courses required to achieve the criteria. Therefore, in this paper, if the learning agent was able to achieve the criteria by the minimum number of courses, we will reward the learning agent.

4.3   Overall Procedure of ACCS with XoL

We show our proposal in the following and its framework in Fig. 3, which we call **ACCS with XoL**.

1. Fetch the required credit numbers for each item.
2. Determine the item number based on the policy that is learned by XoL or RL.
3. Submit the item number to ACCS as shown in the following.
   (a) Matching with syllabus data of the course that is not classified by the sub-committee, determine a set B, $B = \bigcup_{j=1}^{N_t} (myDBc \cap Syl_j)$ where Syl_j is the set of technical terms in the syllabus data of the course j and N_t is the number of courses that is not classified by the sub-committee.
   (b) Select the term that has the highest LL_p in the set B where p is the inputted item number p.
   (c) Output the course that includes the term.
4. Obtain the course derived from ACCS.
5. Consult the sub-committee the classification of the course, and obtain the item number of the course from the sub-committee.
6. Recalculate the number of required credits for each item.
7. Go to step 1.

The learning agent corresponds to XoL or RL. It receives the number of required credits for each item as a sensory input, and selects and outputs an item number based on its policy as an action. ACCS receives the item number as an input, and outputs the course that has the highest value of the lower limit of the confidence interval for the item number to the sub-committee. The sub-committee classifies the course and outputs the item number of the course. We recalculate the number of required credits for each item. The learning agent receives a reward when the criteria have been achieved by the minimum number of courses. If the criteria for all items have been achieved, we can end the judgment of the applicant.

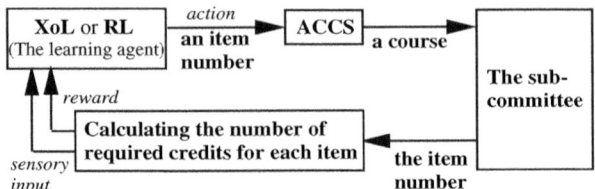

Fig. 3. ACCS with Exploitation-oriented Learning

The learning agent selects an action based on the same probability for every action in every sensory input. This is called *random walk* (RA) and is generally used in XoL and RL, in order to improve a policy. The policy is saved by each individual applicant. If multiple policies are obtained from several applicants, we aim to integrate them by normalizing and accumulating their policies for each applicant's sensory input. The integrated and normalized policy is used to support the course classification of an unknown applicant. On the other hand, known classification results are used as learning data,

The output of the learning agent in Fig. 3 depends on the performance of ACCS. In general, the output of ACCS depends on many factors, such as the threshold value and so on, that are not included in a sensory input. This means that our task has the *perceptual aliasing problem* [1]. Therefore, many RL methods based on *Dynamic Programming* (**DP**) are not suitable for our task because they are designed to resolve *Markov Decision Processes* (**MDPs**).

5   Evaluation of ACCS with XoL

5.1   Learning a Policy by XoL or RL

We show the effectiveness of ACCS with XoL in the information engineering sub-field. It has two steps. First, we aim to learn a policy, that is suitable for the sub-field, by XoL or RL. Next, we evaluate the learned policy combined with ACCS, which is ACCS with XoL.

We must obtain policies for each sub-field in order to use ACCS with XoL. The policy's input is the number of required credits for each item and its output is an item number, as described in Section 4. In this paper, we aim to make a policy that is suitable for the information engineering sub-field, by XoL or RL.

We need to prepare learning data for XoL or RL. We use the syllabuses of the courses that were actually submitted for the information engineering sub-field in fiscal year 2003 and 2004 as learning data to make the policy. They were submitted to NIAD-UE by four applicants independently, and were 48 courses of the science and engineering departments of University M, 28 courses of the engineering department of University K, 56 courses of the engineering science department of University O and 30 courses of the engineering department of University H.

We prepare three types of learning method, that is *Profit Sharing* (**PS**)[3], *Rational Policy Making algorithm* (**RPM**)[4] and *Q-learning* (**QL**)[14]. PS and RPM are XoL methods, and QL is the most representative RL method. PS reinforces the scalar weights on rule sequence at the same time when it receives a reward. The rationality of PS was analyzed in reference [3]. RPM can achieve the same rationality without any scalar weight.

In PS, the initial weight is 0.0 and the reward is 100.0. PS depends on the design of the *reinforcement function* [3] which decides how to distribute a reward. We use the geometric decreasing function whose common ratio is $\frac{1}{4}$ that satisfies the *rationality theorem* in reference [3]. In QL, the initial weight is 10.0, the reward is 100.0, the learning ratio is 0.8, and the discount ratio is 0.9. Those are decided by preliminary experiments.

For each learning method, we continued the learning until the criteria of the information engineering sub-field were satisfied 500,000 times for each applicant, which is equivalent to receiving a reward 2 million times. We only save the optimum policies as a result of learning [2]. The numbers of steps to receive a reward in the optimum policies were 8, 8, 25, and 13 for universities M, K, O and H, respectively.

5.2   Experimental Results

We aim to combine the learned policy in Section 5.1 with ACCS, namely, ACCS with XoL, and evaluate its performance. We use the syllabuses of the courses that were actually submitted from 2005 to 2010 by seven applicants (A to G) to evaluate the policy. We aim to reduce the number of courses required to satisfy the criteria, which thus reduces the number of courses to be judged by the sub-committee.

We show the number of courses required to satisfy the criteria by the three learning methods in Table 2. We conducted 100 trials with a different random seed for each applicant (A to G). Hence, after universities M, K, O and H have been learned by PS, RPM or QL, the courses of the seven applicants (A to G) are evaluated 100 times independently by the policy that has been learned for each learning method. Table 2 shows the average and standard deviation in parentheses for the 100 trials with different random seeds.

In the current classification, the sub-committee usually starts the judgment as to whether the course classification of an applicant is good based on the courses that are classified to item 1 in Table 1 by the applicant. After satisfying the criteria of item 1, the sub-committee selects item 2, and then item 3. Though our target is all specialized courses from item 1 to 4, we do not need to check the criteria of item 4 since it has no condition on the number of credits to be acquired by the applicant. Therefore, the current classification can be restricted to item 1 to 3. We show the results of "123" in Table 2 where the sub-committee has judged courses in order of item number 1,2,3.

---

[2] Note that, in general, there are several types of optimum policies.

Table 2. Results of each method

	A	B	C	D	E	F	G
RPM	6.98 (0.07)	22.00 (0.00)	5.01 (0.04)	6.03 (0.23)	16.33 (0.36)	15.00 (0.00)	6.05 (0.14)
PS	6.94 (0.12)	22.15 (0.36)	5.03 (0.08)	6.43 (0.71)	16.34 (0.76)	15.00 (0.05)	6.21 (0.32)
QL	6.94 (0.36)	22.55 (0.56)	5.15 (0.71)	10.16 (2.73)	16.36 (1.55)	15.00 (0.00)	7.34 (3.39)
123	7.00 (0.00)	22.00 (0.00)	5.00 (0.00)	8.00 (0.00)	17.00 (0.00)	15.00 (0.00)	7.00 (0.00)
321	7.00 (0.00)	22.00 (0.00)	5.00 (0.00)	6.00 (0.00)	15.00 (0.00)	15.00 (0.00)	6.00 (0.00)
gdy	7.00 (0.00)	22.76 (0.88)	5.00 (0.00)	6.54 (0.50)	16.40 (0.92)	15.00 (0.00)	6.00 (0.00)
rnd	9.96 (3.43)	24.19 (2.73)	8.89 (3.12)	15.87 (4.22)	17.06 (2.76)	17.23 (0.61)	14.24 (5.73)

Table 3. Comparison with "321"

	A	B	C	D	E	F	G
RPM	(2.5, 0.0)	(0.0, 0.0)	(0.0, 0.8)	(0.0, 0.7)	(29.8, 60.3)	(0.0, 0.0)	(0.0, 2.4)
PS	(8.9, 1.8)	(0.0, 13.7)	(0.0, 3.3)	(0.0, 27.6)	(27.4, 60.4)	(0.0, 0.5)	(0.0, 13.7)
QL	(6.4, 1.5)	(0.0, 34.5)	(0.0, 6.0)	(0.0, 82.0)	(33.6, 54.6)	(0.0, 0.0)	(0.0, 16.1)

Table 4. Summary of learning results

	A	B	C	D	E	F	G
RPM	>	=+	<	=-	=-	=+	=-
PS	>	=-	<	=-	=-	=+	=-
QL	>	=-	<	<	=-	=+	<

Although we processed all combinations except "123", we only show the results for "321", since it has the best performance among the six patterns "123","132","231","213","312" and "321". In addition, "gdy", which means greedy, is the result when the sub-committee always checks the courses in the item which is furthest from the criteria. Also, "rnd" is the result when the sub-committee always checks the courses in the item decided on the same probability. Note that although "321" has the best performance, we cannot know the order in advance.

Table 3 compares the results with three learning methods and "321" in Table 2. The values on the left of the parentheses in Table 3 are the percentages where the criteria can be achieved by the number of courses less than "321". The values on the right of the parentheses are the percentages where it cannot improve on "321".

The results of Table 2 are summarized in Table 4 where ">" means that it can improve on "321", "=+" means that it is equal to "321", "=-" means that it can improve on the worst case among the six patterns from "123" to "312", and "<" means that it cannot improve on the worst case.

5.3   Discussion

Although the applicant C has "<" for all methods in Table 4, the degree of deterioration is very small as shown in Table 3. Especially, in RPM, it is less than 1%. Therefore, the result of applicant C is not considered to be a serious problem. In contrast, for applicant A, all methods have ">", which means that all methods have cases that can improve on the 7 courses that give the best performance in the current classification. The results are very significant since we can know better results than the known best. Although the same improvement is shown in applicant E, all methods have "=-" since it has a high percentage of deterioration.

QL is the worst among the three learning methods. Especially, applicant D and G show significant performance deterioration in Table 3. Though QL can guarantee the optimality in MDPs, it is not suitable for our task which goes beyond MDPs.

Furthermore, when several policies are combined, QL will receive another negative influence where it is possible to give the maximum weight for an item number that should not be learned. For example, in the sensory input where an insufficient number of courses are 4,4,3 for item 1,2,3, respectively, and the weights for each action 1,2,3 are "0.5, 0.45, 0.05" for University M and "0.2, 0.3, 0.35" for University O, QL averages these weights to "0.7, 0.75, 0.4", that is "0.38, 0.40, 0.22". Therefore, it has a maximum weight for item 2 that should not be learned.

PS and RPM can guarantee rationality in some classes that go beyond MDPs[3]. Since our task belongs to the class, PS and RPM are more suitable for our task than QL. Like QL, PS can receive a negative influence when several policies are combined since it uses scalar weights for learning. However, the effect is less than that of QL, because PS is an XoL method, which strongly enhances the experience.

On the other hand, RPM does not use weights; it evaluates the rule by {0,1} where "0" means bad and "1" means good. Therefore, unlike QL and PS, RPM is not negative influenced and can achieve the best performance.

6   Conclusions

The National Institution for Academic Degrees and University Evaluation in Japan awards awarding academic degrees for applicants based on the accumulation of course credits. An applicant's earned credits must be classified according to pre-determined criteria for his designated disciplinary field. Validation of his classification is carried out by a sub-committee of experts in the syllabus of each course. For assisting this course classification problem, we have already proposed the Active Course Classification Support (ACCS) system.

In this paper, we improved ACCS for course classification based on an Exploitation-oriented Learning mechanism. We presented numerical examples

---

[3] For details of the class refer to reference [4].

for the specific disciplinary field, "Information Engineering" to show the effectiveness of our proposed method.

In the future, we will expand its applicability to various disciplinary fields other than "information engineering", and will improve our system for practical use.

References

1. Chrisman, L.: Reinforcement learning with perceptual aliasing: The Perceptual Distinctions Approach. In: Proc. of the 10th National Conf. on Artificial Intelligence, pp. 183–188 (1992)
2. Claypool, M., Gokhale, A., Miranda, T., Murnikov, P., Netes, D., Sartin, M.: Combining content-based and collaborative filters in an online newspaper. In: Proc. of the ACM SIGIR 1999 Workshop on Recommender Systems: Implementation and Evaluation, Berkeley, California (August 1999)
3. Miyazaki, K., Yamamura, M., Kobayashi, S.: On the Rationality of Profit Sharing in Reinforcement Learning. In: Proc. of 3rd Int. Conf. on Fuzzy Logic, Neural Nets and Soft Computing, pp. 285–288 (1994)
4. Miyazaki, K., Kobayashi, S.: Learning Deterministic Policies in Partially Observable Markov Decision Processes. In: Proc. of 5th Int. Conf. on Intelligent Autonomous System, pp. 250–257 (1998)
5. Miyazaki, K., Ida, M., Yoshikane, F., Nozawa, T., Kita, H.: On development of a course classification support system using syllabus data. Computational Engineering I, 311–318 (2004)
6. Miyazaki, K., Ida, M., Yoshikane, F., Nozawa, T., Kita, H.: Proposal of the Active Course Classification Support System to Support the Classification of Courses at the Degree-Awarding of NIAD-UE. In: Proc. of 6th Int. Symposium on Advanced Intelligent Systems, pp. 685–690 (2005)
7. Miyazaki, K., Kobayashi, S.: Exploitation-oriented Learning PS-r#. J. of Advanced Computational Intelligence and Intelligent Informatics 13(6), 624–630 (2009)
8. Mori, R.: The Credit Transfer System and the Validation Service at the Open University. Research in Academic Degrees (17), 183–198 (2003) (in Japanese)
9. Puirseil, S.: Quality Assurance in Irish Higher Education - The Higher Education and Training Awards Council. Research in Academic Degrees 15, 124–140 (2001)
10. Resnick, P., Varian, H.: Recommender Systems. Introduction to Special Section of Communications of the ACM 40(3) (March 1997)
11. Tachi, T.: A Study on Thomas Edison State College, The External Degree College established by the State of New Jersey. Research in Academic Degrees 10, 73–89 (1999) (in Japanese)
12. Yumoto, H., Mori, T., Nakagawa, H.: Term Extraction Based on Occurrence and Concatenation Frequency. Natural Language Processing 10(1), 27–45 (2003) (in Japanese)
13. Sutton, R.S., Barto, A.G.: Reinforcement Learning: An Introduction. A Bradford Book. MIT Press (1998)
14. Watkins, C.J.H., Dayan, P.: Technical note: Q-learning. Machine Learning 8, 55–68 (1992)

Author Index

GPSR Compliance

The European Union's (EU) General Product Safety Regulation (GPSR)
is a set of rules that requires consumer products to be safe and our
obligations to ensure this.

If you have any concerns about our products, you can contact us on
ProductSafety@springernature.com

In case Publisher is established outside the EU, the EU authorized
representative is:

Springer Nature Customer Service Center GmbH
Europaplatz 3
69115 Heidelberg, Germany

Batch number: 09490872

Printed by Printforce, the Netherlands